D1289940

BATCH PROCESS AUTOMATION

BATCH PROCESS AUTOMATION
Theory and Practice

Howard P. Rosenof

Asish Ghosh

VAN NOSTRAND REINHOLD COMPANY
New York

Disclaimer: The purpose of this work is educational. The information and teachings herein are thus general in nature and are not intended for designing, constructing, or operating a specific facility.

Library of Congress Catalog Card Number: 86-23332
ISBN 0-442-27708-3

Printed in the United States of America

Van Nostrand Reinhold Company Inc.
115 Fifth Avenue
New York, New York 10003

Van Nostrand Reinhold Company Limited
Molly Millars Lane
Wokingham, Berkshire RG11 2PY, England

Van Nostrand Reinhold
480 La Trobe Street
Melbourne, Victoria 3000, Australia

Macmillan of Canada
Division of Canada Publishing Corporation
164 Commander Boulevard
Agincourt, Ontario M1S 3C7, Canada

16 15 14 13 12 11 10 9 8 7 6 5 4 3 2

Library of Congress Cataloging-in-Publication Data
Rosenof, Howard P., 1948-
Batch process automation.
Includes index.
1. Chemical process control—Automation. I. Ghosh, Asish, 1935- . II. Title.
TP155.75.R67 1987 660.2'81 86-23332
ISBN 0-442-27708-3

CONTENTS

PART III: THE PRACTICE OF AUTOMATION

FOREWORD

While continuous processing has always been considered the ideal method of operation of a chemical plant from the economic viewpoint and still remains the ultimate goal of the chemical engineer designing a new plant or unit, it is always amazing to me to see the amount of batch processing that remains in many chemical and petroleum plants. In addition, the more specialized companies such as the chemical specialties makers, the drug companies, and others require even more batch reaction equipment to carry out their production functions successfully. This major need for batch equipment is usually because these latter industries produce very high quality products in quantities too small to justify the economic cost of continuous production equipment. In addition, the use of batch equipment has allowed the production of purities and other quality parameters not readily attainable in continuous equipment.

Batch equipment with its dynamically changing operating conditions during the period of the batch operation has always posed to the control engineer a challenge that has demanded the ultimate in his technical ingenuity. Each of the emerging automatic control technologies has been applied to batch equipment control as it has been developed, with an ever increasing success as control capabilities have advanced. Nevertheless, the batch reactor has continued to challenge the control engineer because of the great individuality and consequent lack of generality shown by each batch reactor control situation.

Major advances have been achieved in batch reactor control with the development of the programmable controller and the minicomputer during the 1970s. These remain the major competitive technologies today since the microprocessor based, distributed control systems have generally not yet shown enough computational ability to supersede the earlier technologies. How long this latter condition will hold in the face of the continued expansion of the capabilities of distributed control systems remains to be seen.

The field of batch reactor control has developed a major literature of technical reports and papers on reactor mathematical models, on process control theory and technology, and on operational applications. The major media for these papers have been in the technical journals, conference proceedings, and related publications of the American Institute of Chemical Engineers, the American Chemical Society, and the Instrument Society of America. The technical journals, *Instruments and Control Systems* and *Control Engineering*, have also published an extensive set of papers in this area.

Despite the many papers, very few books about batch reactor control have been published. Thus *Batch Process Automation: Theory and Practice*, by Rosenof and Ghosh, fills a long felt need for the profession. These authors have covered all aspects of the field from the basic definitions of the field to discussions of such advanced topics as: optimal scheduling of batch reactor systems, proper use of plant safety interlocks, some of the more detailed aspects of project justification, and methods of testing and proving the final systems design for the project. They have provided important material for all classes of readers from those requiring an elementary treatment to those needing the full technical details of specific examples of process equipment.

The examples chosen by the authors, while perhaps unconventional at first consideration, very descriptively and pointedly illustrate the principles involved and thus provide a very good learning experience for the reader.

I expect this book to be an important addition to the literature of the field of batch process control and one of which Messrs. Rosenof and Ghosh should be very proud. I am happy to be able to contribute to it in this small way.

THEODORE J. WILLIAMS
Professor of Engineering
Director, Purdue Laboratory
for Applied
Industrial Control
Purdue University

PREFACE

A well-designed control system can go a long way toward making a batch-process plant reliable, safe, and easy to operate. It helps achieve manufacture within specification and allows the flexibility to make different grades of product, thus helping the process achieve financial success. On the other hand, a poorly designed system can be a money loser. Its products may be unrepeatable or even unsalable, with operator attention required continuously to second-guess the system's decisions.

We believe that the difference between success and failure is in choosing the right hardware and software and in properly applying them. Control system vendors now provide not only the hardware but also comprehensive software packages for controlling the batch process. Choosing the right combination for an application requires considerable insight into the functions of the control system along with detailed process knowledge. This allows the user to generate an appropriate specification of the requirements.

Choosing the right hardware and software combination is only the first step toward success. This must be followed by the arduous task of design and application unique to the process to be controlled. Over the years, we have developed an approach to planning and implementing batch control systems. We have seen first-time batch automation users learn the intricacies of this business, take an active role in determining system requirements, and take full responsibility for system operation. In contrast, we have met people who experienced substantial disappointment in trying to automate batch processes. In many cases they were able to compare their planning approach with ours and discover the cause of their difficulties. Our employer, The Foxboro Company, has permitted and encouraged us to place our method in the public domain. We have done so by writing articles for major control and user industry publications and by making presentations at professional meetings. This book represents the accumulation of this work and would be useful to anyone interested in automating a batch process.

The book is divided into three parts. Part I (Chapters 1-4) is an introduction to the batch process and the batch control system. It deals with batch and sequential processes and familiarizes the reader with common terms (Chapter 1). It also deals with the historical aspects of batch control and gives an introduction to the equipment used today (Chapters 2 and 3). Finally, the benefits and problems associated with batch-process automation are discussed (Chapter 4). This part will prove invaluable to those unfamiliar with this field. Because standard terms for batch control are yet to be universally accepted, experienced engineers will also find this part useful in orienting themselves toward the terminology used in the rest of the book.

Part II (Chapters 5 and 6) elaborates on the functional requirements of a batch control system and then describes how the typical control system packages available today meet these requirements. This part will be useful in gaining enough appreciation of the batch control systems to allow the reader to make intelligent choices from among the available hardware and software packages. This part will be useful to those unfamiliar or with limited familiarity with batch control and will act as the stepping stone to the next part.

Part III (Chapters 7-14) deals with the practice of automation, which includes the specification, design, implementation, and testing of batch control systems. Chapter 7 gives the essence of our planning method, called design by levels, which divides a batch plant's controls into a hierarchical system. Chapter 8 covers the higher (executive) levels, Chapter 9 the sequence levels, Chapter 10 the regulatory levels, and Chapter 11 the interlock levels. Chapter 8 includes a brief discussion of plantwide batch control, including optimal scheduling. The interested reader should study the literature in operations research. We have assumed that our readers are knowledgeable in continuous-control systems, and we have included, in Chapter 10, only a short discussion of the differences between regulatory controls for continuous and batch processes. Comprehensive discussions on plant safety interlocks are provided in Chapter 11. Chapter 12 has a brief discussion on the possible economic returns for automation at different levels. Hardware architecture and reliability issues are given in Chapter 13. Finally, the importance of testing the designed system along with an outline of the types of tests are given in Chapter 14. This part should prove very useful to experienced engineers. Beginners should thoroughly study the first two parts before reading the third.

Batch processes use and manufacture materials that are toxic, carcinogenic, flammable, explosive, and otherwise dangerous. This book is intended to be used by practicing engineers who have the background to determine which parts of this book are and are not applicable to the specific processes under consideration. Naturally, responsibility for the safety of a process plant remains with those responsible for its design, construction, and operation.

ACKNOWLEDGMENTS

We are grateful to many Foxboro personnel and clients who helped us in developing the implementation methods described in this book. Our special thanks go to William Saunders and Robert Mick, respectively former and present Foxboro colleagues. Bill Saunders developed a highly sophisticated master-sequence recipe application that helped lead to the discussion of this recipe type in Chapters 6 and 8. Bob Mick helped develop the concepts of structured English for documenting sequence logic, as described in Chapter 9.

We also acknowledge the support of The Foxboro Company in preparing this book, and of the many individual management, professional, and technical personnel who helped us in a wide variety of ways. Tom Schonbach, Charlie Hall, Ed Bristol, Ray Sawyer, Al Storti, and Leo Johnson helped establish the Company's commitment to our effort. Dick Sherman provided major assistance with the publication arrangements as well as welcome editorial support. Paulette Bosquette and Mary Nettle provided the word processing support and were tolerant of indecipherable manuscripts and frequent changes of mind. Helen Stevens and the staff of Foxboro's RD & E Library helped us immensely in our research of previous work in this field. Several persons helped us by reviewing our work and offering constructive criticisms: in particular Ed Bristol, Peter Martin, Dave Noon, Mo Prasad, Jim Verhulst, and Jack Wu. In addition to writing the Foreword, Theodore J. Williams reviewed the entire manuscript and made a number of useful suggestions.

HOWARD P. ROSENOF
ASISH GHOSH

BATCH PROCESS AUTOMATION

PART I

INTRODUCTION TO DISCONTINUOUS PROCESSES AND THEIR CONTROL

CHAPTER

1

BATCH AND SEQUENTIAL PROCESSES

Although continuous processes are more widely discussed, batch processes are very common. We cook food by batch processes. We prepare dishes by measuring and mixing ingredients, heating the ingredients, mixing in other ingredients at predefined intervals, testing as required (to check quality), and so on. When the cooking is finished, the food is removed from the heat source and served hot or allowed to cool. Thus, a typical cooking process involves a list of ingredients and step-by-step instructions for mixing, heating, frying, roasting, and so on. It also involves testing the food at certain intervals to check for taste and texture and to take the appropriate actions, such as adding spices. Though a procedure might appear straightforward, it may be modified, based on the experience of the cook or the results of the tests carried out during the cooking process.

Figure 1-1 is a recipe for a fruit cake. The procedure can be divided into three broad phases: preparation, cooking, and cooling and storing (Fig. 1-2).

The steps in each phase must be followed in the proper sequence to make a good cake. Some steps, such as the beating of the egg yolks and whites, may be done in parallel with the main activity and therefore are not documented explicitly (they may be shown as separate phases or as part of the same phase). The procedure also involves setting a timer, which possibly generates an audible message (ringing) when the set time of 2½ h has elapsed. Manual intervention is then necessary. The cook must test the cake to determine if it is done or if further baking is needed.

The procedure assumes normal operation with no unexpected happenings. But suppose the oven temperature controller or the timer breaks down, causing

Fruit Cake

1½ cups butter	3½ cups flour
2 cups sugar	½ teaspoon salt
6 eggs, separated	2 cups golden raisins
½ teaspoon cream of tartar	1 cup chopped, dried apricots
1 cup milk	½ cup chopped candied orange peel
1 teaspoon brandy extract (or ¾ cup milk, ¼ cup brandy)	1 cup coarsely chopped walnuts
	1 teaspoon vanilla

Beat butter until light and creamy. Gradually add sugar, beating until smooth. Beat egg yolks together lightly and add. Combine milk, brandy and vanilla. Mix flour and salt together. Alternately add milk mix and flour mix to butter mix. Fold in fruits and nuts. Beat egg whites with cream of tartar until stiff but not dry. Fold whites into batter gently but thoroughly. Pour mixture into two buttered and floured six-cup molds. Bake at 275 degrees for 2½ hours or until they test done. Cool. Unmold.

Figure 1-1. A typical cooking process, for fruit cake (California Raisin Advisory Board, 1983).

the batch to burn. The actions to be taken in such an event are left to the experience and judgment of the cook.

We can find other examples of batch and sequential processes: washing and drying clothes, cleaning dishes, and showering. In contrast, the control of water supply pressure to the house is a continuous process because the supply is maintained at a constant pressure (within limits) even when there are large variations in demand.

In industry, batch and sequential processes are used in many ways. For example, in

Chemical processes, such as manufacturing of polyvinyl chloride (PVC)
Dyeing in the textile industry
The batch digester in the pulp and paper industry
Starting power plants and other basically continuous plants
Manufacturing corn syrup and beer in the food industry
Manufacturing drugs and fine chemicals
Solid material handling, such as iron ore concentration

Even in areas normally associated with continuous processes (e.g., refining petroleum and the manufacturing of petrochemicals), unit start-up and shutdown, along with such functions as wax extraction and catalyst regeneration, are generally sequential. A typical batch process requires continuous control for a limited period of time—for example, maintaining temperature and pressure to specified points in a reactor during a reaction.

```
Preparation Phase

    Beat specified amount of butter until light and creamy
    Add specified amount of sugar and continue beating
    Add beaten egg yolks
    Add the milk mix and flour mix alternately
    Fold in fruits and nuts
    Fold in beaten egg white mix
    Pour the batter in buttered and floured molds
    Go to cooking phase
End of Phase

Cooking Phase

    Set oven temperature to 275°F
    Wait until the oven temperature is greater than or equal to 275°F
    Put the molds with batter in the oven
    Set timer for 2½ hrs.
    Wait for the timer to time out
    If baking not done
    | Then repeat-until baking is done
    | |       |       Set timer for 15 minutes
    | |       |       Wait for the timer to time out
    | |       End repeat-until
    | Else continue
    Go to cooling and storing phase
End of phase

Cooling and Storing Phase
    Switch off the oven
    Take the molds out with the cakes
    Wait until the cakes are cool
    Unmold the cakes and store
End of Phase
```

Figure 1-2. The three-phase procedure for making fruit cake.

DISTINCTIONS BETWEEN PROCESSES

Some engineers argue that all processes are essentially batch: Continuous processes are really batch processes with long holding or operating times. They reason that since all continuous processes need sequential operations during start-up and shutdown, these processes are not truly continuous. This is true, but that argument does not make them batch processes either. During normal operation a batch

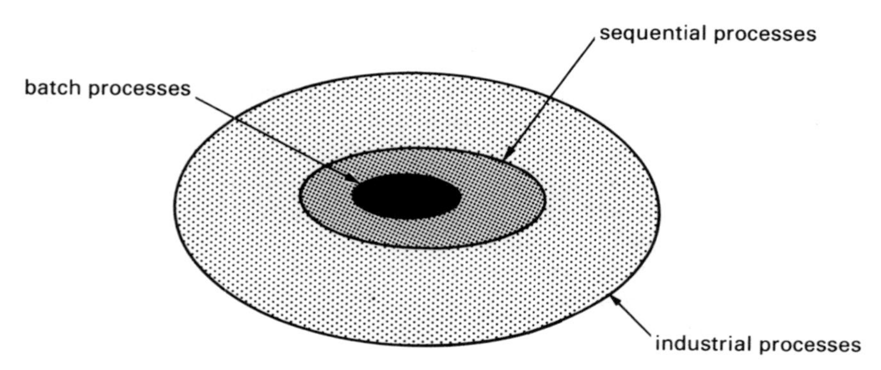

Figure 1-3. Among all processes some are considered sequential, and some of the sequential processes are truly batch processes.

process does not deliver its product continuously, but in batches, whereas a continuous process delivers its product almost continuously. The test of a process lies in the procedure for doubling the amount of product: A batch process must produce twice as many batches of a given size; a continuous process simply must be run twice as long. A batch process is largely sequential, though it may use continuous control functions during certain periods of manufacture. Although continuous processes might need some sequential control, especially during start-up and shutdown, its products are generally manufactured continuously during normal operations. Processes that include batch and continuous operations are discussed later.

So far, we have used the terms "batch process" and "sequential process" rather loosely. They might appear synonymous, but they are distinct: All batch processes are largely sequential, but the converse is not always true. For example, the manufacture of polyvinyl chloride (PVC) or corn syrup, which are batch processes, consists primarily of sequential operations. But the start-up of a steam turbine or the regeneration of the catalyst in a refinery's catalytic cracker unit, though sequential in operation, can hardly be called batch processes. The reason is that neither produces any product in batches. Thus, batch processes are those that manufacture products in batches and are subsets of sequential processes (Fig. 1-3).

CONTROLLING PROCESSES

Controlling batch and sequential processes is quite different from controlling continuous processes. The essential requirement for controlling a continuous process during normal operation is the ability to control to desired set points (within limits) in spite of process and load disturbances. For example, when controlling the crude-oil temperature at the furnace outlet of an oil refinery distillation unit (Fig. 1-4), a controller measures the outlet temperature, compares it with that of a preset value, and generates an output signal. This output signal is

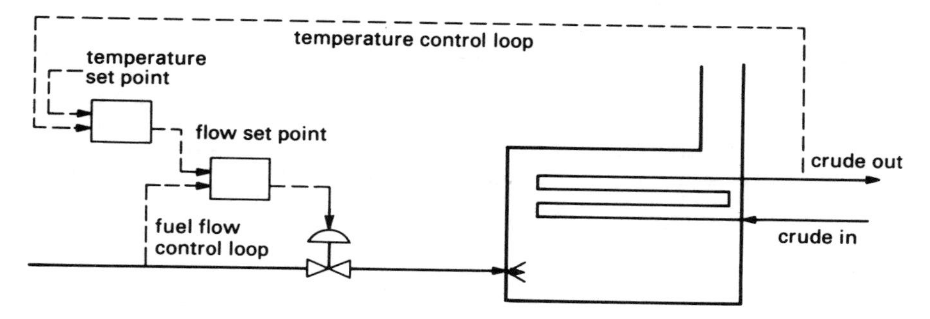

Figure 1-4. Crude-oil temperature control.

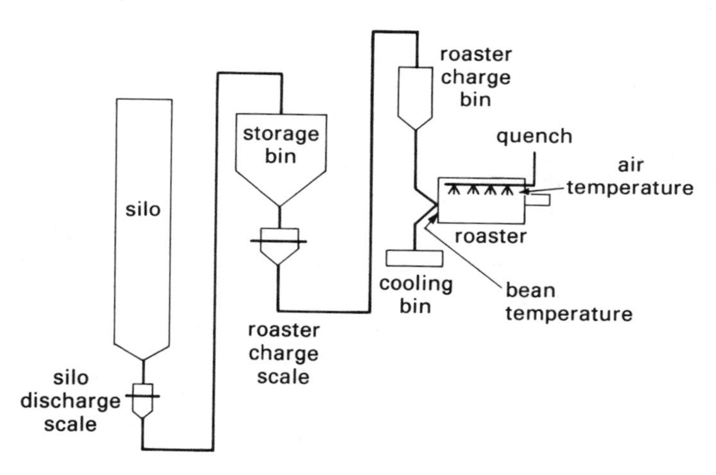

Figure 1-5. Coffee roasting control. *(After Kafer, 1982; copyright © 1982 by the Instrument Society of America.)*

fed as a set point to the secondary control loop, which controls the fuel flow to the burners. Thus, the control system tries to maintain a steady outlet temperature in the face of disturbances such as the changing crude-oil flow, the changing fuel supply pressure, and the changing heating value of the fuel. The controller constants for the primary and secondary loops are adjusted for the best possible control in the face of these disturbances. Once adjusted, the controls are seldom changed.

Batch and sequential controls, on the other hand, usually are based on a predetermined set of functions to carry out a production process. The process is checked frequently and, if necessary, modified. The process usually ends with a finished product or with a controlled system in the desired state.

In a typical roasting process (Fig. 1-5) the coffee beans from a storage bin are weighed into individual batches and sent to charge bins above the roasters (Kafer,

Table 1-1. Example of a Recipe

	Variable Name		
	Grade A	*Grade B*	*Grade C*
Wt. of coffee beans (lb)	200	150	300
Hot air temp. (°F)	325	300	400
Final roasting temp. (°F)	280	280	320
Cooling temp. (°F)	75	75	80

1982). At the start of each roasting, the roaster is charged from its bin. The burner is ignited, and the air temperature is governed by controlling the fuel flow to the burners. The roasting process is continued until the bean temperature reaches a preset value. The flame is then shut off, and the water quench is activated. When the beans have cooled sufficiently, they are discharged to bins for further cooling, storage, and grinding. The roasting of the next batch of beans may then be started.

Controlling the roasting process involves a set of sequential operations that includes measuring the beans, transferring the materials, starting and switching off the burners at appropriate times, and so forth. The process also requires continuous control of the temperature for a limited period of time. Thus, controlling this typical batch process requires sequential operations as well as starting and stopping continuous control functions and manipulating continuous control parameters such as set points and alarm limits.

For controlling the roasting operation, we must include in the operating sequence the steps to be performed if there is a deviation from the normal conditions (e.g., the failure of burners or temperature detectors). Steps that control a process during normal operations are called *normal logic;* those that control contingencies are called *service logic*. The term *hold logic* is used for a service logic operation that stops the process.

Sequential operations may remain essentially the same between batches in one roaster, and between different but similar roasters. However, the weight of beans per batch, the hot air temperature, the final bean temperature, and other variables may vary from batch to batch to produce different grades of coffee or to accommodate different bean types. One set of these variables can then be associated with a particular variation and is then generally called a recipe (Table 1-1).

As previously explained, there is a significant difference between the terms "batch process" and "sequential process." Batch processes are subsets of sequential processes, but the control of batch and nonbatch sequential processes is generally similar. Obviously, both processes need sequential operations to control. Batch processes frequently need recipes to manufacture different grades or products. For nonbatch sequential processes such as regenerating catalysts in a catalytic cracker in an oil refinery or starting or shutting down a material handling system, recipes generally are not required. Thus, except for recipes, controlling batch and nonbatch sequential processes is usually similar.

CHAPTER

2

HISTORICAL APPROACHES AND THEIR PROBLEMS

Before industrialization, batch production of essentials such as food, drink, and clothing was routine. Simple devices for measuring weight, volume, and time were supplemented by the human sight, taste, and touch during manufacturing. Products were delivered in small batches, and because standardization of quality was no great issue, minute measurement deviations were of no great consequence.

Industrial-scale processing introduced instruments for accurately measuring process variables, such as temperature, pressure, and chemical composition, as well as quantity and time. Thus, although batch processes in the early industrial era were largely manual, products of more uniform quality were manufactured. Techniques for controlling these variables in closed-loop fashion were then developed, reducing the need for human effort. These feedback control techniques, which allowed flow, temperature, pressure, chemical composition, and other variables to be controlled precisely to preset values, were ideally suited for controlling processes that would run continuously at the same preset values instead of in small batches. Hence, continuous processes turned out to be generally more economical than their batch counterparts, so process industries were strongly motivated to transform batch processes to continuous operations. However, some processes are not easily amenable to continuous operations. For example:

Processes with feedstock and/or products that cannot be handled efficiently in a continuous fashion, such as solids and highly viscous materials (e.g., those found in iron ore enrichment and separation of wax)

Processes in which the reactions are slow, requiring the reactants to be held in process vessels for a long time (e.g., fermentation for beer and wine)

Processes in which only small quantities of products and/or different grades of the same product are required in limited quantities (e.g., dyestuffs and specialty chemicals)

Processes that need precise control of raw materials and production along with detailed historical documentation (e.g., drug manufacturing)

Processes in which raw materials are available for a limited period during the year (e.g., making sugar and fruit preserves)

Processes that by nature are discontinuous, such as regenerating a catalyst in a catalytic cracker (because of the catalyst's fixed position in the bed)

From the early 1900s to the mid-1960s continuous-process control improved considerably. Improved sensors and controlling devices along with pneumatic signal transmission systems made continuous-process automation a reality, whereas, except for regulatory control functions, batch-process control saw little improvement. Automation of sequential control operations in batch processes became practical only after the introduction of electrical and electronic control devices and signal transmission systems. Electromechanical cam switches and relay logic circuits eventually gave way to microprocessor-based programmable controllers and digital computer systems. Modern tools for controlling batch processes are just as sophisticated as those for continuous processes. However, it is worth discussing the simple tools and systems first (some are still used extensively) before examining microprocessor-based control systems.

MANUALLY OPERATED BATCH PROCESSES

Although the sequencing operation is done by hand in manually operated batch processes, sophisticated instruments may be used to measure process variables. Where needed, regulatory control (e.g., for controlling reaction temperature) may be implemented by a feedback control loop similar to that in a continuous process. Sequencing operations, such as charging raw materials, starting and stopping agitation, enabling and disabling a temperature control loop, and transferring a finished product, are all initiated manually.

In a typical process the quantities of ingredients to be used, the sequencing steps to be followed, and the tests to be made are usually documented in an instruction book. Trained operators follow these instructions, factoring in their knowledge and experience, where required, to carry on the manufacture of a batch. Actions required to correct problems with the plant or the process are not always specified, and the experienced operator must rely on his or her knowledge and judgment.

The accuracy to which quantities of raw materials are measured and to which other process variables are controlled (or set) depends a great deal on the operator, and inconsistencies between operators could cause considerable variation in product quality. The batch record is essentially the log kept by the operator(s), possibly supplemented by recorder charts of process variables.

When batches are manufactured in parallel, the demand of common resources such as process vessels, supply lines, and common pumps could cause problems. A plant manager or supervisor can resolve these problems ahead of time by manually drawing vessel allocation charts. Or the contentions can be resolved as they occur.

A manually controlled batch process has flexible sequencing operations and, due to human judgment, less need for specific instructions for dealing with emergencies. But its drawbacks include the following:

High labor content.

Considerable chance of human error in controlling the quantity of ingredients and other process variables, causing variation in the quality of end products.

Batch records are not always accurate enough to allow determination of the causes for out-of-specification batches.

EARLY AUTOMATIC SEQUENCERS

One of the earliest devices for automatic sequencing is the electromechanical cam switch. The cams are rotated at a fixed speed by an electric motor (Fig. 2-1). The output switches are set on or turned off by the position of protrusions in the cams.

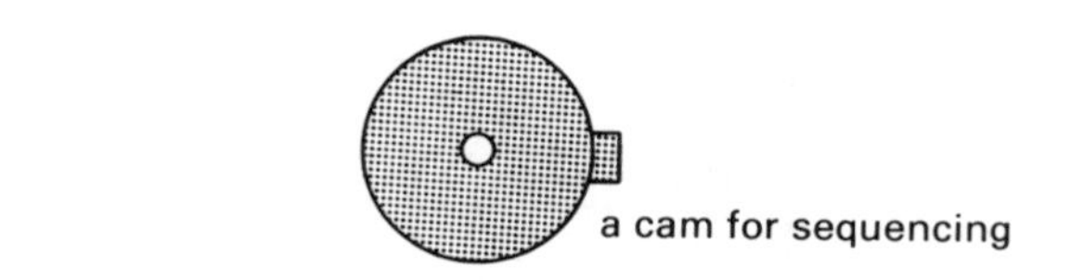

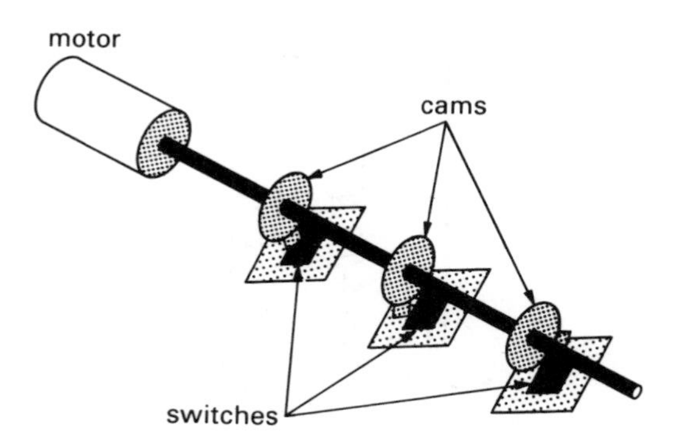

Figure 2-1. A cam sequencer.

The repeatability of this arrangement is extremely high, and the sequencing can be altered by repositioning the cams relative to one another and/or by replacing them by cams with protrusions of different sizes. Where sequencing requirements are relatively simple and require little flexibility, cam sequencers are still favored. They are commonly found in washing machines and dishwashers.

Reprogramming is easier with a perforated drum (Fig. 2-2), where one drum replaces a set of cams. Sequencing is arranged by inserting plugs in the appropriate perforations, which serve the same function as the protrusions in a cam. The drum is rotated by an electric motor, and the activating switches are mounted on a bar adjacent to the drum.

Cam and drum sequencers are limited in the number of sequencing steps they can perform. The limitations vary with the size and rotation speed of the cams or drums. We can increase the number of steps by increasing the diameters of the cams

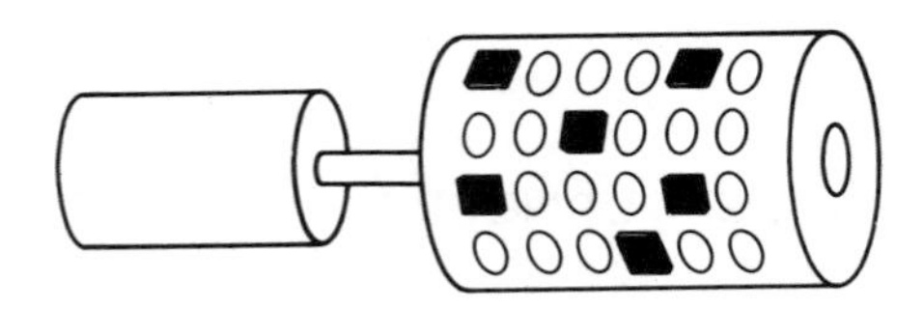

Figure 2-2. A drum sequencer with movable plugs.

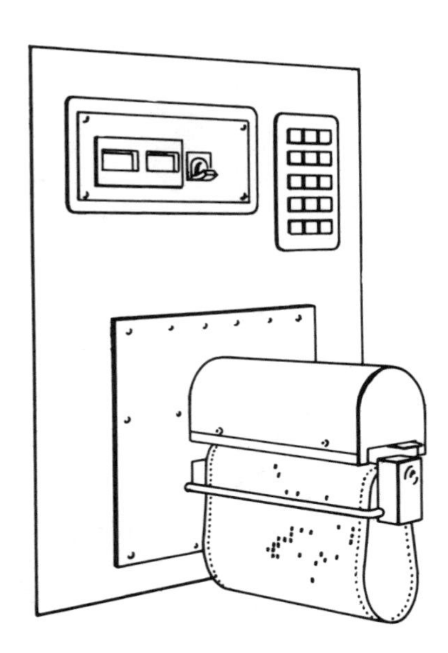

Figure 2-3. A paper tape sequencer.

or drums, or, in a much greater way, by using perforated tapes (Fig. 2-3). The perforations replace the protrusions or plugs and specify the state of the output switches. Another way is by using a clear plastic card (Fig. 2-4) moving at constant speed through an optical reader. Typically, the optical reader consists of a light source and a set of sensors. An output switch connected to a sensor changes state according to whether light impinges on it. Masking tapes are pasted at appropriate positions on the card according to the required states of the output switches. Flexibility is greatly increased by this arrangement because masking tapes can easily be applied or removed. We can use different cards, each with a unique program, to store the required sequence steps for manufacturing different products.

During manufacture most batch processes need, in addition to sequencing, the control of analog process variables. For example, reactants might need to be heated at a certain rate, held at a preset temperature during reaction, and then cooled at a predefined rate. These functions are achieved with an analog controller, where the set point of the controlled variable is adjusted by an arm following the contour of a

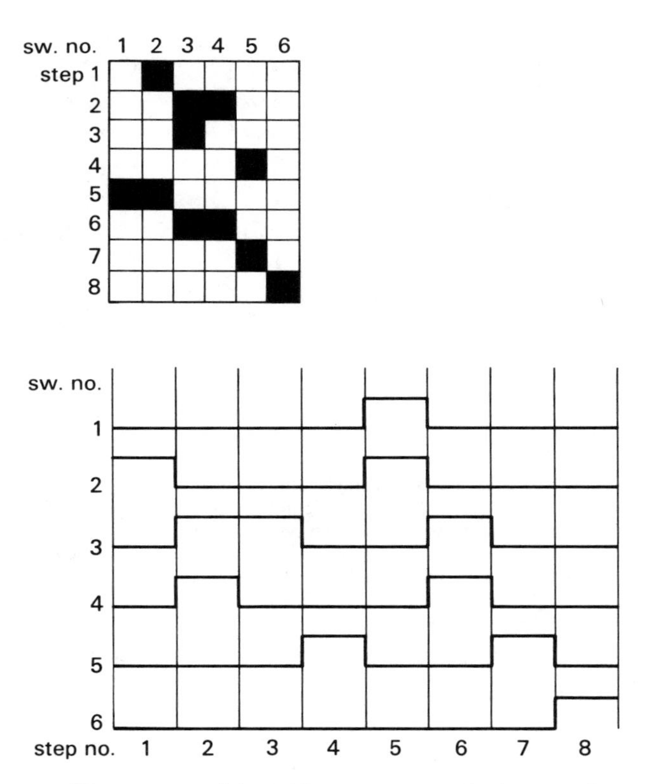

Figure 2-4. Adjustable sequencing with plastic card.

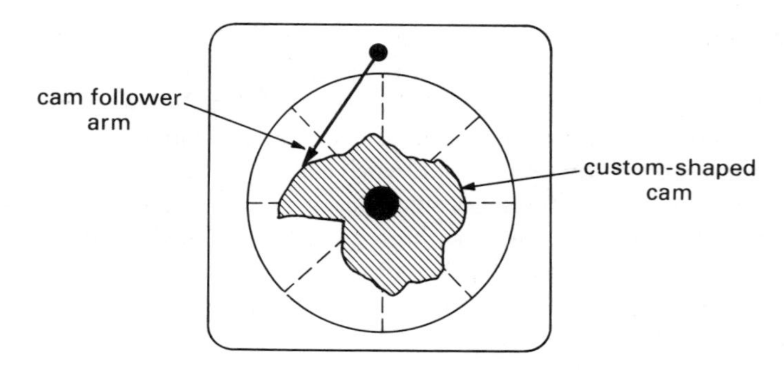

Figure 2-5. A cam set-point controller.

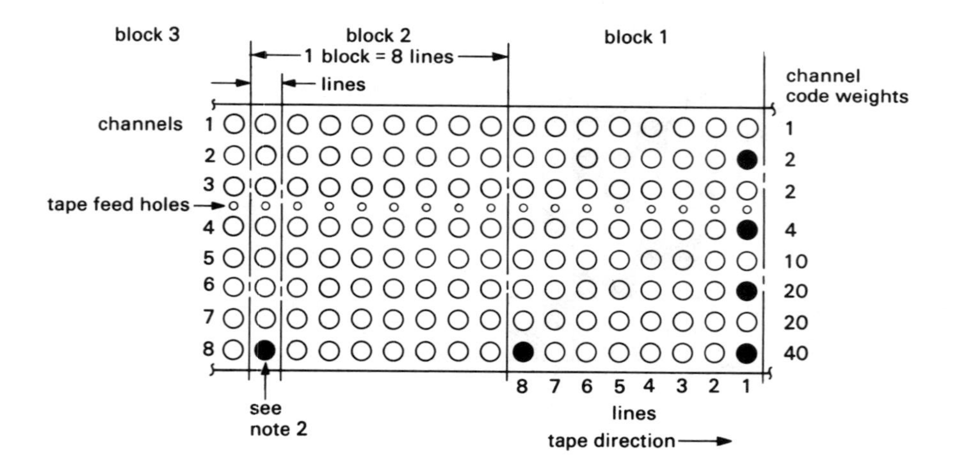

Notes

1. A typical tape code arrangement, where
   line 1 = temperature set point 0–100% in 1% increments.
   line 2 = rate of rise of the temperature;
   0–10% of scale/min, 0–1%/min increments.
   lines 3 and 4 = total batch flow 0–9999 gal.
   lines 5 and 6 = total step time 0–9999 min.
   lines 7 and 8 = 15 on/off type operations.
2. Channel 8 line 8 will signify proper block alignment.
3. Each block represents one step in the process.

Figure 2-6. Typical layout of a block of tape. (*Source: Functional Description of Punched Tape Programming System DR-358; copyright © 1963 by The Foxboro Company, reproduced by permission.)*

specially shaped cam (Fig. 2-5). Unlike the protruded cams, which are used for sequencing, these special cams are cut into shapes conforming to the required set points for the controlled variables.

Automatic sequencing devices, such as cam switches, drums, and perforated or transparent cards or tapes when combined with analog controllers with cam followers are capable of sophisticated batch automation. Lamps on a control panel indicate the states of the output switches and/or the sequence step. In advanced punched tape control systems, the tape reader reads a block of perforations rather than one line of tape. A block is usually eight lines of punched code, with each line specifying the value for an analog variable or a set of digital outputs (Fig. 2-6). These variables are then fed to appropriate converters and controllers to effect the control (Fig. 2-7). The only feedback provided to such a system is a signal that allows the next block of tape to be read. Depending on the application, this "permissive" may be tied to an analog variable reaching a certain value or to a contact input changing its state. This type of arrangement is still used for controlling batch processes in certain industrial environments, notably textile dyeing.

Automatic sequencing arrangements offer considerable advantages over the manually controlled procedures described earlier. They are reduced human effort and increased repeatability of sequencing operations between batches. However, automatic sequencing generates little documentation for batch records, other than perhaps charts of certain analog variables. The biggest drawback of automatic sequencing arrangements is their inability to modify the sequencing steps after feedback from the process. When something goes wrong in the plant or in the controlled process, the sequencing either halts or carries on as usual without taking corrective action. The operator must take the necessary recovery actions manually.

ALARMING AND SAFETY INTERLOCKING

Visual and audible alarms are often provided for manually operated and automatically sequenced batch processes. They are usually independent of the automatic sequencing arrangements. Many analog indicating and controlling devices allow alarm limit setting: When such a limit is exceeded by the process input, an alarm lamp or buzzer is actuated. Safety interlocks independent of the controlling mechanisms are also provided. They are active in the event of a controller malfunction. Interlocking generally refers to establishing a relationship between pieces of equipment in order to maintain safety. A common example of a safety interlock is the float and valve arrangement (Fig. 2-8), which relates tank level and inlet flow.

Alarming and safety interlocking devices used commonly in batch processes include pressure, flow, level, and temperature switches. These switches are simple mechanical or electromechanical devices that on detection of hazardous conditions activate on/off valves, motors, and other plant equipment to bring the process

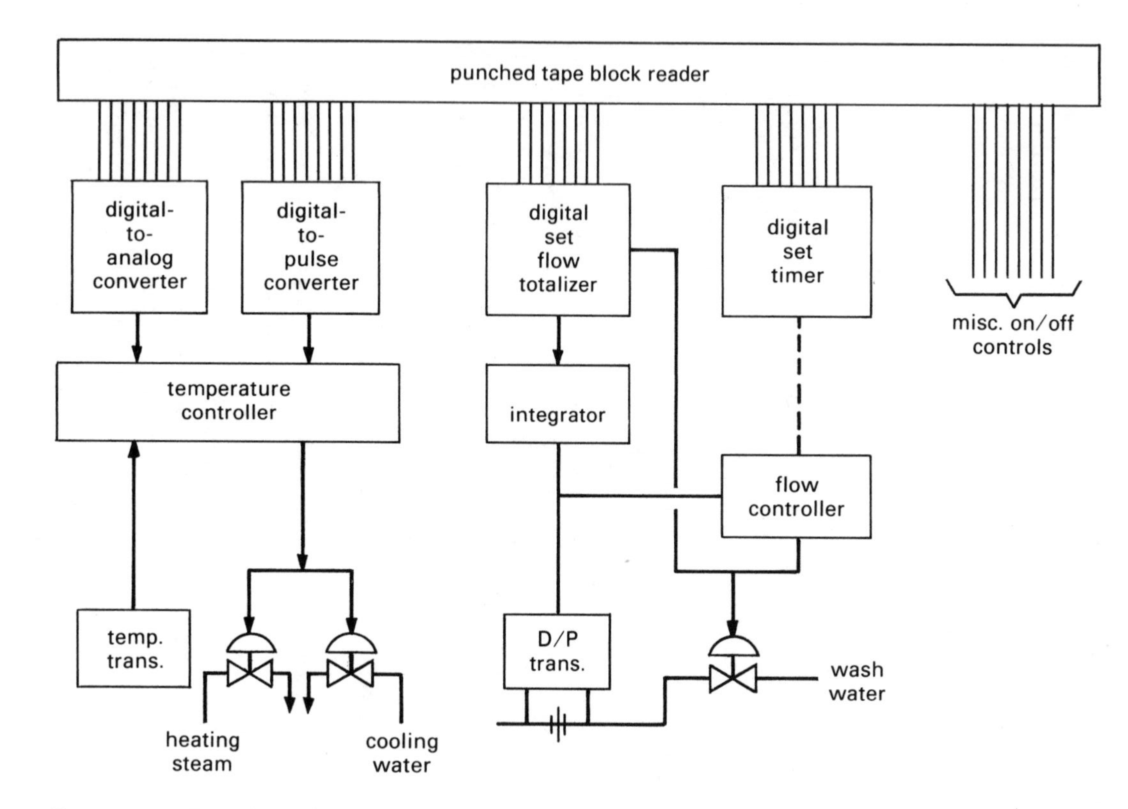

Figure 2-7. Typical punched tape control system layout. *(Source: Functional Description of a Punched Tape Programming System DR-358; copyright © 1963 by The Foxboro Company, reproduced by permission.)*

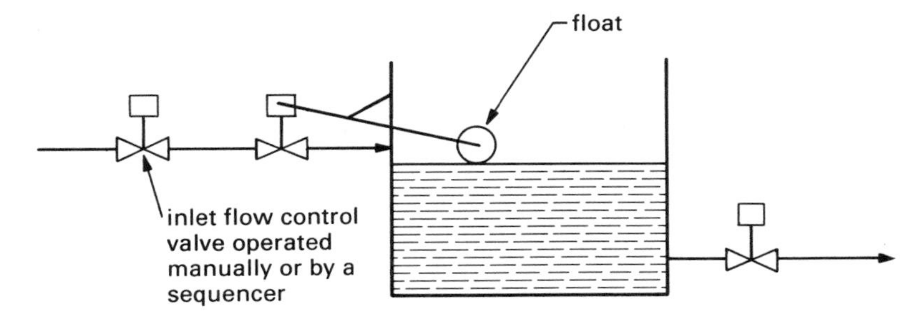

Figure 2-8. Mechanical interlock to prevent tank overflow.

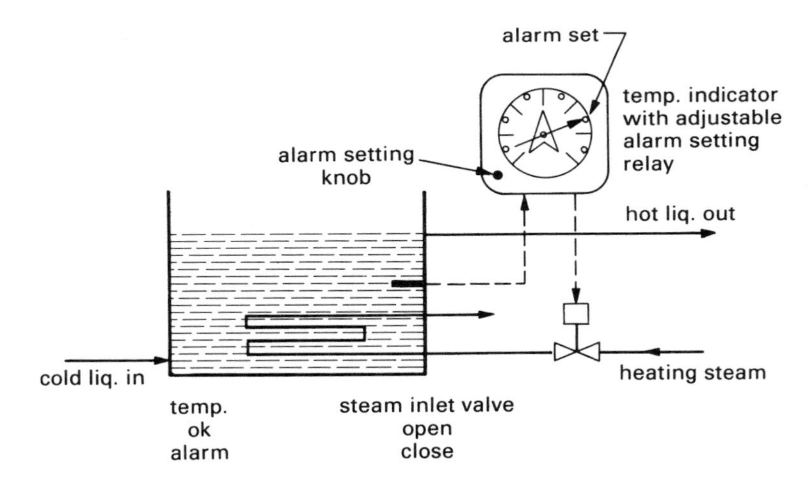

Figure 2-9. Electrical safety interlock to prevent overheating.

back to a safe state. They also may activate visible or audible alarms to draw an operator's attention.

In a typical arrangement (Fig. 2-9) the heating steam in a reactor is cut off automatically before the temperature can reach a dangerous level. Here, a temperature indicator is provided with an adjustable alarm limit setting. If the temperature reaches the alarm limit, the alarm signal triggers a relay that shuts off the steam inlet valve. Other safety interlocks include automatic motor trip on overload, and not starting the pump when the outlet valve is closed.

These alarming and interlocking schemes are critical to the safety of a manufacturing process. As stated earlier, they are generally independent of the sequencing and other controlling arrangements. This setup is desirable because it guarantees that the interlocking system will work after the control system fails, thus preventing dangerous conditions. These schemes are adequate where simple actions, such as simultaneously opening and/or closing single or multiple outputs are required. However, when a more elaborate arrangement is needed (e.g., opening and closing valves, pumps, or agitators, in sequence), these arrangements are generally inadequate because there is little communication between alarming and safety interlocks and the sequencing operations.

CHAPTER
3

COMMON EQUIPMENT FOR BATCH AUTOMATION

The most common devices for controlling batch processes are programmable controllers, digital computers, and distributed batch controllers. All are microprocessor-based digital systems that are quite similar in their inner workings, but they differ considerably in their methods of programming and presenting information for human use. Because of different packaging, the environmental requirements for the normal operation of a digital computer usually differ from those of a programmable controller or a distributed batch controller. This chapter discusses briefly the hardware and the working principles of programmable controllers, digital computers, and distributed batch controllers.

PROGRAMMABLE CONTROLLERS

A *programmable controller* (PC) is a solid-state control system mainly used for interlocking and controlling sequential operations. It can readily replace mechanical logic controllers, such as cams and drum programmers, and electromechanical controllers, such as relay logic circuits. The PC is designed and packaged for high reliability and long operating life in industrial environments with ambient temperatures of 0°C to 60°C and relative humidity up to 95%, in addition to mechanical vibration and electrical noise.

Programmable controllers were originally designed as replacements for electromechanical relays for sequential operations in manufacturing industries. Controllers are now capable of purely sequential control, mathematical and logical operations, and regulatory control of analog variables. This capability has made PCs more useful in process applications where sequential operations predominate.

Historical Background

In the late sixties, General Motors prepared a specification for a reprogrammable system capable of performing simple sequential operations, which were then performed by hardwired electromechanical relay circuits. The primary objective was to reduce the expense of wiring the relays and their rewiring whenever sequential actions were changed. However, this new system had to retain the programming interface based on the relay ladder logic familiar to electrical and control personnel. They also expected the programmable system to operate on the plant floor during wide variations in ambient temperatures, electrical noise, and mechanical vibrations.

Thus, early PCs were little more than relay replacements, but they offered considerable advantages because they were easier to install, they required less space, their diagnostic facilities made troubleshooting easier, and they were reusable for different applications. Naturally, these devices found quick acceptance for controlling repetitive sequential tasks in manufacturing industries.

Since their inception, PCs have been improved considerably. They now include

- Sophisticated operator interfaces based on CRT displays and keyboards, allowing the display of multiple lines of relay equivalent logic simultaneously
- The ability to save and reload logic programs using magnetic tape recorders
- The addition of arithmetic functions and improved instruction sets
- Analog control capabilities
- Variable memory sizes, allowing the user to choose the correct size for each application

Now, what is the distinction between a PC and a process control computer? The National Electrical Manufacturers Association (NEMA) Standard ICS3-1983, part ICS3-304, defines a PC as

> a digitally operating electronic apparatus which uses a programmable memory for the internal storage of instructions for implementing specific functions such as logic, sequencing, timing, counting and arithmetic to control, through digital or analog input/output modules, various types of machines or processes. A digital computer which is used to perform the functions of a programmable controller is considered to be within this scope. Excluded are drum and similar mechanical type sequencing controllers.

Based on this definition, all PCs are computers, but not all computers are PCs. The difference lies in the packaging and programming. Recent improvements in PC functions are reducing some of these differences.

Some recent improvements include:

The ability to program in a high-level language, in addition to or in lieu of traditional ladder logic.

Improved storage of data and reporting capabilities such as batch reports, scheduling, and diagnostic reports.

Improved operator's interface allowing live display of plant data in graphical and tabular formats.

A remote input/output subsystem, allowing the multiplexing of many inputs and outputs via a single pair of twisted wires, thus reducing wiring costs.

The ability to connect PCs to a high-speed communications network, allowing them to communicate with each other and with operators' consoles, printers, and other computers. This ability gives plantwide automation with each section being controlled by an individual programmable controller (distributed control).

Hardware Components

The basic configuration for a PC includes the processor, a power supply, input/output (I/O) panels, and a portable programming panel (Fig. 3-1). The processor

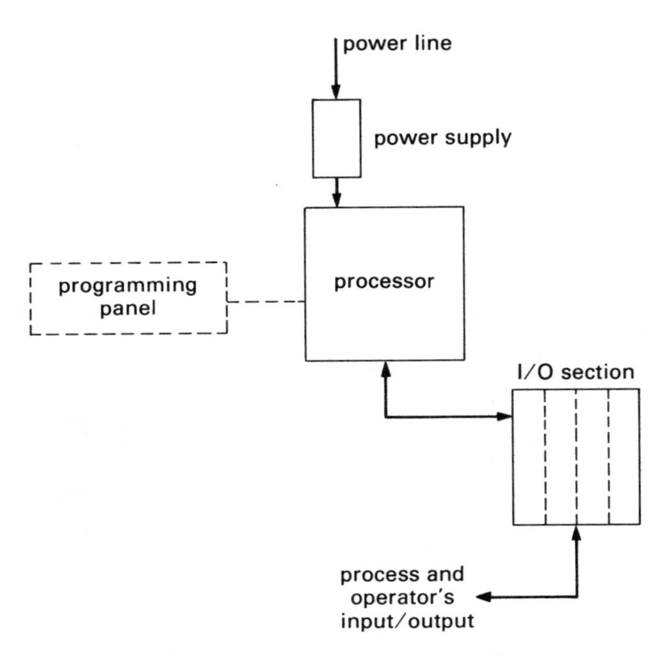

Figure 3-1. Basic programmable controller system.

contains the executive and the user memory (Fig. 3-2). The memory for the application logic and data is normally random access memory (RAM) with read and write capabilities. The maximum user memory capacity in a programmable controller varies enormously between models. For most, it is possible to have less memory than the maximum wherever this is appropriate.

The input and output systems provide the connections between the outside world and the processor, including field devices along with operator inputs and outputs (e.g., switches, thumbwheels, and lights). In early PCs the field devices were only contact inputs and outputs. Now allowable inputs and outputs include binary coded decimal (BCD) and analog inputs of various ranges. Some PCs allow their I/O racks to be located remotely from the processor. These remote racks may contain their own power supplies to drive the I/O and the communications with the processors.

For entering the application-specific, sequence logic program, a portable programming panel with pushbuttons is sometimes used (Fig. 3-3). In recent years,

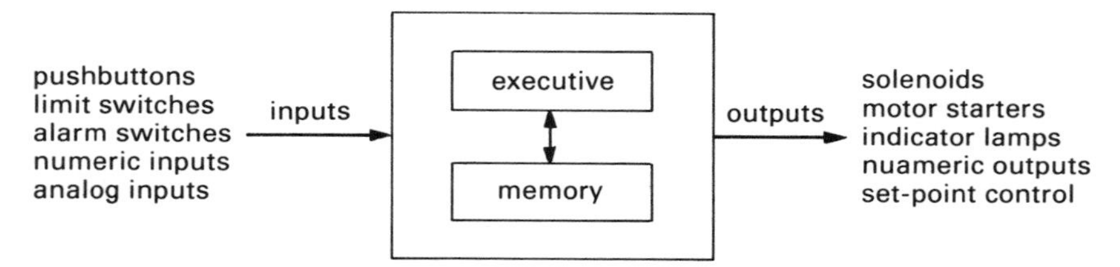

Figure 3-2. Processor in a programmable controller.

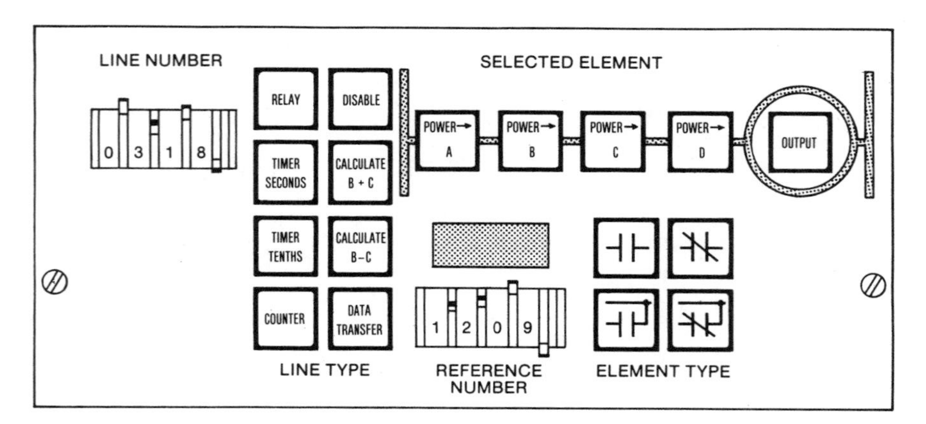

Figure 3-3. Typical programming panel for a programmable controller. (*Source: Cycle Logic System Product Specification SP53002CL-B; copyright © 1975 by the Foxboro Company, reproduced by permission.)*

cathode ray tube (CRT) consoles with keyboards or small personal computers have become common for this task. With these consoles, it is easier to insert and display multiple logic steps.

Additional input and output devices include

Tape recorders (for storing and loading programs)
Operator data entry and output devices such as thumbwheel switches, lights, alphanumeric displays, CRT consoles, printers
Communications devices including telephone and high-speed data link interfaces

Principles of Operation

A PC processor can be divided functionally into four parts (Fig. 3-4): executive, memory registers, I/O status area, and user logic. The executive is a collection of permanently stored programs supplied by the vendor with the hardware system that are generally not accessible to the user. The functions of an executive include

Scanning and storing process inputs in the I/O status area after data conversion, where required
Executing user-defined logic (ladder logic or other programs), including service logic functions, servicing timers, performing arithmetic calculations, and moving data
Storing historical data
Generating control outputs
Generating alarms and process reports
Communicating with other processors and with operator and programming interfaces
Diagnostic checks

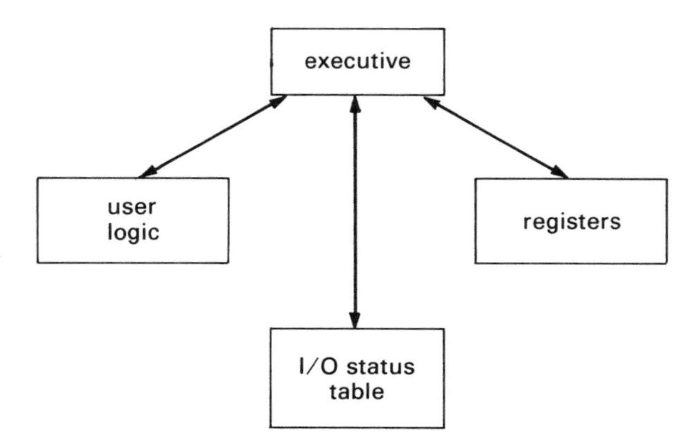

Figure 3-4. Programmable controller processor functions.

Registers are temporary storage devices for various types of data. A register has a fixed number of bits (usually 16) and can represent analog data, the contents of a timer, or a set of intermediate values.

The I/O status area may also be considered as a set of registers containing the status of field contacts, analog inputs, operator inputs, field outputs, and the operator outputs.

The user logic area consists of user developed application programs, written in ladder logic or other programming languages, and the addresses of process and internal inputs and outputs.

During normal operation, the executive interprets the user logic along with the available data in the I/O status area and in the registers to generate the required output functions. Generally, the executive in a programmable controller executes the entire user logic once every cycle, which differs from the normal operation of a general-purpose computer where multiple tasks (programs) may be running in parallel but asynchronously.

The Ladder Logic

Ladder logic or the ladder diagram is the most commonly used format for specifying the control functions in a PC. This section gives a brief introduction to the ladder logic format. Detailed vendor supplied instructions should be used when designing logic for a particular PC.

In a typical ladder diagram the power source is indicated by the vertical uprights of the ladder. The horizontal rungs indicate the relative positions of the switches to control the power (Fig. 3-5). The most common symbols are the switches, which are normally open or closed, and the coils, which represent internal storage or outputs (Table. 3-1). The switches may be arranged in series and in parallel to generate different logic functions (Fig.3-6).

In a ladder diagram each switch is referenced by an address whose content is used to evaluate the output when the logic is executed. These addresses may refer to real process inputs and outputs or to internal flags and registers. Different address ranges ordinarily are used for the different types of variables. Each horizontal rung in a ladder diagram may have no more than one coil (or output) and a limited number of switches. The maximum number of switches varies between systems but is usually between 4 and 16. However, this is not a serious limitation because, when more contacts are required in series, the coil in one line may control a switch in another.

Earlier PCs were limited to performing the relay equivalent functions discussed so far. However, the need for more capability and greater flexibility has given rise to considerable extensions in the types of components in a ladder diagram. Their extent and format vary with PCs of different vendors. Figure 3-7 shows typical examples. The ladder logic program may be entered in the memory of a PC by

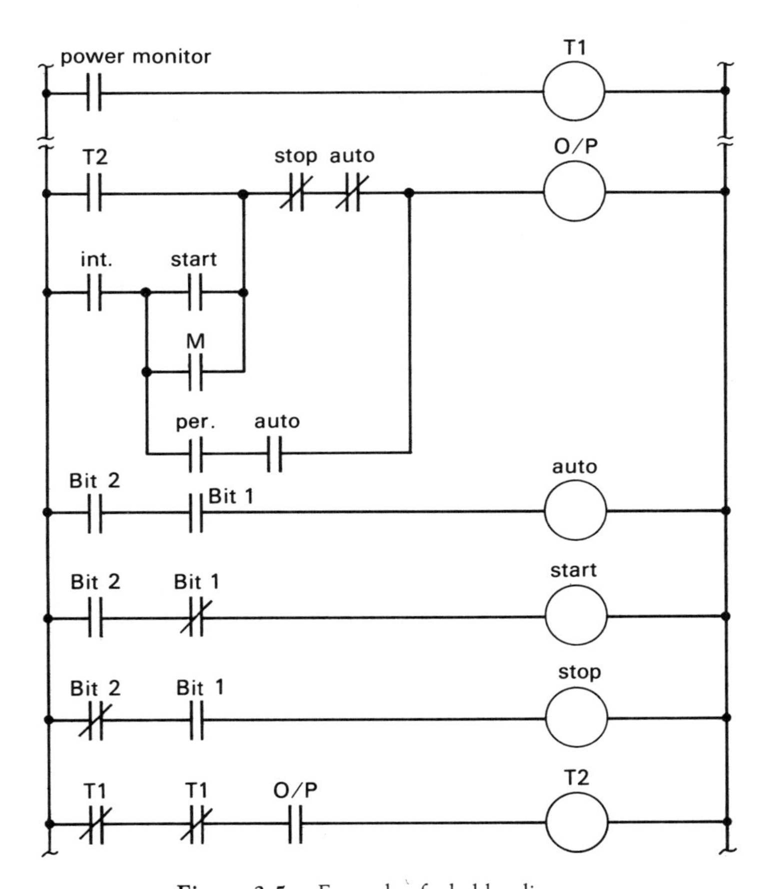

Figure 3-5. Example of a ladder diagram.

using a portable programming panel (Fig. 3-3). A programming panel of this type can enter or examine only one line of a ladder. The line number is first entered by using the thumbwheel; then the appropriate relay contact types are entered for that line.

A programming panel can also be a valuable tool for checking the operation of a ladder logic. It usually allows the logic program to be isolated from field inputs and outputs. Once isolated, the inputs can be manipulated and the outputs can be checked to ensure the proper working of the logic.

More recently, programming panels consisting of a CRT display and keyboard have become more common. They are more user friendly, since they allow multiple lines of logic to be displayed simultaneously. Hence, system checkout and maintenance is rapid and easy. Nowadays personal computers are often used for CRT display and keyboard functions.

Table 3-1. Basic Ladder Logic Symbols

Functions	*Ladder Logic Symbols*
Series Normally Open	
Series Normally Closed	
Parallel Normally Open	
Parallel Normally Closed	
Coil (Output Or Internal Data Storage)	

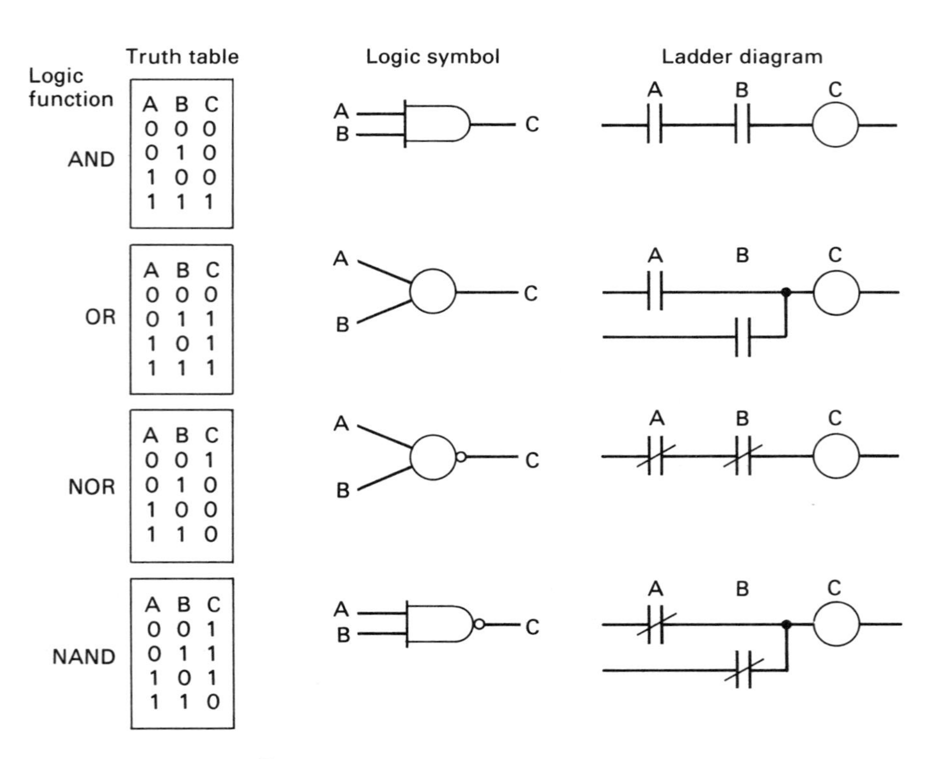

Figure 3-6. Examples of logic functions.

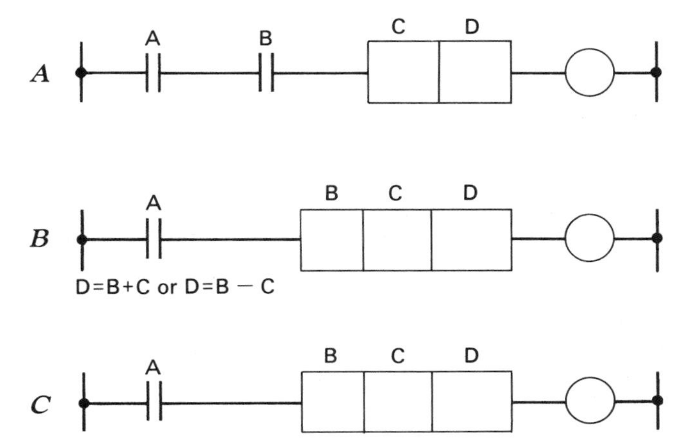

Figure 3-7. Typical extensions to ladder logic. ***A:*** A timer line. The timer is activated when both A and B elements are closed. Element B resets the timer. Element C contains the preset time, and element D the elapsed time. When elements C and D are equal, the coil is energized. ***B:*** An addition/subtraction line. This performs a new calculation at each run cycle when contact A is closed. Depending on programming, either the contents of B and C are added or the contents of C are subtracted from B. In both cases the result is stored in D. The coil is generally energized when the sum exceeds a maximum value or the subtraction produces a negative value. ***C:*** A function line. In a typical function line B contains the address of a source register or the first of a group of source registers. The function code is stored in C, and D contains the address of the destination register or the first of a group of destination registers. Depending on the code specified in C, such a line can perform mathematical calculations, register to register transfer, PID control, and so on. The specified function is executed once per cycle, only when contact A is closed.

High-Level Programming Language

Although a ladder diagram is the most commonly used language for programming a PC, high-level languages are becoming more available (Fig. 3-8). The concept of high-level programming languages is derived from those used in general-purpose digital computers. The statements and the format of high-level languages vary among PCs of different manufacturers, but they are usually similar to the BASIC language, which is common to digital computers.

As they are more process oriented, high-level languages are generally easier to use than ladder logic for specifying sequencing operations. They are especially useful when PCs are used to perform mathematical functions, achieve supervisory control, and perform complex data handling. Modularizing control tasks, and hence their documentation, is also easier. Depending on the system, the high-level language can be interpretive or compiled. Execution is generally much faster with a compiled language.

```
CYCLE BATCH
 LET CRT_PORT = 2
 SELECT CRT_PORT
 LET CRT_FORMAT = 136
 LET STATUS_REG = ASCI_STATUS
 SET INPUT_ENABLE ON
 PRINT 196:C1
 REPEAT EVERY 25 MSEC
   PRINT 196:C1, 197:C1, 6:C1, 2:C1, 'FOXBORO PROGRAMMABLE CONTROLLER'
   DELAY 5 SEC
   PRINT 196:C1, 197:C1, 6:C1, 4:C1, 'TWO ELEMENT BATCH CONTROL'
   DELAY 4 SEC
   SET VALVE_A ON
   PRINT 196:C1, 197:C1, 2:C1, 2:C1, '1. VALVE "A" OPEN'
   DELAY 6 SEC
   SET VALVE_B ON
   SET VALVE_A OFF
   PRINT 197:C1, 4:C1, 2:C1, '2. VALVE "B" OPEN, VALVE "A" CLOSED'
   DELAY 4 SEC
   SET VALVE_B OFF
   SET AGITATOR ON
   PRINT 197:C1, 6:C1, 2:C1, '3. VALVE "B" CLOSED, AGITATOR MOTOR ON'
   DELAY 4 SEC
```

Figure 3-8. Example of a program in a high-level language for a programmable controller. *(Source: Fox Programmable Controller Programming Manual MI 824-150; copyright © 1983 by the Foxboro Company, reproduced by permission.)*

Normally only one programming language is available with a particular model of PC. Thus, the user has little choice of ladder logic or high-level language once the hardware has been selected. When choosing the type and size of PC suitable for an application, the user should also take the programming language into consideration. He or she must take into account the persons responsible for implementing the language as well as those responsible for maintaining it. Admittedly, the ladder diagram is easier to use by electricians and others familiar with it. Computer-type high-level languages seem easier to those not oriented to ladder diagrams. Computer language literacy is rising rapidly due to increasing use of personal computers in all areas of life.

Communications and Networking

With the increased complexity of control systems, the requirements for effective communication between PCs are becoming more and more important. A large plant may be controlled by several PCs, so communication between them is necessary to coordinate their functions.

In the early days the only available method of communication involved connecting a process output of one PC to a process input of another controller. This method was inefficient because it provided only one bit of information for each connection. Many PCs now have interfaces for connecting to local area networks (LANs). These networks support transmission speeds of up to 10 million bits per second (bps). Other methods of communications include RS-232 interfaces with dedicated links or telephone lines (RS-232 is a communications standard published by the Electronic Industry Association). Usually the transmission rate for RS-232 interfaces varies between 300 bps and 9600 bps.

Advantages over Relay Logic

For very small applications (up to 10 inputs and outputs) the initial cost of a PC would probably be higher than that of hardwired relay logic. However, the relay logic would not offer the same degree of flexibility as a PC, which is a reprogrammable device. The electromechanical relays also have long-term reliability problems, since the mechanical parts are subject to deterioration. The testing and diagnostic facilities available with a PC usually allow quicker checkout and debugging of the system than would otherwise be possible. Thus, a PC usually turns out to be economical in the long run, even for smaller applications.

For a large application, even the initial cost of a PC turns out to be less than that for an equivalent relay logic system. Additionally, the user reaps all the advantages of flexibility and reliability associated with a PC.

DIGITAL COMPUTERS

Early digital computers were bulky machines that used vacuum tubes as active elements. Attempts to build digital computers were made even before vacuum tubes were invented. Notable among them is the nineteenth-century analytical engine by Charles Babbage, which used mechanical rotary shafts, and the automatic sequence controlled calculator, by Howard Aiken (around 1940) which used mechanical and electromechanical devices.

Vacuum tube computers needed a lot of electric power, were expensive and slow compared to modern devices, and were not especially reliable. They were used mainly for mathematical calculations and scientific research where continuous operation for a long time was not required. As transistors replaced vacuum tubes, computers became less bulky and more reliable.

Few industrial processes were controlled by digital computers before the transistor era. Also integrated circuits (ICs) and large-scale integration (LSI) have made computers more powerful, less expensive, and smaller than the earlier machines using discrete components. This paved the way for minicomputers, which, unlike their predecessors, required no special room, could be used in industrial control room environments, and made digital process control affordable for smaller processes.

In the early days of process control, digital computer users functioned mostly on their own. Little or no software was available, except perhaps the operating system and a compiler or interpreter for converting instructions in high-level language to machine code. This meant all the necessary application programs had to be written from scratch, which required considerable skill of the user and was quite expensive. Therefore only very large industrial organizations could afford to maintain the nucleus of such a skill to exploit the full benefits of digital process control.

Application to Batch Processes

The first process control packages available with digital computers were used for regulatory control functions. This kind of package allowed a user to specify control requirements as sets of data instead of as sets of programming instructions (Fig. 3-9). The package used these data to set up the required regulatory control functions. Now users could concentrate on designing control strategies rather than on detailed programming issues.

For batch control applications, however, no package was then available. With much effort, users had to design their own when required. Digital computers began to be widely used for batch processes in Europe and in North America in the late 1960s and early 1970s. A notable application was by Imperial Chemical Industries (ICI) in England for manufacturing dyestuffs (Bowen et al., 1975), where a vendor-supplied sequencing language with extensions was used to carry out lower-level batch functions (opening and closing of values, starting motor, etc.). For higher-level functions (i.e., coordination of lower-level functions and supplying recipe values) the user had to design and implement the necessary software.

Early software packages for the control of batch processes include the TABL Language by Taylor Instruments, PCP88 Batch Control System by The Foxboro Company, and PROSEL Language by Kent Automation Systems Ltd. (now Kent Brown Boveri) in England. These batch packages are general purpose and provide the framework (i.e., coordinating the execution of logic, high-level, sequence language; operators' communications, scanning, and driving I/Os, etc.) for designing and implementing particular applications. Significant reductions in the price of computer hardware and the availability of sophisticated batch control packages since then have made computerized batch process control viable for an increasing number of processes.

The Hardware Components

The basic architecture of a digital computer, whether used for process control, data processing, or other applications, remains essentially unchanged (Fig. 3-10). The

PID (PROPORTIONAL-INTEGRAL-DERIVATIVE) BLOCK

Block name	NAME: _ _ _ _ _ _
Block type	TYPE: P I D
Scan period (1 through 255 seconds)	PER: _ _ _
Block index integer (-32 768 to 32 767) PMDS range (1 - 2049; 0 = no action)	BLKIND: _ _ _ _ _ _
Annunciator index integer (1 through 64) 0 if no annunciator. Links blocks to annunciators. System supports up to 15 entries per annunciator	ANCIND: _ _
Measurement or controlled variable	MEAS: _ _ _ _ _ _
Set point	SP: _ _ _ _ _ _
Reset feedback input (PID output is valid)	RSTFBK: _ _ _ _ _ _
Proportional band scaling factor + sign = INC/DEC control direction - sign = INC/INC control direction	PBFACT: _ _ _ _ _ _
Proportional band	PB: _ _ _ _ _ _
Integral (reset) time in minutes	RT: _ _ _ _ _ _
Derivative (rate) time in minutes	DT: _ _ _ _ _ _
Enable reset? Y = Yes N = No	RESET?: _
Output type M = Measurement R = Reference N or <> = Null	SIGTYP: _

Figure 3-9. A typical data base form for a continuous-control package. *(Source: Fox 300 Foxboro Control Package (FCP) MI 802-226; copyright © 1985 by the Foxboro Company, reproduced by permission.)*

main hardware components are input and output units, main memory, auxiliary memory units, arithmetic and logic unit, and control unit.

The input and output units are the interface for the computer to communicate with the outside world. They include keyboard/printers or CRT display units for communications with plant personnel and I/O multiplexer racks for interfacing with the process. Other input and output interfaces include card and tape readers, card punches, and serial data interfaces.

The main and auxiliary memories contain all the programs and necessary data (systems with no auxiliary memory need to have all programs and data in the main

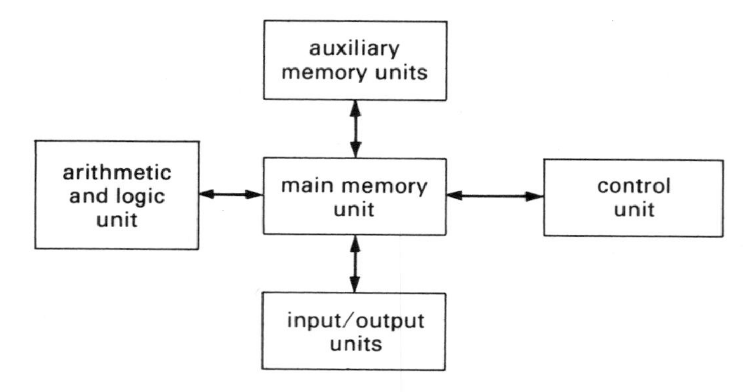

Figure 3-10. Major information flows in a digital computer.

memory). A program must be in the main memory before it is executed. When a program is too big to fit in the main memory, it is broken into sections and each section transferred to the same area of the main memory for execution. The main memory usually consists of solid-state devices. Auxiliary memory units can be fixed- and moving-head discs or drums, solid-state devices, magnetic tape units, and optical storage devices, among other forms. The arithmetic and logic unit handles all arithmetic calculations, logic functions (e.g., AND, OR, NOT), and comparison of variables.

The control unit synchronizes the functions of all these components in a unified manner and transfers data to and from the main memory. It also sends control signals to every other component.

Although the basic architecture of computers used for data processing and those used for process control are essentially similar, there are significant differences between them in terms of peripherals and the way they are programmed to respond to changes in their inputs. In a data processing application, such as payroll preparation, the program execution time is of economic interest but is not critical to the process. All the necessary programs and data are first stored in the computer memory, and the output is produced as a series of paychecks and/or statements. But in a process control application, the states of the process must be scanned at fixed intervals and there must be immediate response to changes in the inputs. In addition, the architecture in a process control computer must allow the orderly suspension of a task currently being performed so that a more important (higher priority) task can be executed before the original task is resumed.

A process control computer, unlike a data processing computer, needs process input and output interfaces, usually in the form of I/O racks with multiplexers. For operator interface a process control computer needs interactive graphic displays, alarm annunciators, process trend displays, and other functions to query and change the real time process variables. Hence the interface for a process control computer is generally different from that for a typical data processing computer.

There is little difference in hardware between a process control computer controlling a continuous process and one controlling a batch process, but there could be major differences in software. Generally there are distinct software packages for regulatory control and for batch control. In applications where only continuous processes are controlled, there is often no need for a batch control package. For controlling a batch process, both continuous control and batch control packages are usually required because batch processes ordinarily need at least some regulatory control functions.

Figure 3-11 shows a typical configuration of a process control computer where all the peripherals are connected directly to the computer. However, in recent years, most vendors have designed systems where computers and peripherals are interconnected by using one or more data highways. A network of one or more computers and their peripherals and other control devices, such as programmable controllers, can then be configured easily (Fig. 3-12). Networks take various shapes and forms (topologies) and allow much greater flexibility for configuring many control devices than would otherwise have been possible.

Batch Application Specification

General-purpose, high-level languages such as BASIC, FORTRAN, and PASCAL, sometimes with extensions, are used in many systems to specify sequence logic (Fig.

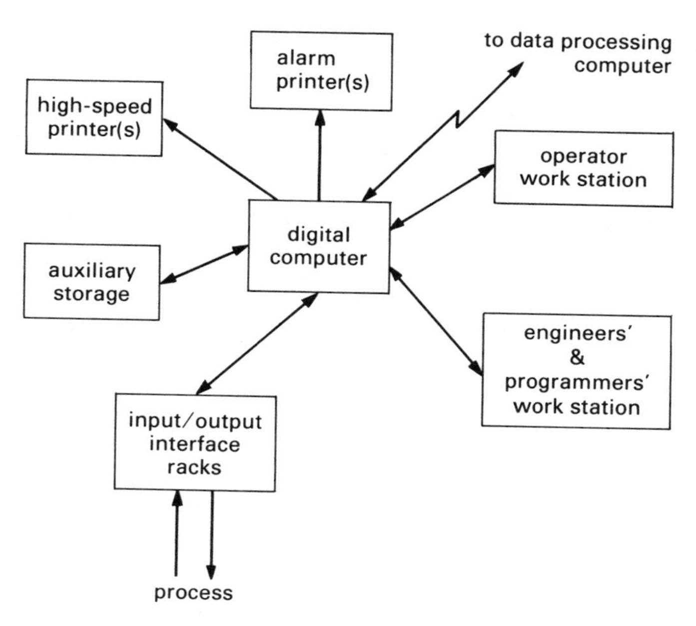

Figure 3-11. Typical process control computer configuration.

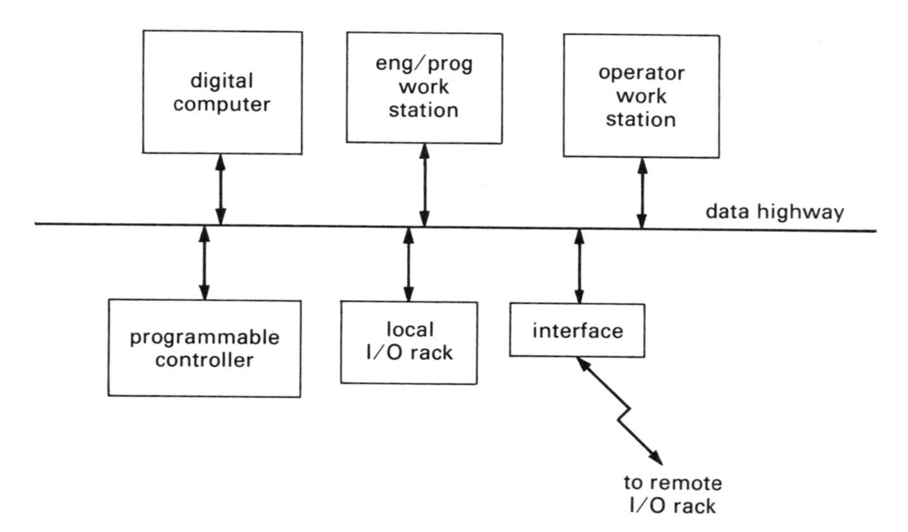

Figure 3-12. Computer process control system in a network.

```
              C STEP #12 - BOIL THE GLUNK
              C
0039            200        TARGET =996.28
              C
              C  SET UP TRACE ARRAY - 'BOILING-01'
              C
0040                       DO 201 I=1,5
0041                       TRACE(I)=STEPB(I)
0042            201        CONTINUE
0043                 CALL  SETPNT (PRESS2,TARGET,IVAL)
0044                       IF (IVAL.NE.0) GO TO 999
0045                       STEP=14
0046            209 CALL   BEXIT (209)
              C
              C STEP #13 - A DUMMY
              C
0047            300        GO TO 999
              C
              C STEP #14 - MORE TESTS
              C
0048            400 CALL   MEASUR (FLOW1,VAL2,IND)
0049                       IF (IND.NE.0) GO TO 999
0050                       ASSIGN 410 TO RENTER
0051            409 CALL   BEXIT (409)
```

Figure 3-13. Batch sequence logic in FORTRAN.

3-13). However, special-purpose, high-level languages specially designed for batch control applications are becoming more common (Fig. 3-14). These languages allow the sequencing steps to be specified in a more process-oriented format than would otherwise be possible. They also allow indirect addressing of process inputs and outputs and other variables so that common logic may be used for multiple process units requiring similar sequencing operations.

The process I/O definitions and other pertinent information for a process unit are stored in areas of main memory or in files in the auxiliary memory area. This area is generally called the *batch data base.* Typically, a data base is generated by entering data in a fixed format (Fig. 3-15) or through a question-and-answer session. A set of programs called "data base generator" reads the entered data, checks validity, and, after appropriate transformation, stores them in data base files.

As stated previously, the sequence logic programs in a computer, unlike in most PCs, do not run continuously. At the start of a batch, the operator usually initiates the sequence logic for the first unit from the operator's console. The sequence logic for the unit started and for all other units started subsequently are time shared,

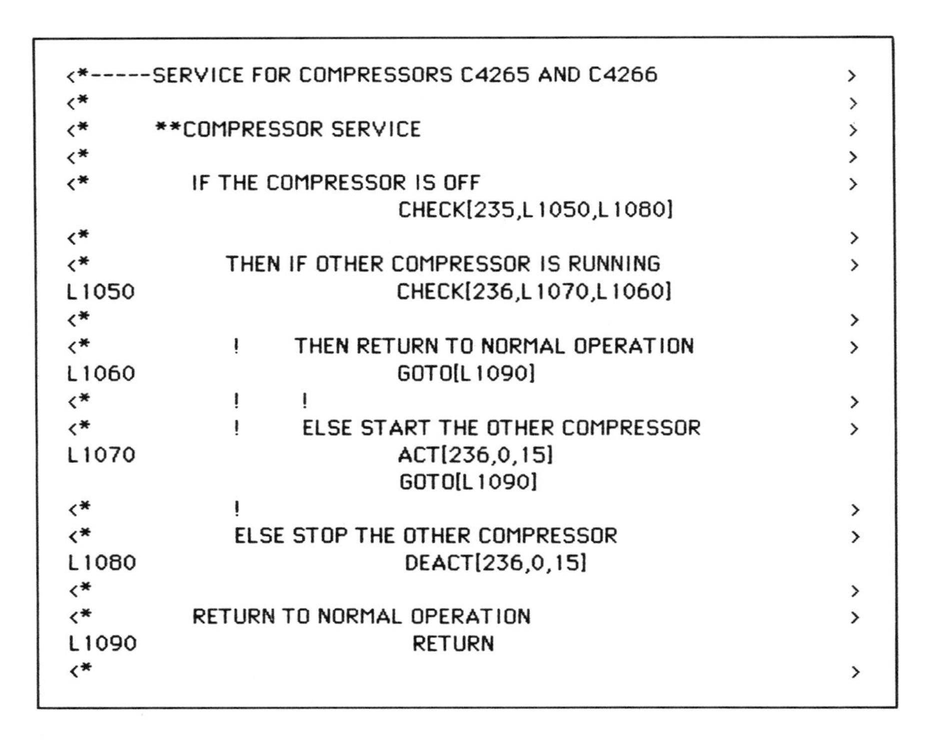

```
<*-----SERVICE FOR COMPRESSORS C4265 AND C4266                        >
<*                                                                    >
<*       **COMPRESSOR SERVICE                                         >
<*                                                                    >
<*          IF THE COMPRESSOR IS OFF                                  >
                              CHECK[235,L1050,L1080]
<*                                                                    >
<*             THEN IF OTHER COMPRESSOR IS RUNNING                    >
L1050                         CHECK[236,L1070,L1060]
<*                                                                    >
<*             !      THEN RETURN TO NORMAL OPERATION                 >
L1060                         GOTO[L1090]
<*             !      !                                               >
<*             !      ELSE START THE OTHER COMPRESSOR                 >
L1070                         ACT[236,0,15]
                              GOTO[L1090]
<*             !                                                      >
<*             ELSE STOP THE OTHER COMPRESSOR                         >
L1080                         DEACT[236,0,15]
<*                                                                    >
<*          RETURN TO NORMAL OPERATION                                >
L1090                         RETURN
<*                                                                    >
```

Figure 3-14. Batch sequence logic in a high-level language.

Figure 3-15. Example of a batch data base form (*Source: Fox 1/A Batch Data Base Generation MI 854-220; copyright © 1984 by the Foxboro Company, reproduced by permission.*)

```
BATCH DATA FORM                                          SWITCH INPUT INFORMATION

IDENTIFICATION

    STYPE  = S /_1   SWITCH INPUT
S W ___ = ____________ ID
    SMESG  = ______________________________
             SMESG = DESCRIPTION OR ID OF SWITCH WITH SAME DESCRIPTION
    S      = _  IS SMESG ID OF SWITCH WITH SAME DESCRIPTION (1 = YES)
    UNIT   = ___   ID OF UNIT IN WHICH SWITCH RESIDES
    STA    = _  STATUS ONLY? (1 = YES)
    KEYLOK = _  BYPASS INHIBIT (1 = YES)
    OSTATE = ______   OPEN STATE MNEMONIC
    CSTATE = ______   CLOSED STATE MNEMONIC
    BLANKS = _  DISPLAY BLANKS IF NOT IN ALARM OR BYPASS? (1 = YES)
```

INPUT SOURCE INFORMATION

```
    INTYP    = __   INPUT TYPE (0 = HARDWARE, 1 = LOGIC, 2 = IMPAC, 3 = SHARED)
IMPAC LIMIT SWITCH (INTYP 2)
    BNAME    = __ __ __ __ __ __ __ __ __ __  SCAN BLOCK ID
    LIMIT    = __   L = LOW, H = HIGH, B = BOTH
SHARED SWITCH ID (INTYP 3)
    SSNAM    = __ __ __ __ __ __ __ __ __ __
```

PHASE INITIALIZATION DATA

```
    SAMAS    = __ __ __ __ __ __ __ __ __ __  ID OF SWITCH WITH SAME DATA

               DES  ALM  SVC  HLD  HDS
P H __ __ __ = __   __   __   __   __    DES = DESIRED STATE
                                               (0 = OPEN, 1 = CLOSED)
P H __ __ __ = __   __   __   __   __    ALM = MESSAGE AND HORN?
                                               (1 = YES)
P H __ __ __ = __   __   __   __   __    SVC = SERVICE LOGIC?
                                               (1 = YES)
P H __ __ __ = __   __   __   __   __    HLD = GO TO HOLD?
                                               (1 = YES)
P H __ __ __ = __   __   __   __   __    HDS = HOLD DESIRED STATE?
                                               (0 = OPEN, 1 = CLOSED,
                                                X = AS IS)
P H __ __ __ = __   __   __   __   __
P H __ __ __ = __   __   __   __   __
P H __ __ __ = __   __   __   __   __
P H __ __ __ = __   __   __   __   __
P H __ __ __ = __   __   __   __   __
```

thus allowing more than one unit to be controlled at the same time. The sequence logics for inactive units are not executed.

A batch package generally contains facilities to specify sequence logic in a high-level language and to generate the batch data base as well as recipe capabilities and basic human-process interface programs; but the user must apply considerable design and application effort specific to his or her process needs. Major application effort is needed in specifying the data base, sequence logic, graphic displays, and for historical data recording and reporting. In the area of sequence logic, the user not only must specify procedures for the normal operation but must also address responses to failure conditions and possibly other asynchronous events.

DISTRIBUTED BATCH CONTROLLERS

A *distributed batch controller* (DBC) is essentially a digital computer specifically designed to control a small batch process or a part of a larger process (Fig. 3-16). Depending on the software packages and programs, a digital computer can perform many different functions and give the user great flexibility in carrying out the control functions. A DBC, on the other hand, allows the user little choice of software packages and programming facilities. All application-related information,

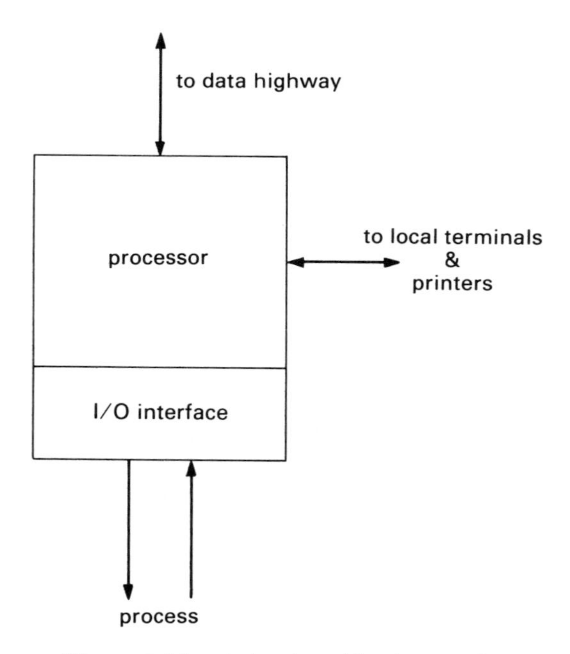

Figure 3-16. A distributed batch controller.

except for sequence logic steps, are entered as preformatted data, and the user has little choice in specifying the display and report formats. This makes the system less flexible, but it makes the application easier and demands less sophistication from the user.

A DBC is usually packaged in a single cabinet with its own process input and output interfaces. Its main memory is limited and it lacks bulk storage capability, so it scans and controls fewer inputs and outputs and sequence logic steps than a digital computer does. However, a DBC can readily interface with other such controllers, host computers, and intelligent operator consoles via networks and data highways. Thus, for controlling large processes DBCs may be configured in a network with intelligent operator stations, host computers, printers, and other such devices (Fig. 3-17). The DBCs are usually installed near their respective process areas. The operators' stations and host computers are located in central control and computer facilities.

The size and capabilities of DBCs vary considerably between products of different vendors. A DBC has its own data base for batch and regulatory controls. The data base and sequence logic steps are either specified locally, by using a directly connected terminal, or down-loaded from a host computer or a remote

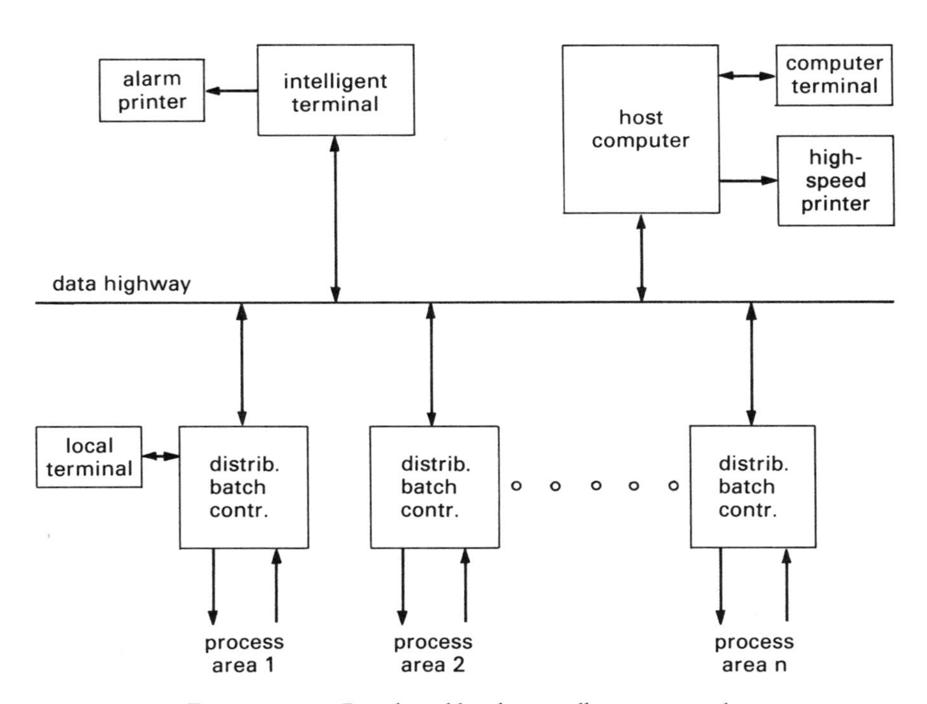

Figure 3-17. Distributed batch controller in a network.

intelligent terminal. There is usually a high-level language for specifying the sequence logic steps.

Distributing the control of a large process over several controllers increases the overall security of the system, since the failure of a single controller would directly affect only part of the total process. Also, most vendors provide options for redundant controllers and a communications network with automatic switchover capabilities, thus increasing system security. The distributed control scheme also increases the modularity and expansion capability of the system. The user may initially configure a small system with fewer controllers that may be expanded as the need arises.

A DBC is generally adequate for safety interlocks, regulatory control, and sequential control for a process area. However, it is not usually capable of process optimization and sophisticated communication with operators and plant management. Intelligent terminals with their own processing capabilities are usually required for sophisticated communication functions. Host computers are generally required for process optimization and management and for collecting large amounts of historical data. A designer has to be careful in not overloading the communications network when designing a distributed control system. When closely related process areas are controlled by multiple controllers, the communication requirements increase substantially. Proper design of the process and the control system (for details see Chapter 13) will minimize this increase.

PROGRAMMABLE CONTROLLERS VERSUS DIGITAL COMPUTERS

As stated before, PCs and digital computers, including DBCs, are microprocessor-based digital devices and, as such, are very similar in their inner structure and working principles. In many cases common IC chips are used. What sets them apart are their packaging, their human-process interfacing, and their programming methods.

The advantages of PCs over digital computers include the PCs ruggedness and suitability for operating in harsh industrial environments. Many digital computers need an air conditioned atmosphere and reasonable freedom from electrical noise and mechanical vibration. (This does not imply that PCs are of higher "quality" than computers—only that the packaging is different.) A PC is also somewhat easier to program and maintain, especially by personnel familiar with the ladder logic, and generally more economical than a digital computer for smaller applications.

A digital computer, on the other hand, is basically a general-purpose machine providing substantially more flexibility in the way it can be programmed. Thus, it is generally possible to use several programming languages in the same system: one for specifying the sequence control; another for doing involved calculations; and yet another for collecting historical data and printing reports. A digital computer also has a distinct advantage over a PC in the area of storing historical data for reporting and trending purposes. The high-level-language feature in a digital computer allows better modularity of the sequence structure by way of main tasks and subtasks and common subroutines. This makes it easier to maintain specification and design discipline for large applications. Digital computer systems also generally provide better operator interfaces than PCs. User-defined graphic displays, with live process values and the facilities for initiating functions and entering data, are commonly provided with digital computer systems.

In conclusion, the digital computer system for batch control provides more flexibility but also demands greater application expertise from the users. The computer also has economic advantages in many medium and large systems. Programmable controllers are generally easier to program and are sometimes more economical for smaller applications. In simpler applications requiring little operator intervention, PCs, or stand-alone DBCs, may be used advantageously. However, for large and complex processes and those requiring change of recipe and/or frequent operator intervention, digital computers, with or without DBCs offer considerable advantages over PCs.

With data highways and networks, control systems incorporating computers, DBCs, and PCs can now be designed. Each component can then be used to its maximum advantage. In such a system the PCs usually maintain safety interlocks; the DBCs usually carry out lower-level sequencing operations for individual units; the computer(s) provides the operator interface, reporting, recipe handling, and coordination of the different units in the process.

CHAPTER

4

AUTOMATING BATCH PROCESSES: BENEFITS AND PROBLEMS

Chapters 1 through 3 dealt with the batch process in general and the kinds of equipment generally used to control it. This chapter outlines generic (that is, common to many batch processes) benefits and problems.

BENEFITS

Numerous benefits derived from automating batch processes are similar to those associated with continuous processes. For example, routine operator tasks are reduced, which in turn reduces the number of operators needed and allows them to concentrate on exceptions, rather than routine, conditions. Batch process automation allows process variables to be controlled within narrower bounds, resulting in more consistent product quality from batch to batch. Because human-process communication interfaces in an automated process are centralized, an operator can obtain all process information more easily and perform necessary supervisory tasks more efficiently.

The automation of a batch process also offers unique benefits. One of them is the shortening of batch cycle time achieved by the reduction of time delays between process steps and by the deployment of process equipment more efficiently. Such a reduction, which effectively increases the capacity of the capital equipment, can itself justify automating many batch processes.

Consistency in product quality between batches is a significant consideration in batch process automation. In a manually controlled process, where relatively small changes in process variables cause readily measurable changes in product quality, consistency is obtained only by blending several batches. Automating these processes gives more consistent product quality, which reduces energy consumption, raw material usage, and capital outlays for storage tanks and blending equipment.

An automated batch process not only promotes consistency and repeatability in controlling the process between batches but also maintains the flexibility of manufacturing different products, or various grades of the same product, by allowing changes in sequence logic and/or process parameters. Thus, in a programmable controller (PC) the process parameters may be changed by modifying a table, and the sequences may be changed by loading an alternative set of ladder logic to change a product. Similar product changes are easily accomplished when computers are used for controlling batch processes.

One problem in the manual control of a batch process concerns responding to contingency situations. An emergency can arise from such equipment malfunctions as the failure of a pump or the sticking of a valve. Similar situations may also arise when the product is off specification or when a plant item or raw material is not available when required. The actions required in such situations range from simple to elaborate. Much training and quick reaction from the operator of a manually controlled process may be required to cope with them. In an automated process, however, most of these predictable emergency situations can be resolved by prespecified corrective actions. Where specific actions are not possible or have not been specified, the automated control system may put the process in a safe state (hold) before asking the operator to take appropriate action. Thus, the actions taken by an automated batch control system are normally equal to or better than those of the best-trained operators and show noticeably faster reaction time.

Another important benefit of automating a batch system derives from improved reporting. The report for a batch may consist of (Fig. 4-1) the amounts of raw materials added, the changes made to process parameters, a list of alarm conditions, and other items. Accurate batch reports are particularly important when analysis is required for batches out of specification. In a manually controlled system the batch report is essentially what an operator writes down as log entries together with the chart records of measuring instruments, if any. The quality of reporting in these cases could vary notably between operators. Also, in an emergency situation an operator's priority is in handling it rather than in making records in the logs. In a fully automated batch system controlled by a computer, accurate and extensive reporting by display or printout is readily obtainable. For other control systems the reporting capabilities vary between these two extremes.

PROBLEMS

On the surface, batch process control might appear simple. All that is needed is a control system with regulatory control functions along with the requisite enhancements to do sequencing for opening and closing valves and starting and stopping motors,

```
BATCH DIGESTER BLOW REPORT-----  5-10-85  , TIME-16:12
--------------------------------


    BLOW NUMBER-----678
    DIGESTER--------3
    SPECIES---------SW


    ACTUAL TIME -----        60# STM      175# IND      175 DIRT      BLOW TIME
                             13:36        14:10         14:46         16:12


    STEAM AND COOK TIME-------156
    SPACING-------------------25.0


    TOTAL STEAM  USAGE-----
    60 #-----  20.478  (KLBS)    175 #----- 32.812  (KLBS)


    CHEMICAL USAGE-----
    WHITE LIQ---- 14517.0  (GALS) BLACK LIQ---- 9049.5  (GALS)


    OPERATING CONDITION-----
                                TARGET            ACTUAL
    DRY WOOD (TONS) -----                         31.62
    MOISTURE (PER CENT)-----                      51.0223
    AA (PER CENT) -----         20.46             20.4766
    LIQ/WOOD -----               4.4               4.45896
    H FACTOR -----             1503.0            1509.0
    WHITE LIQ CONC (#/GAL)---                       .85025
    K NO                        23.0
```

Figure 4-1. Example of a batch report.

conveyor belts, agitators, and so on. This may indeed be true when a batch process consists of a single unit (e.g., a reactor, a simple conveyor system, or a mixing tank) with a simple set of sequential operations and no parallel functions. Moreover, the single unit operates in virtual isolation from extraneous processes. This single unit also requires an uncomplicated procedure for handling all alarm conditions, which may simply involve halting the process and alerting the operator.

However, in reality very few batch processes can be handled so simply. Most have multiple units running in parallel but asynchronously (Fig. 4-2). Some units might be feeding others, and some of the equipment and services may be shared by several units. These units may be producing the same product or dissimilar products at the same time, so the facility to easily change sequence logic steps and their parameters (recipes) is needed. Generally, close coupling is required between units when the output of one unit feeds another. In addition, when two units are running in parallel, interaction between them is generally required for accessing common services and shared plant equipment.

A batch control system is required to continuously monitor the states of the plant equipment and to take appropriate actions on alarm conditions. For continuous control, where the process remains at the same state under normal operating conditions, an alarm condition response may be a single set of corrective actions. But since a batch process is multistate by nature, the same alarm condition at different states in a batch might have different implications and require different corrective actions. The failure of a reactor agitator during the reaction will have a very different implication from that for a similar failure during the cleanout. Thus complexities arise from many possible states in a batch process and, consequently, the number of different sets of corrective actions that may be required to take care of all possible malfunctions (Appendix 1).

In a batch control system the quality of the operator's interface is crucial. In an emergency requiring manual intervention, the operator not only requires the values of relevant process variables but also needs to know precisely the state of the process in order to take corrective action. Sophisticated reporting is also required, as stated earlier, for analyzing an unsatisfactory batch and for various management functions, such as production reporting, process unit scheduling, and energy management. Operator interface is also required for changing the state of a process unit from manual to automatic control, and vice versa, when the system must perform the necessary initialization functions. A sophisticated backup system may be required where control failure is not tolerable. For an effective takeover a backup system needs accurate knowledge of the process state at the time of the failure of the primary controller.

Another problem when putting batch process control into effect lies in the application design documentation. The documentation needs to spell out clearly, among other things, the sequencing requirements for a process, along with the process variables and the interunit communication requirements. This documen-

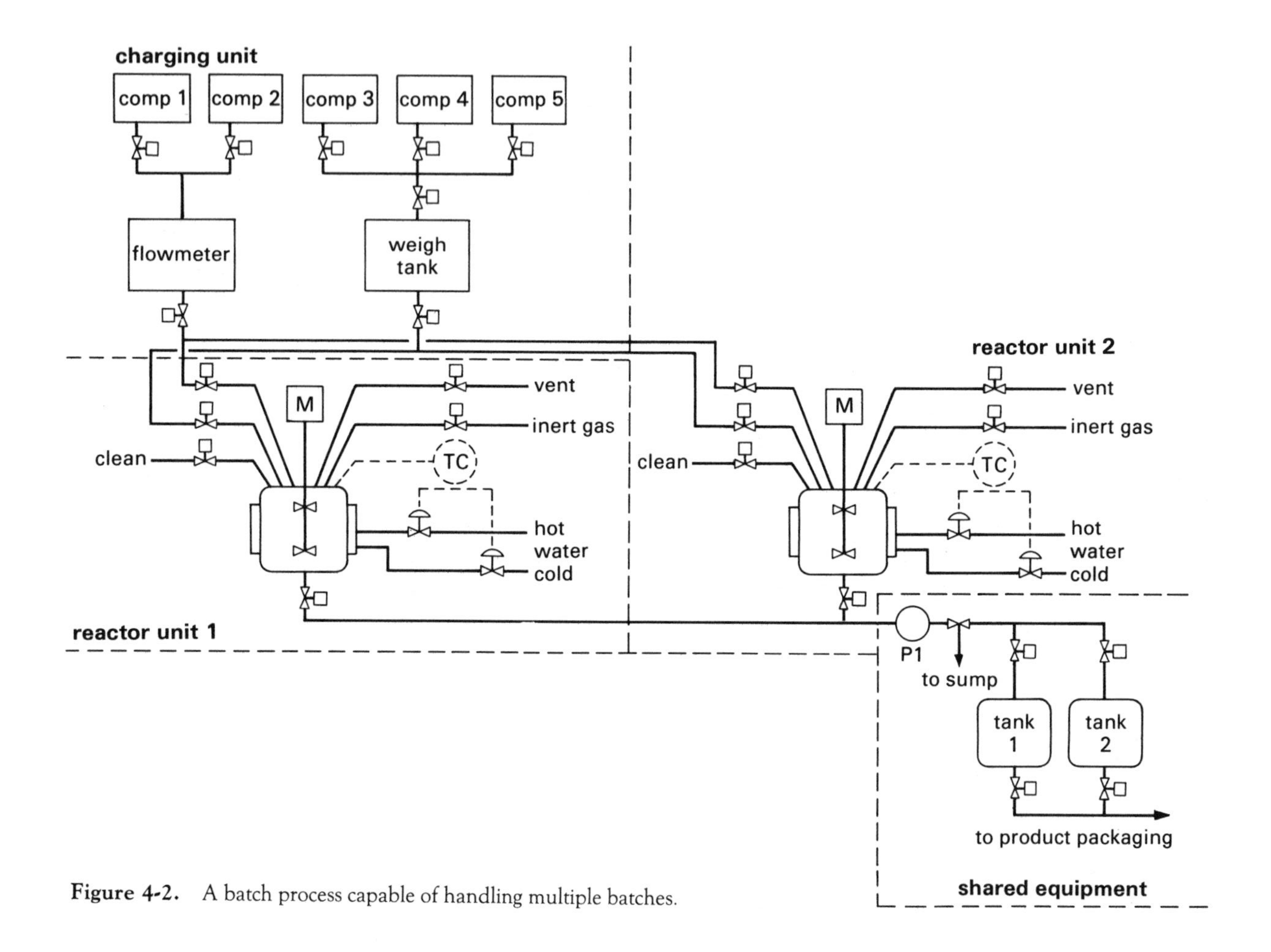

Figure 4-2. A batch process capable of handling multiple batches.

tation requirement is far more extensive than that for continuous control of a plant of similar size. Industry standard terms (e.g., PID controller, cascade loop, etc.) and standard ways of specifying control schemes (e.g., P & I diagrams) exist for continuous control. However, for batch control such standards have yet to be universally accepted. Chapter 9 proposes such standards, see also Appendix 5.

The main design problem for a batch control system concerns integrating all the tasks into a reliable system. Thus designing and implementing batch control systems are considerably more complicated and time consuming than for a continuous-control system for a process of equivalent size. However, both technical and economic rewards are ample, as outlined in the section on benefits in this chapter.

To sum up, batch control problems cannot be solved merely by selecting and using suitable control systems. They may be solved only by using suitable strategies in conjunction with the appropriate systems. Parts II and III essentially deal with these issues.

PART II

AUTOMATED CONTROL OF BATCH AND SEQUENTIAL PROCESSES

CHAPTER
5

BATCH AUTOMATION FUNCTIONS

Consideration of batch-process automation requires an examination of different aspects of batch control, such as (Fig. 5-1)

Handling field signals
Interlock functions
Executing sequence logic functions
Interfacing sequential and regulatory control functions
Flexibility with repeatability
Interfacing with the operating staff and the plant management

These functions are generally necessary in addition to the regulatory control capabilities required in most batch control systems. The need for some of these functions and their relative importance will vary with the complexity and the type of process to be controlled. The ways in which these functions are implemented also differ considerably with the type of control system used.

HANDLING FIELD SIGNALS

An automated batch control system needs an effective means for routine handling and alarm detection of field signals. Field signals include contact inputs and outputs, analog inputs and outputs, and sometimes pulse signals. Since analog input and outputs are generally handled by the regulatory control systems, they will not be considered here in detail. However, a proper interface with the regulatory

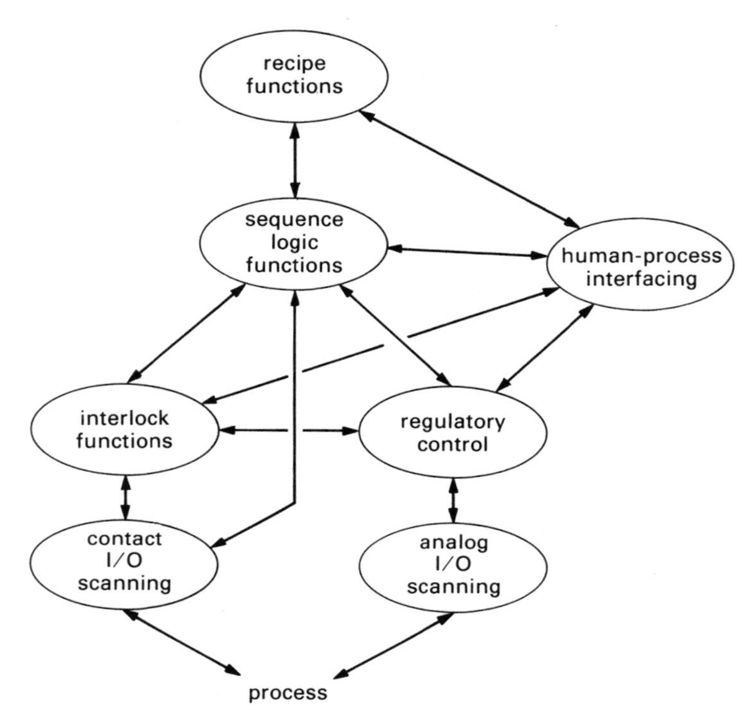

Figure 5-1. Automatic batch control functions. Linkages show flow of data.

control system is vital to any batch control system and will be addressed later. Handling contact input signals requires

Scanning and storing in real time the contact input states
Comparing the current states of these contacts with their expected states
Taking actions, if any, on a change of state (e.g., printing a message)
Taking necessary actions to resolve discrepancies between the current and the expected states.

The scanned contact input states specify the current condition of the relevant items in a plant or process, such as an open/closed valve, a pump (on or off), a tank (empty or not), and so on. These input states may be read either on demand or at predefined intervals. In most automated systems they are read at regular intervals. An interval may be common for all contact inputs, or it may be defined individually. In systems where they are read not at regular intervals but on demand, the requests are generally made by the sequence logic steps. This puts an additional burden on the logic functions, which is not generally favored.

Once these field signals have been scanned, they are compared with their expected states. No action is generally taken when they are found to be the same. An automated batch process may take one or more of the following actions in case of a discrepancy between the actual and expected states of a contact input:

No action (ignore the discrepancy).
Generate an alarm to draw the attention of the operator.
Suspend any further execution of the batch until the discrepancy is removed or until the operator allows it to continue.
Automatically take the corrective actions necessary, such as starting a spare pump when the primary pump has failed.
Put the process in a safe state (i.e., cool down a reactor or stop the conveyor system after the load has been discharged).

Appropriate actions depend on both the process and the criticality of the item in discrepancy at the time of the occurrence. For example, the failure of a cooling water pump during batch cooling might not be as critical as a similar failure when an exothermic reaction is taking place. In the first case, an alarm drawing the operator's attention might suffice, but in the latter case an immediate and automatic corrective action might be required—for example, suspending further addition of reactants and/or adding inhibitors to stop the reaction. Thus, in automated batch control the actions needed on discrepancy must be predefined and have the capability of being changed as a batch operation progresses.

Whereas the contact inputs are generally scanned at regular intervals, the contact outputs ordinarily are driven only when required. The requirements may reflect:

Initialization at the start of a batch or a phase
Normal execution of the logic
Action required in the event of an abnormal condition as detected by the contact input checking routine
Action required due to alarm from the continuous control area

Initialization of the contact output states at the start of a batch is a definite requirement in an automated system. This insures that all the plant items are in their proper states at the start of a batch. Thus, if a valve is supposed to be closed or a pump running at the start of a batch, then it should be driven to that state regardless of its earlier status. Under automatic control the initialization actions when a phase ends and a new phase is started might not always be necessary, since the states of the outputs at the end of a phase are generally the same as those required at the start of the next phase.

However, if the automated system lets individual phases be completed manually, it is possible to start the automatic control of a batch at the beginning of any phase.

In that case the initialization function might be necessary at the beginning of these phases. Thus, when there is a transition from manual control to automatic control, it might be necessary to initialize contact output states. The required initialization states may, however, vary from one phase to another.

Normal batch-process operation requires the manipulation of such plant items as opening and closing valves, starting and stopping pumps, and similar actions. Each of these functions requires the driving of one or more contact outputs.

For abnormal conditions detected by the contact input scanner or by the analog control system, the contact outputs might need to be manipulated. Actual instructions for manipulating the contact outputs may be generated by interlock functions or by exception logic (service/hold) functions.

The operator also might need to drive contact outputs and can do so by manipulating a valve or a pump from the operator's console. This function is legitimate when a process is not in automatic control. However, when the process is under automatic control, the issue of operator manipulation of the outputs requires careful consideration. Such manipulation may cause conflict situations where the operator and the automatic controller try to drive the same hardware in two different states simultaneously.

So far, we have explored the need for scanning and checking the contact inputs and driving the contact outputs. It is possible however, in some control systems to read back the state of the contact outputs. These read backs indicate the state of the outputs at the interface of the control system, but they do not show the actual states of the equipment. Thus, when a contact output is closed to start a pump, the read back of the contact output will ensure that the output signal has reached the interface correctly. However, this will give only scanty information concerning the malfunction of the pump itself. The primary value of the read-back capability occurs in a network system, where it might be possible for more than one controller in the network to drive the same contact output. Read back alerts the control system to contentions so that appropriate actions may be taken.

The most reliable way to ensure correct manipulation of a plant item is to check appropriate feedback inputs. A feedback input can directly indicate the state of a plant item, such as a limit switch on an open/close valve. The state of a plant item can be indicated indirectly by a process parameter—for example, for a pump, where the presence or absence of flow or the outlet pressure will show if the pump is running.

INTERLOCK FUNCTIONS

In a simple example, a device like an open/close valve can be opened or closed by opening and closing a contact output (Fig. 5-2). The valve has two limit switches indicating the state of the valve (open or closed). If, for example, the valve normally takes 5 s to travel from fully open to fully closed, then the relationship between the

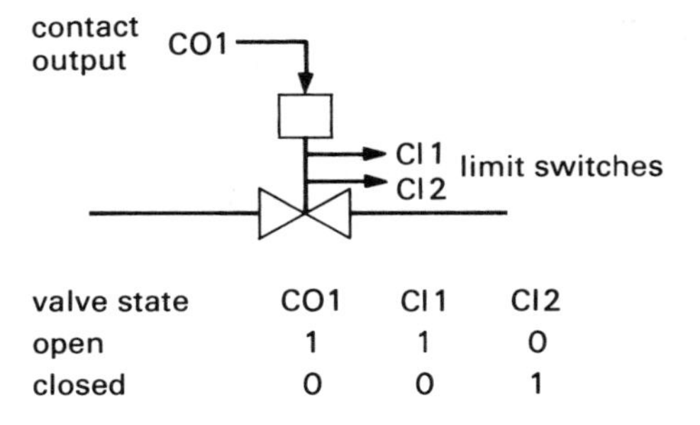

valve state	CO1	CI 1	CI 2
open	1	1	0
closed	0	0	1

travel time 5 s

Figure 5-2. Truth table for an on/off valve.

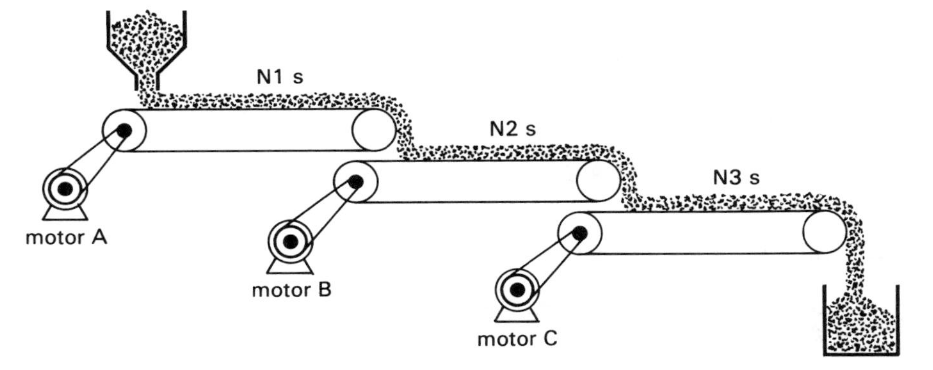

Figure 5-3. A conveyor system.

contact inputs and the output may be monitored by an interlock function. The valve is open when the contact output is closed (1), the contact input 1 is closed (1), and the contact input 2 is open (0). When closed, the states of these inputs and outputs are reversed. When traveling, the contact output could be opened or closed, depending on the direction of the travel; and both the contact inputs may be open. However, the traveling state does not normally exceed a fixed period of time, 5 s in this example. If it does, that indicates some problem with the valve or limit switches, or in the connection between the control system and the valve. The operator or the control system will have to attend to this problem; if it is critical, it could require immediate automatic action, such as shutting down the unit.

More complicated interlock functions may define the relationships among several plant items. For example, in a conveyor system where the conveyors feed to one another, starting and shutting down each conveyor needs to be synchronized (Fig. 5-3). This hypothetical example has three constant-speed motors, each driving one conveyor belt. Motor B should be started later than motor A, the start time depending on the speed and the length of conveyor A. Similarly, motor C must be started after motor B. When the conveyor system must be shut off, the motors need to be switched off in the same sequence to ensure that no material is left on the conveyor belts. There should also be provisions for the emergency shutdown of the conveyor system on operator's request or when one of the conveyor motors fails. The logic for starting and stopping the conveyor system can be described as follows:

Starting the conveyor

```
START MOTOR A
WAIT UNTIL MOTOR A IS RUNNING
OPEN HOPPER DISCHARGE VALVE
WAIT N1 SEC
START MOTOR B
WAIT UNTIL MOTOR B IS RUNNING
WAIT N2 SEC
START MOTOR C
WAIT UNTIL MOTOR C IS RUNNING
INDICATE THE CONVEYOR IS RUNNING
```

Stopping the conveyor

```
SHUT HOPPER DISCHARGE VALVE
WAIT N1 SEC
STOP MOTOR A
WAIT UNTIL MOTOR A HAS STOPPED
WAIT N2 SEC
STOP MOTOR B
WAIT UNTIL MOTOR B HAS STOPPED
WAIT N3 SEC
STOP MOTOR C
WAIT UNTIL MOTOR C HAS STOPPED
INDICATE THE CONVEYOR HAS STOPPED
```

Emergency stop

```
SHUT HOPPER DISCHARGE VALVE
STOP MOTOR A
STOP MOTOR B
STOP MOTOR C
INDICATE THE CONVEYOR HAS STOPPED
```

Alarm actions
Motor A stopped

```
TURN OFF POWER TO MOTOR A
SHUT HOPPER DISCHARGE VALVE
WAIT N2 SEC
STOP MOTOR B
WAIT UNTIL MOTOR B HAS STOPPED
WAIT N3 SEC
STOP MOTOR C
WAIT UNTIL MOTOR C HAS STOPPED
INDICATE THE CONVEYOR HAS STOPPED
```

Motor B stopped

```
TURN OFF POWER TO MOTOR B
SHUT HOPPER DISCHARGE VALVE
STOP MOTOR A
WAIT N3 SEC
STOP MOTOR C
INDICATE THE CONVEYOR HAS STOPPED
```

Motor C stopped

```
TURN OFF POWER TO MOTOR C
SHUT HOPPER DISCHARGE VALVE
STOP MOTOR A
STOP MOTOR B
INDICATE THE CONVEYOR HAS STOPPED
```

Thus, in this example, the interlock functions can be defined to ensure the proper starting, running, and shutdown of the conveyor system. All of these actions can be implemented by combinational logic (Figs. 5-4 and 5-5), but this gets quite complicated for nontrivial applications.

In the examples for the on/off valve or the conveyor system, the interlock functions are independent of the process or end product. Thus, interlock functions generally depend on the inherent physical characteristics or the safety aspects of the plant or equipment rather than on the manufacturing process or the product. Interlock functions may specify the states of contact outputs and the expected states of contact inputs at a given time. They may also inhibit their checking for a limited time during changes. Scanning of contacts and alarming are generally implemented by the I/O handling system (see *Handling Field Signals*). Corrective actions may be defined in the interlock function itself or addressed by the sequence logic functions (see the next section). For complicated interlock functions with

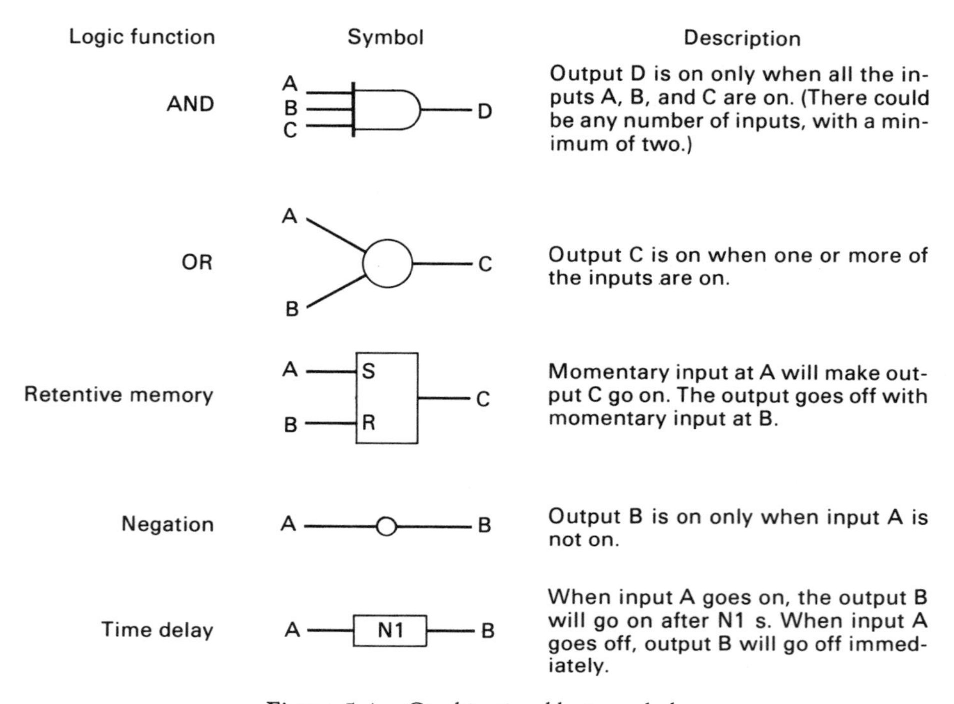

Figure 5-4. Combinational logic symbols.

numerous steps and time delays, sequence logic functions are simpler to use. However, when safety and/or fast corrective actions are required, interlock functions are generally preferred because they are active all the time and are "closer" to the process.

EXECUTION OF SEQUENCE LOGIC FUNCTIONS

The execution of sequence logic functions is central to the control of most batch processes. Sequence logic functions specify how and in what sequence plant operations are to be carried out to deliver the desired product. Examples of sequence logic functions include opening or closing valves, starting and stopping pumps, and setting and changing alarm limits for a temperature control loop. Not all sequence logic functions address control functions in the same way. They vary from a single action, such as opening or closing a contact output, to a series of actions to accomplish a specific function. One contact closure could be required to start an agitator, for example, whereas charging a specified material to a reactor might involve opening and closing valves in a predefined sequence, along with metering the material being charged.

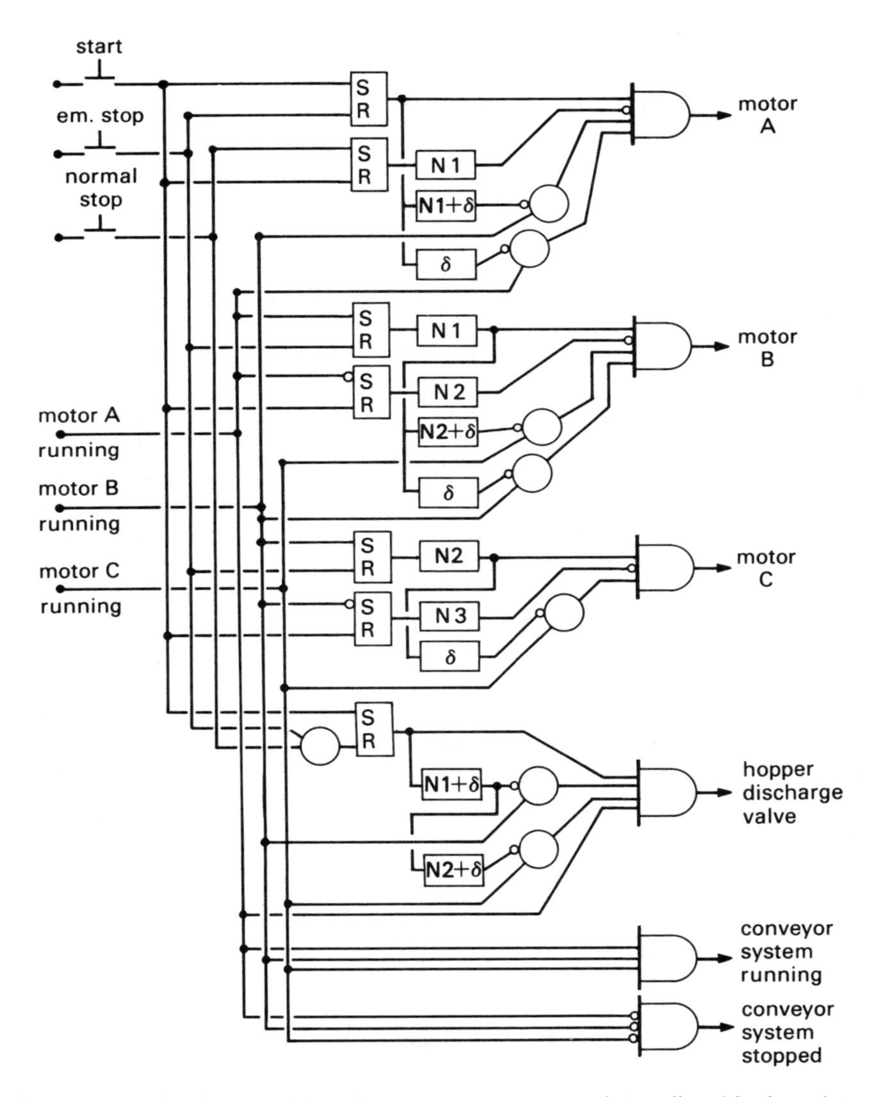

Figure 5-5. Combinational logic for conveyor system control. $\delta =$ allowable elapsed time for a motor to be fully running after the power has been applied to it.

A sequence logic step can directly manipulate plant items via the field I/O handler, but this function is often accomplished by interlock or regulatory controls. Thus, a pump can be started directly by a sequence logic instruction to close the output switch. If, however, starting the pump involves conducting a series of checks, such as the status of the pump motor and delayed read back of the pump outlet pressure, then that can be accomplished by enabling the appropriate interlock function. An analog function such as a temperature control loop is initiated through regulatory controls.

Sequence logic functions required for controlling a batch process must include the necessary steps for responding to abnormal plant or process conditions. Abnormal conditions occur due to

Equipment malfunction (e.g., pump failure, valve sticking)
Plant emergency (e.g., fire)
Product problems (e.g., product off specification or excessive foaming)
Unavailability of plant item or raw material (i.e., a common discharge line in use by another reactor, air supply pressure low, or product storage tank full)

When each of these abnormal conditions occurs, the normal sequence logic functions must be suspended, and service logic steps (Fig. 5-6) must be invoked, in addition to any interlock action that might have taken place. The service logic may vary from one process to another and may differ for various phases of the same process. In the simplest cases the service logic may involve suspending further action and calling the operator's attention to the problem. Normal sequence operations will then be resumed only on the operator's instruction. However, in most automated batch control systems, just drawing the operator's attention may not always suffice. The control system is expected to take care of most routine failures. When a pump fails, the backup might have to be started immediately without manual intervention. A cooling water supply failure during an exothermic reaction might require adding inhibitors. Thus, different sets of service logic might be required to resolve the different types of failures, and in all but the simplest processes the service logic requirements tend to be much greater than the normal logic requirements.

The normal and service logics required to control a process will vary with the process. For example, the logic for controlling the process for dyeing cloth would

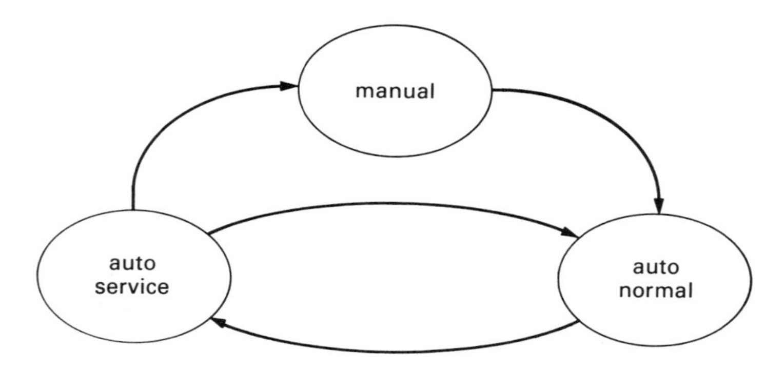

Figure 5-6. The process unit states.

differ significantly from the logic for filling milk cartons. However, when two or more similar processes run in parallel, their sequences would be very similar, even as they address different sets of plant items such as valves, pumps, and reactors. In batch control systems where the inputs/outputs are addressed directly, as in a programmable controller, multiple sets of logic generally will be required to control several similar processes. Thus, to control two similar reactors running in parallel, two sets of sequence control logic are required. In other batch control systems where indirect addressing of the inputs and outputs is possible, as is typical with a computer, the same logic could be used to control two similar reactors by addressing different sets of data for the two reactors.

INTERFACING WITH THE REGULATORY CONTROL FUNCTIONS

Most batch control systems require regulatory control as well as sequential control. In a typical batch process in the chemical industry, the reactants must be heated to a specific temperature and maintained at that temperature for a certain period until the reaction is complete; then the process is cooled. The flow rate of the reactants and finished products might need controlling. They are examples of regulatory control functions.

Regulatory control functions must be enabled and disabled at appropriate times during the manufacture of a batch. Regulatory function parameters also must be inspected and adjusted when they are enabled and also when they are active. Parameters requiring inspection and manipulation include

Measured values of analog variables
Set points
Alarm limits
Such control constants as proportional band and integral time
Closing and opening control loop cascades

Interfaces are also required for the regulatory control function to notify the batch control system of a violation of alarm or other limits so that necessary sequential or interlock actions can be initiated. For example, if the reactor temperature during an exothermic reaction is too high, sequential control actions to slow or stop the reaction will probably be initiated.

Still another example is where the contents of a local raw material tank (day tank) are measured by an analog measuring instrument; opening and closing the inlet valve could then be initiated by sequence logic. Here the sequence logic, rather than periodically checking the level, can set regulatory control alarm limits for empty and full conditions so that the regulatory function can inform the sequence function when to open or close the inlet valve to the tank.

THE REQUIREMENTS OF FLEXIBILITY WITH REPEATABILITY

Repeatability is an essential requirement in an automated batch control system (Chapter 4). When executing sequence logic functions, repeatability ensures the consistency of the end products and represents one of the principal advantages of automated batch control over manually operated batch plants. Without repeatability, controlling the quality of the products will be more difficult, thus increasing rejection rates and the need for blending.

However, along with repeatability comes the independent requirement for flexibility of manufacture of different grades of the same product. For example, a plant might manufacture different grades of polyvinyl chloride (PVC). The manufacturing process for each grade is essentially the same, but the processes might differ in the proportions of raw materials and catalysts and in process parameters such as temperature, pressure, and reaction time. Different sets of sequence logic steps can be specified to manufacture each grade of product, but it would be a waste of engineering effort and control system resources. It is better to specify a common set of sequence logic functions for all grades and then list those variables that differ for each grade of the product. Each set of these variables is called a recipe, and each variable in a recipe is called a *recipe item* (Table 5-1).

In a set of simple recipes, the items are ordered identically for all the recipes. Thus, the third item for recipe 1 will be for the same variable as for recipes 2 or 3. For each recipe item a pair of limits (upper and lower) should be specified. The control system could then prevent the entry of grossly erroneous recipe item values.

Recipe items generally include process variables (temperature, pressure, amount of material, time for reaction, etc.) and plant items (storage and raw material, tank numbers, etc.). A recipe is generally selected at the start of a batch. In more complicated recipes different grades of products might have different orderings of the recipe items. Recipes may specify changes in the order in which the sequence logic may be executed for different grades of products. Some of these advanced aspects of the recipe are considered in Part III.

Recipes are generally not required in nonbatch sequential processes such as catalyst reactivation in a catalytic reformer or starting and shutting down a conveyor system. Recipes also are not required for batch processes intended to manu-

Table 5-1. Part of a Recipe

		Limits		*Recipe No.*		
Item No.	*Description*	*Low*	*High*	*1*	*2*	*3*
1	Amount of reactants (lb)	500	1,000	600	750	820
2	Reaction temperature (°F)	170	208	188	192	207
3	Temperature ramp rate (°F/S)	1.0	3.0	2.5	2.5	2.8

facture the same grade of product all the time; where used, they do facilitate engineering changes.

INTERFACING REQUIREMENTS

A batch control system interfacing with the operating staff and the plant management must provide adequate information about plant and process performances and the capability of changing process parameters and strategies. Ideally, the information presented by a control system should include enough detail for its stated purpose without an abundance of extras that might otherwise distract or overburden. Thus, in a multiple alarm condition the control system should be able to draw the operator's attention to the most critical situations first, preferably accompanied by the recommended operator actions. Safeguards are also required to make the system foolproof so that commands that are unacceptable for safety or process reasons are rejected. Again, these checks must not seriously reduce system flexibility. Thus, the facilities offered by a control system interface in accessing information and manipulating the process exerts a great influence not only on the effectiveness of the control system but also on the general operation of a process.

The human-process interfacing requirements in a batch control system tend to be considerably more extensive than the requirements for a typical continuous-control system. Additional requirements for the operating staff include

Status of a unit or part of a unit in terms of phase and steps along with the display of alarm conditions
Status of individual contact inputs and outputs and devices displayed in tabular format or with process graphic displays
Bar charts displaying the progress of individual batches, occupation of plant items, critical paths, and so on
Trend plots of analog and contact variables

The interactive displays for operating staff include

Facility for the starting of a batch along with the assigning of appropriate recipes
Facilities for putting a batch on hold, shutting down a process, and restarting it
Acknowledging alarm conditions
Displaying of sequence logic functions and the facilities for changing them
Displaying of sequence logic messages and the facility for entering data when required
Facility for the manipulation of the states of contact outputs, devices, and so on, with safeguards to prevent contention problems between operator and control system

The reporting and logging facilities for the operating staff and the plant management should include reports of alarms, of principal events in a batch, and of operator actions. These reports may be generated as they occur, on demand, or at regular intervals (i.e., for each shift or each day).

In addition to operator requirements, plant management needs information concerning the past and present performance, production, and efficiency of the plant, the usages of raw materials, the efficient utilization of equipment, and so on. This information is generally derived from data generated in the course of controlling the plant, and it should be presented in a compact, self-explanatory form. A manager also requires the ability to change and review items defining the longer-term aspects of control and production planning. Some examples of needed reports are

Inventory
Energy usage
Product quality
Quantity and types of effluents for environmental considerations
Preventive maintenance of plant items

In general, the operator and management interface should be clear and extensive, making all the necessary data for the management and control of the plant and process easy to access and modify. The interface should also be sufficiently flexible to allow the user to specify and modify the formats for data input and output.

CHAPTER

6

AUTOMATION SYSTEMS

The batch automation functions described in Chapter 5 can be implemented through many different hardware systems and configurations. This chapter addresses the functional aspects of the three most commonly used hardware types—programmable controllers (PCs), digital computers, and distributed batch controllers (DBCs)—and discusses the ways in which they meet those requirements.

PROGRAMMABLE CONTROLLER SYSTEMS

Programmable controllers are used extensively to control small-to-medium-sized sequential and batch processes. Multiple controllers, along with digital computer(s), may be used for controlling larger processes. Programmable controllers are appealing as control devices because they are easy to program and require little sophistication from the user. Although ladder logic may not be the best way to represent sequential operations, it is straightforward for interlocks and simpler sequential operations.

A PC comes with all the necessary facilities for application programming. The user needs to specify all the application-related functions—interlocks, sequencing, and regulatory control—in ladder logic with appropriate extensions. Functions such as recipes, custom displays, and reports are available in a PC in rudimentary fashion or not at all.

Contact Scanning

Although details vary significantly among manufacturers, in principle the scanning techniques in PCs are similar to one another and to scanning techniques in computers.

Some computers use networked input and output (I/O) interfaces with enough "intelligence" to function as network devices, but as far as the authors are aware all PCs use I/O circuits directly connected to the controller's central processor. This is effectively true even for "remote I/O." Where redundant processors are used, there is generally some switch to determine which one is in control. A PC uses one or more I/O channels. Each channel functions more or less independently, so the overall speed of I/O processing can be maximized.

For most applications, and for most controllers, I/O communication can be considered transparent. Output values are usually read from a buffer table in the main memory, so these values are available to the logic even before they have been communicated over the I/O channel to the output interfaces and, subsequently, to the process. In some cases this communication time can be important. The division in time of each channel's subgroupings and the detailed timing relationship between this scanning and central processor logic solving also might be important in some applications. Input and output communications are normally fast: some programmable controllers can exceed 10 scans and output drives per second.

Interlock Functions

The PC excels at interlock functions, since most are programmed in the language of electrical interlocks—ladder logic. OR logic is shown as parallel contacts, and AND logic is shown as contacts in series. An advantage of PCs in interlock applications is that a contact (input or output, including outputs not actually connected to the process) can be used in logic equations any number of times, whereas in conventional relay logic the number of uses is limited by the available relay circuits. Supporting the simple combinatorial AND and OR logic functions (along with simple logical inversion invoked by showing contacts as normally closed in the logic) are a variety of memory, timing, and counting functions.

Sequence Logic Functions

Sequential control functions can be established by building up combinations of the basic logic function types, but generally there are easier ways. Most PCs provide some sort of sequencer function, equivalent to a drum driven by a stepping motor. The drum, equipped with removable pins or similar mechanism, changes the state of one or more switches along its length as it rotates. Compared to the drums they are modeled after, PC sequencers can usually be combined as needed to make them "longer" (more switched circuits) or to increase their "diameters" (more steps). They can usually be driven to a "rest" state, or to any state, without activating intermediate steps even momentarily.

Regulatory Control Functions

Many PCs offered for process control use have some regulatory control (P, PI, PID) capability. This capability is not as comprehensive as that normally available in process control computers and distributed controllers, but it is adequate for many applications. When one of these feedback controllers is built, various parameters (i.e., set point) are assigned addresses within the PC's memory. Data may be read from and written to these addresses so that a sequencer can, for example, write new set points to the controller as required. Similarly, the sequence may be programmed to wait until some condition, typically the achievement of a desired operating point by the loop's measured variable, is reached. Programmable controllers are generally limited to fixed-point representation of analog values.

Flexibility with Repeatability

Programmable controllers can control processes with a high degree of repeatability. As long as the various measuring instruments it uses show proper repeatability, a PC can ensure specification integrity batch after batch. However, they are not very flexible. Whereas interlock logic is readily understood and changed, and hence flexible, changing a sequential control strategy might well require the services of an engineer familiar with the strategy's details. Programmable controllers have comparatively primitive recipe-handling functions that are suitable for multigrade (different amounts of the same material), but not multiproduct, plants. The PC's inflexibility is minimized when it is used in conjunction with a computer that has good recipe-handling capabilities.

Other inflexible aspects of a PC, compared to a computer, are that report formats—when reports are available at all—are difficult to change for various products; display capability is limited; and because PC addresses are based on hardware, hardware assignments cannot usually be changed without changing control software.

Interfacing with Operating Staff and Plant Management

Traditionally, human interface for PCs could communicate only through I/O devices that could be interfaced through field I/O connections. These were typically contact-driven outputs, such as indicator lamps and numerical displays, and contact input devices, such as pushbutton, toggle, and thumbwheel switches. Various PC and third-party manufacturers offer small operator-access panels that enable the operator familiar with PC addressing to read and/or change variable values. More recently, manufacturers have made available CRT-based operator terminals similar to those normally associated with computers. These terminals are generally limited to display and modification of parameter values, along with graphics, although certain other computerlike functions, such as trending, are sometimes available. Some of these terminals are implemented through common personal computers.

DIGITAL COMPUTER SYSTEMS

Because of their inherent flexibility arising from a variety of programming languages and utility programs, general-purpose digital computers offer the greatest opportunity for designing batch control systems. Scanning and driving field inputs and outputs, interlocking and executing sequence logic functions for multiple batches in parallel, and interfacing with regulatory control functions, operating staff, and plant management can all be performed by a properly designed, computer-based batch control system that guarantees flexibility and repeatability. However, the design of a batch system from scratch, with all these functions, requires an enormous engineering design and programming effort. Computer control of batch processes would be prohibitively expensive if every batch control system had to be designed this way. Fortunately, it is now possible to buy, along with a process control computing hardware system, a comprehensive batch control package that can provide the framework for accomplishing all of these functions. The user then may need only to define the requirements for a particular application, which in most cases can readily be incorporated in the standard package.

The following sections discuss the various functions of automated batch process control and how they can be provided by standard facilities in a typical batch control package.

Handling Field Signals

In a computer-controlled batch system the field signals may be fed directly to the control computer via I/O interface racks, or they may come from other intelligent devices, such as PCs, tank gauging systems, other computers, and so on. A routine that reads and manipulates contact inputs and outputs is called a *scanner/driver*. Usually, different scanner/driver routines are provided, one for each type of hardware interface. These scanner/driver routines can be chained or run asynchronously.

Typically, a scanner/driver reads the states of the contact inputs and reads back contact output states, where such facilities are provided, and stores them in an array (table) in a predefined order. Different scanner/drivers might store the input states in different arrays or in different parts of the same array. The array(s) for input states might contain not only the actual states of hardware contact inputs and read back of contact outputs but also the states of analog variables (in or out of alarm), flags used internally within the computer system, states of other control equipment, and so on. Once the states are represented in their arrays, they are handled the same way regardless of their source.

Table 6-1 shows a case where the scanner/driver stores the actual states of contact inputs. Each of these actual states is compared with its predefined desired state by the contact input checking function, and necessary action is taken according to the alarm and other actions specified. The predefined desired states are established by the operator or by the computer logic. Thus, in this example, for valve 1, the actual state is 0 (open) and the desired state is also 0 (open), so there is no discrepancy and no action is required. For reactor pressure the actual state is 1 (in alarm), but the desired state is 0 (no alarm); thus there is a discrepancy. The alarm and call service actions are required, so that an alarm message is generated and the service logic is called to respond with appropriate action. For tank level there is a discrepancy between the actual state and the expected state and the alarm is required, but a call to service logic is not necessary. Additionally, for reactor pressure driving the unit to hold is required on discrepancy. In this case the contact input checking routine will send out an alarm message and will invoke the service/hold logic with the instruction to drive the unit to hold. The functions of the service/hold logic are specified by the user as part of the sequence logic, since they vary between applications. The details of the specification of service/hold logic are discussed in the section *Sequence Logic Functions*. Specified by the user in the database are the initial values of expected states, alarm message requirement,

Table 6-1. Switch Input Handling

		State		*On Discrepancy*[b]		
No.	*Description*	*Actual*[a]	*Desired*[b]	*Alarm*	*Call Service*	*Drive to Hold*
1	Valve 1	0	0	1	0	0
2	Valve 2	1	1	0	0	0
3	Reactor pressure	1	0	1	1	1
4	Tank level high	0	1	1	0	0
5	Disch. pump	1	1	1	0	0
6	Valve 1 output read back	0	0	1	0	0

[a]Set by I/O scanner, regulatory control system, and so on
[b]Initialized as per data base and changed by sequence logic

call to service logic and the requirements to drive the unit to hold. They are modified by sequence logic as required.

In this example, three levels of priority have been specified. The least critical need alarming only. The more critical need calls to the service logic to take corrective actions, but immediate action to put the unit in a safe state (hold) is not required. For the most critical cases, calling the service logic and placing the unit in hold must be done. The number of priority levels may differ among systems.

The contact outputs are also driven by the scanner/driver when specified by the interlock function, sequence logic function, or operator. A typical arrangement consists of (Table 6-2) a set of arrays (tables): one array defines the states to which outputs are to be driven (desired state), and another defines the states to which they were last driven. Whenever an interlock or a sequence logic function wants to drive a contact output, it sets the state of the contact in the array for the new desired state. The scanner/driver, on finding a discrepancy between the desired output state and the state to which it had last been driven, will drive that particular contact and will change the value in the array for the last driven state. The contact output will not be driven again until a requested change is stored in the array for the desired state. In this example, a third array for contact output handling defines the states desired when the corresponding batch unit is on hold. When a unit has to be put on hold or shut down due to an alarm condition, the outputs will be driven to the states specified in this "hold desired" array. The definition of the hold desired states allows all the contact outputs to be driven to specified safe states by a single command.

The scanner/driver for contact inputs and outputs along with the checking of contact inputs is usually part of the standard control system software supplied by a vendor. However, the user has to specify the data base for a particular system, which may include

Table 6-2. Switch Output Handling

		State		
No.	*Description*	*Desired*[a]	*Last Driven*[b]	*Hold Desired*[a]
1	Valve 1	0	0	0
2	Valve 2	1	1	0
3	Reactor press releas.	0	0	0
4	Disch. pump	1	1	1
5	Emg. cooling water	1	0	0
6	Tank 1 inlet	0	0	0

[a]Initialized as per data base and changed by sequence logic, interlock functions or operator command
[b]Set by I/O scanner/driver

Switch tag names (e.g., PSH-101A)
Switch descriptions (e.g., reactor pressure alarm)
Mnemonics for the open and closed states of the switches (e.g., OK, alarm)
Switch types (source of information, e.g., field I/O)
Hardware termination addresses, regulatory control block names, and so on, as required for the switch types used

The data base should also include the phase initialization data for each input and output. They specify the expected states, alarm requirements, and emergency actions at the start of each phase (Tables 6-1 and 6-2). However, these are changed by the sequence logic commands, as required, during the execution of a phase.

In some batch control systems the data base specifies devices instead of defining individual contact inputs and outputs. Each device may consist of multiple contact inputs and outputs. The details of individual device inputs and outputs are sometimes specified with the device definition. In other systems they might need to be specified earlier. Many interlock functions can be embedded in the device definition.

Interlock Functions

In computer-controlled batch processes the procedures for handling interlock functions vary greatly among control systems. Most systems allow the interlock functions to be established through the sequence logic language (see next section). In addition, many systems provide the facility for specifying them through Boolean equations. In a Boolean equation the relationships between switch inputs and outputs may be specified as in the example (Fig. 6-1). Here, the valve at the

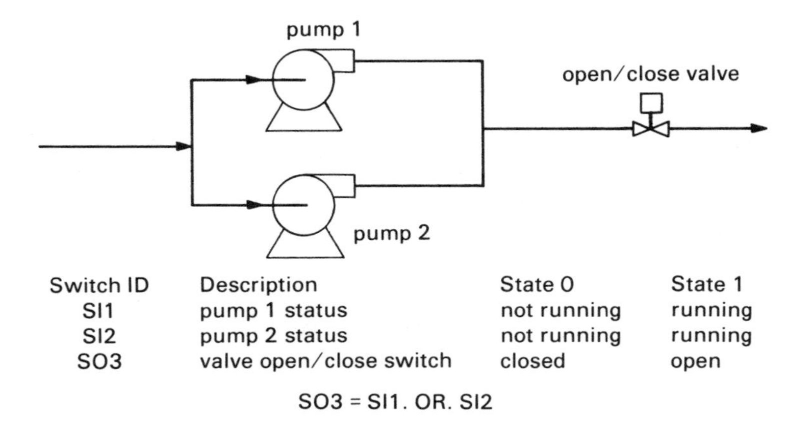

Figure 6-1. Boolean equation for specifying interlock function between the valve and the two pumps.

common outlet manifold for two pumps must be open when one or both pumps are running; the valve should be closed when neither pump is running. This condition may be specified by the simple equation

$$SO1 = SI1 \text{ .OR. } SI2$$

Similar equations may be specified with other Boolean functions, such as AND, NAND, XOR, NOR, XNOR, and others. A switch input or output in a Boolean equation is not limited to indicating the state of a plant item. It may also refer to other two-state devices or functions, such as an indicating lamp (on or off), analog input (in alarm or not), or internal flag. Thus, a Boolean equation may be used to set an alarm when a certain combination of plant or process conditions occur.

Boolean equations, however, are not well suited for specifying interlock functions involving time delays. Time delays are usually easier to specify in the sequence logic language. A user is more likely to use sequence logic or a hybrid of sequence logic and Boolean equations to specify such functions. A sequence language statement for specifying time delay may be as simple as

```
WAIT 10
```

for a 10-S delay. Multiple time delays, in combination with other functions, are also easy to specify in sequence logic language. When a batch control system includes hardwired relay logic circuits or the like in addition to control computers, the interlock functions are usually taken care of by those devices rather than by the computer. Thus, in the example of multiple conveyors (Fig. 5-3), the commands for starting and stopping the conveyor system can be generated by the sequence logic in the computer. But the interlock functions involved in starting and stopping the system and monitoring and responding to alarms can be performed by a relay logic circuit, a PC, or another, similar, device.

As stated earlier, interlock functions generally take care of the inherent physical characteristics or the safety aspects of a plant or process. In a computer-controlled batch system the interlock functions are executed at regular intervals, usually once a second or faster, and they run independently of sequence logic and other functions. However, the interlock functions communicate with other control functions such as scanning and driving contact I/O, regulatory control, and sequence logic functions to provide integrated control.

A standard batch control package usually provides facilities for interlock functions as a separate entity or as part of the sequence logic commands. The user must specify the application-specific requirements.

Sequence Logic Functions

Executing sequence logic functions is central to controlling most batch-type processes. Sequence logic functions interface directly with the contact and analog I/O scanner/

drivers and via interlock and regulatory control functions to manipulate a process. Sequence logic also interfaces with the recipe functions and operator and management interface routines.

Dissimilar and Similar Units

A typical process consists of several units running asynchronously. Thus, independent sets of sequence logic functions are required to specify the sequence logic for different units. For example, if a chemical process plant consists of raw material preparation, reaction and storage units, then each of these units needs separate sets of sequence control logic or plant production will be constrained. Each sequence control logic set should be able to run asynchronously and concurrently so that while the reactor is running for one batch the raw material preparation unit can weigh and mix material for the next batch.

Although different sets of logic must be specified for dissimilar units because their operations are different, a common sequence logic can be specified for similar units. Thus, if a process plant has a number of similar reaction units, their sequence logic functions will probably be similar. However, logic for these reaction units will be running asynchronously and will be addressing different sets of inputs and outputs.

In a computer-controlled batch system only one set of sequence logic functions may be specified for all units with the same requirements. Then in the sequence logic steps the plant inputs and outputs and other variables are addressed indirectly by common names or numbers. Thus, the indirect addresses for the discharge valves are identical for all similar reactors. However, the data base for each reactor unit contains the actual address of the discharge valve for that unit, which is obtained at the time of sequence logic execution (Fig. 6-2).

Since these units may run asynchronously, executing the sequence logic for these units must be monitored individually. One method is unit status tables (USTs), one for each active unit. More on these status tables later.

Unit States

The states of a computer-controlled batch unit may be divided broadly in two categories: manual and automatic (auto). There is another state, which may be termed "off-line," that occurs when the control computer is not up and running or has no account of the unit in question. The "off-line" state will not be addressed here.

When a unit is in manual state, the computer control system does not execute any sequence logic functions. However, it may scan the states of inputs and outputs for display purposes and allow the manipulation of the outputs by an operator from the console. In the auto state the computer control system controls the unit by executing prespecified sequence logic functions. Under normal conditions the

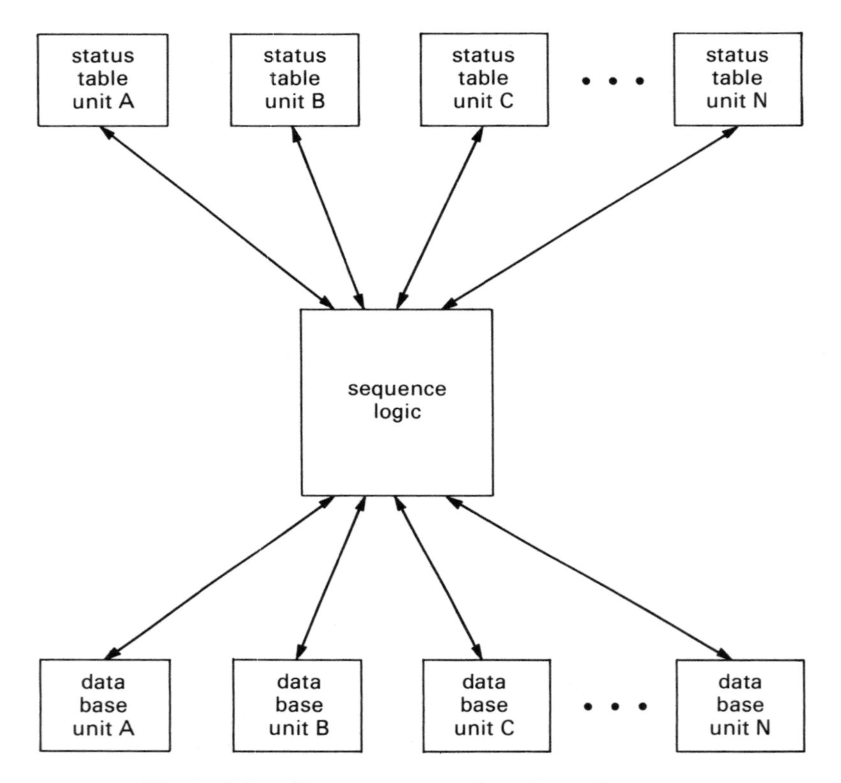

Figure 6-2. Common sequence logic for similar units.

control system will execute sequence logic functions specified for normal operations, but under abnormal conditions, appropriate predefined service logic will be executed. Depending on the situation, the service logic may successfully resolve the problem, in which case normal operation will be resumed. If the problem cannot be resolved automatically and requires manual intervention, then the service logic will drive the unit to a safe state, defined here as *HOLD*. Depending on the nature of the abnormal condition, interlock logic, independent of service logic, might also come into play. In that case the action taken will be a combination of service logic and interlock logic.

Figure 6-3 illustrates the manual and automatic states of a unit. The auto state is further subdivided into normal, service (service logic being executed), and hold (driven to a safe state, with no logic executing). In a typical computer-controlled batch system, the transition from manual to auto always requires manual intervention—that is, an operator must enter commands from the console to start a unit. Operator action is also required to bring it back to the manual state. (A unit may go to idle state under automatic control when a batch is completed normally.) The transition from the normal state to the service state in a unit may occur automatically due to alarm conditions. A unit also can be driven to the

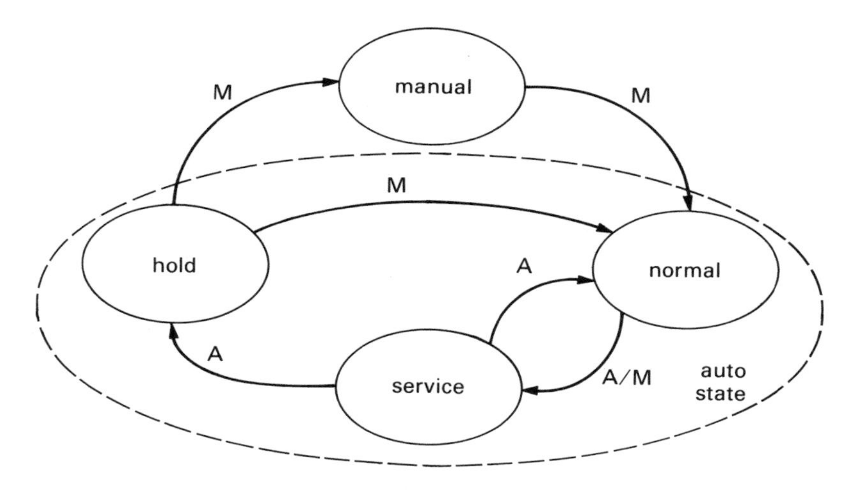

Figure 6-3. States of a computer-controlled unit. M = Manual intervention required to change state. A = Change of state may happen automatically.

service state by an operator command. When in service, the recovery to normal state may happen automatically if specified process conditions are met. If recovery is not possible, then the unit is driven to the hold state, where operator intervention alone can restore it to normal operation or to the manual state.

Thus, the sequence logic functions for unit control should specify the steps for normal operation as well as for executing service logic and transferring to hold. The service logic may need to take different actions to deal with the various alarm conditions. Therefore a typical service logic consists of a collection of sequence logic steps, where each set of steps takes care of one or more similar alarm/failure conditions.

Sequence Logic Phases

The sequence logic for a unit is generally divided into multiple phases. For example, a reactor unit may have three associated phases: charge, reaction, and cooling. Depending on the control system, separate service logic can be specified for each phase (Fig. 6-4), or a common service logic can be specified for all phases (Fig. 6-5).

As stated earlier, a UST keeps track of the sequence logic steps executed for a unit. It also carries other information pertaining to the current status of the unit. A typical status table might contain information on the

State of the unit (manual, normal, in service, or hold)
Current phase
Pointer to the next instruction in the sequence logic to be executed

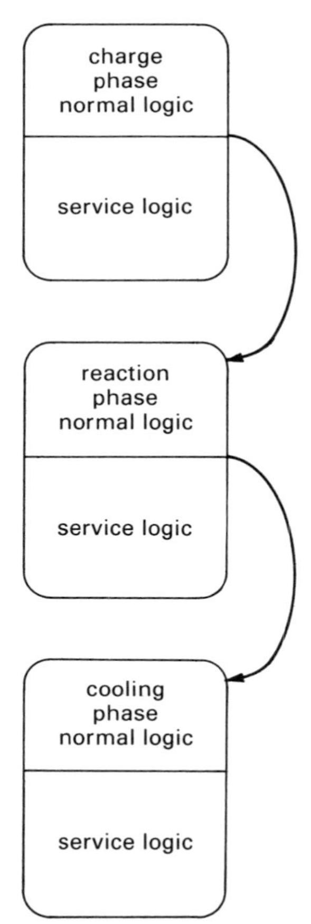

Figure 6-4. A set of phases with service logic for each phase.

Suspension of logic pending manual entry of data or change to a specified process condition
Service logic entry address

Batch product manufacturing usually involves executing multiple phases spread over several units. The order in which these phases are to be executed can be specified in two ways. The first is *direct linking,* where each phase specifies the next phase to be executed, if any. Thus, in Figure 6-5 the charge phase will specify reaction phase as the next phase to be started. Similarly, the reaction phase will specify cooling as the next phase. In the last phase for product manufacture, no phase will be specified, or, if the manufacture of the next batch is to be started automatically, the first phase will be specified (Fig. 6-6). The other common method

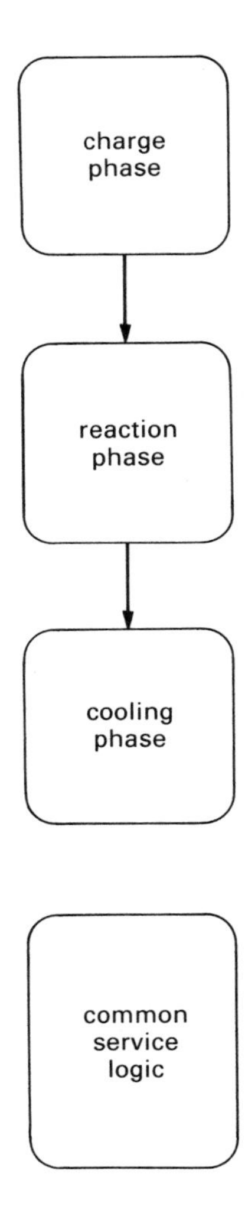

Figure 6-5. A set of phases with common service logic.

of linking is by specifying the order of the phases in a tabular format (Fig. 6-7), commonly called a *procedure*. There the phases are specified in the order of execution. Optionally, one or more process parameters required by the phase logic are specified.

When a phase has to be repeated, specifying the phase sequence in a procedure with multiple entries of the repeated phase is easier than using direct linking. With direct linking it is also difficult to specify process-dependent variables for a set of phases. Evidently the procedure function, in addition to its ability to pass parameters, is a more flexible method of specifying phase execution. However, the alternative still has some advantages as discussed in Chapter 9.

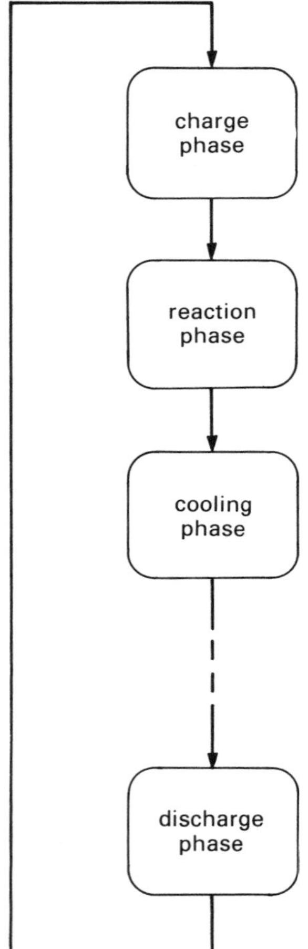

Figure 6-6. Direct linking of phases with automatic restart of next batch.

Product - Blue Paint

No.	Phase name	Para 1	Para 2	Para 3
1	charge	200.0 lb	30.5 lb	23.5 lb
2	reaction	273.5°F	5.5 h	—
3	cooling	102.2°F/s	—	—
	•	•	•	•
	•	•	•	•
	•	•	•	•

Figure 6-7. Example of a procedure.

Sequence Logic Language

In a computer-controlled batch system the steps in sequence logic functions are usually specified in a general-purpose, high-level language or a special-purpose batch language. The general-purpose, high-level language could be FORTRAN, BASIC, or PASCAL, for example, preferably with some enhancements for real-time and batch functions. But, more often, a special-purpose language is provided by the vendor as part of the batch package. A general purpose, high-level language may be familiar to users. A special-purpose batch language, on the other hand, is more likely to be unfamiliar to the user; but because this language is designed only for specifying sequence logic functions, its execution speed is faster and computer memory requirements are usually less. Also, the special-purpose batch language mirrors process requirements, so its use demands only minimal programming knowledge. A good compromise is a batch system that provides a special-purpose batch language along with the facility for writing parts of the phases in a general-purpose, high-level language. This facility accommodates the best of both worlds—the efficiency of the special-purpose language and the power of a general-purpose language.

Another aspect of the language worth considering is whether it is interpretive or whether it is compiled to generate executable code. An *interpretive* language is one in which the sequence language steps are stored and executed as they have been specified, as in most versions of BASIC (compiled BASIC is also available), or they may be translated in some form of data, as with some macro languages. The source language or the data in these cases are interpreted when the sequence logic steps are executed. Noninterpretive languages are those in which the sequence logic steps, once written, are compiled into executable code. A compiler does the translation and checks for syntactic errors. Once compiled without errors, a set of sequence logic may be run when required. Most high-level languages like FORTRAN, PASCAL, and many special-purpose languages are in this category. The advantages of a compiled language are its execution speed and reduced storage requirements. On the other hand, a compiled program is relatively difficult to modify. Any modification to the sequence logic, however small, could involve recompiling the module (a phase or a subroutine), whereas for sequence logic written in an interpretive language, on-line modification is usually easier. Neither compiling or linking is then required. Incrementally compiled languages act as interpretive languages in this case.

Whatever the language for specifying sequence logic steps, the commands generally fall under the following five headings (Appendix 2):

Commands that manipulate contact outputs, contact input desired states, devices, and flags
Commands that interface with the regulatory control functions
Commands that perform arithmetic calculations
Commands that communicate with operators, supervisors, and plant managers
Commands that make decisions and control the direction and timing of the sequence logic steps

Commands in the first category open and close contact outputs, change the desired states of contact inputs, and manipulate devices and flags. While changing the states of contact outputs and devices, a command of this type might set up interlocks to inhibit the alarming for expected periods when devices are changing from one state to another (such as the travel time of a valve from fully open to fully closed). Other commands in this category include reading and testing the states of contact inputs and outputs, and flags and devices, setting and clearing alarming functions for inputs and devices, and setting the criticality level of inputs and flags.

Commands that interface with the regulatory control functions include those that read and modify continuous-control parameters, such as set points, outputs, alarm limits, timing and control constants, and the status of blocks and loops, such as block on and off, and opening and closing of control loop cascades.

Most sequence logic functions must be able to perform numerical calculations. Many sequence logic languages provide commands for such simple numerical functions as addition, subtraction, multiplication, and division. Since general-purpose, high-level languages (BASIC, FORTRAN, PASCAL, etc.) have more powerful capabilities for numerical calculations, subroutines in these languages are generally used whenever complicated calculations are required.

Commands that communicate with plant operators, supervisors, and managers do so by printing or displaying messages or activating lamps and annunciators. Included in this category are commands that ask operators to perform manual functions or enter process-related information to allow the sequence logic to proceed in the right path. Those commands that store information in files for generating future reports for the operators and the plant management are also of this type.

Commands in the last category are those that make control decisions: conditional and unconditional jumps, waiting for a specified period or until certain external conditions are satisfied, calling and returning from subroutines, specifying the start and end of phases, setting and clearing counters and timers, and so on.

All of these commands are usually standard with any batch control language. However, the commands and their parameters vary widely between different systems and languages.

Interfacing with Regulatory Control Functions

As stated earlier, most batch systems require regulatory control functions for controlling analog process variables at specified values for given periods. In most computer-controlled batch systems, regulatory control functions are provided as separate packages or as part of batch control packages. Since several existing textbooks address regulatory control functions, no discussion of control theory will be given here. However, aspects of regulatory control unique to batch systems will be discussed later. The proper interaction between batch and regulatory control functions is vital to the control of most batch systems. They include

Enabling and disabling control loops and blocks and specifying the set points and initial output values
Opening and closing cascade control loops
Setting, clearing, and changing alarm and other limits
Setting and changing controller constants
Reading process variables, such as controller constants, alarm limits, set points, and controller status.

All of these functions are usually handled by the standard sequence logic commands. A set of sequence logic commands might be required to specify the set point and the alarm limits before enabling control blocks to regulate a process variable. The regulatory control function normally continues controlling until the sequence logic function either changes or disables the control blocks at some subsequent stage in the execution of the batch (Fig. 6-8). Thus, the direct interaction between the sequence logic functions and the regulatory functions takes place when specific commands are executed.

All communication between the batch and the regulatory control functions specified so far is initiated by the sequence logic functions. The regulatory control function, however, must communicate with the batch control system on alarm and other limit violations of analog process variables when they occur. This communicating is usually done by specifying these limits as switches (flags) in the contact input data base. Thus, rather than indicating the states of plant hardware, these switches indicate the statuses of these limits. When a variable violates a limit, the state of the appropriate switch is changed and the contact I/O check routine then takes appropriate actions such as alarming and/or invoking sequence logic functions, depending on the criticality of the variable, as specified in the data base. Here the contact I/O check routine makes no distinction as to whether a switch repre-

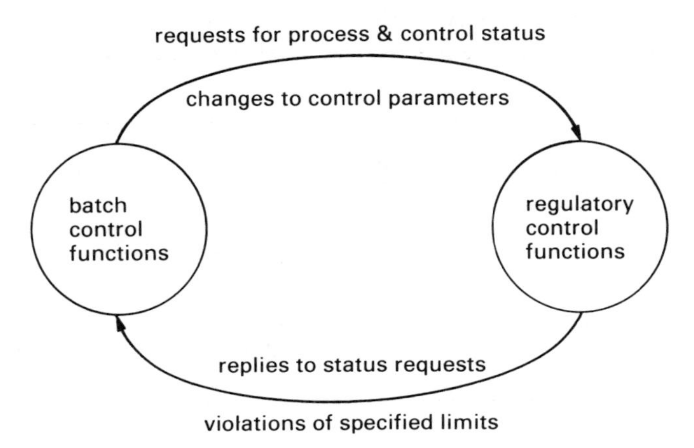

Figure 6-8. Interface with regulatory control functions.

sents the state of a piece of plant hardware or the state of a process variable limit. This feature is used for process alarm limit violations and as a convenient way to inform the batch control system of process values reaching predetermined points.

Flexibility with Repeatability

Flexibility in manufacturing different grades of a product with different sets of recipes was outlined in Chapter 5. The recipes are generally easy to implement in computer-controlled batch systems. In simple recipes the recipe items specify process variables such as temperature, pressure, amount of material to be added, and reaction time, which vary from one grade of product to another. Additionally, when different grades of products have to be stored in different product tanks, recipe items may specify tank numbers or contain other identifiers. A sequence logic step on reading an identifier will invoke the appropriate routine for routing the product to the right storage tank. This concept of specifying a numerical value in the recipe can be extended to other decision-making situations, where an appropriate section of sequence logic can be executed depending on the numerical value. In some applications the scope of recipe items has been extended to include alphanumeric data.

Recipe Logic Locations

So far, we have considered recipes structured as simple tables. Another approach is to use the concept of *logic location* (Chapter 8), where a recipe item is identified by a particular step (logic location) in the sequence logic (Fig. 6-9) where that recipe item is required. A recipe item may also include an identifier to specify the type of variable (temperature, pressure, etc.) or whether a material is to be added to the source tank or the ingredient name. Here the sequence logic, rather than accessing a recipe item by its fixed table position, will read all the items associated with the

Ser. No.	Logic Location	Ident.	Value
1	5	Cat. A	3.75
2	12	Deg. C	157.5
3	15	Mat. X	764.0
4	15	Mat. Y	360.5
5	17	Gal/Min	24.5
6	20	Mat. Z	520.0

Figure 6-9. A recipe with logic locations.

current logic location. Thus, when logic location 15 is reached during the execution of the sequence logic, recipe items 3 and 4 are read and acted upon by sequence logic functions.

Recipes of this type are advantageous where the number of recipe items vary from one grade of product to another. Thus, when many materials need to be added at the same time, the same logic location number may be used. For another grade of product where not all the materials are used, the recipe will contain items only for required materials. The same function can be realized with conventional recipes, where the recipe items not required may be set to zero. Recipes with logic locations are more compact and are useful where a widely varying number of recipe items is required for different grades of products. As computers are equipped with more and more memory, this advantage will become less important.

Variable-Sequence Recipes

The recipe types considered so far are suitable where the essential manufacturing process varies from one grade of product to another only in predetermined ways. Common steps in the sequence logic specify the manufacturing process for the different grades, and the recipe items specify the variation between them. The execution of sequence logic steps can be varied to a limited extent by specifying in a recipe a particular path in a multiple-choice situation. This method might not be suitable when the manufacturing of different product grades requires considerable alterations in the order in which the sequence logic steps must be carried out. An alternative is to specify different sets of sequence logic for different product grades. But when a plant produces many different grades of products, specifying numerous sets of sequence logic steps is tedious and time consuming. The best solution is to design recipes that specify the process variables and the order of execution of the sequence logic steps. Such recipes are called *variable-sequence recipes*, as opposed to the *fixed-sequence recipes* discussed earlier.

Variable-sequence recipes have been successfully implemented in different ways. In one method (Chapter 8) the recipe is written in a tabular format (Fig. 6-10)

Step	Value 1	Value 2	Value 3
1 (Change temperature)	122.4 (Temp. SP, °F)	1.15 (Ramp, °F/min)	
2 (Wait for temperature)	110.0 (Temp. °F)		60 (Max. time, min)
3 (Charge ingredient)	522.0 (Gal.)		3 (Ingredient No)
2 (Wait for temperature)	122.4 (Temp. °F)		15 (Max. time, min)
4 (Take sample)			1 (Sample tag)

Figure 6-10. Example of a variable-sequence recipe in tabular format. *(After Rosenof, 1982b; reproduced by permission of the American Institute of Chemical Engineers.)*

```
              :DYESTUFF YELLOW X-Y97 - 22.8.73.
1.0   23/100  DIAZOTISATION
1.1   23/110  Set AGITATOR to 80%
1.2           CHARGE FROM BP 23/110 with transfer valve 100% open and
              pump speed 1 (800 lb/min amine slurry)                    40
1.3           Message (Inform BP operator that 23/110 is released).
1.4           Set SAMPLE RECIRCULATION at 40%
1.5           Manual (Open valves to pH system: close bypass and
              open caustic isoln valve : check pH reading)              10
1.6           TEMP ICE to 20 deg C alarm limits 10 deg C and 25 deg C 10
```

Figure 6-11. Variable-sequence recipe in free format. *(After Bowen et al., 1975; reproduced by permission of the Institution of Electrical Engineers.)*

where each recipe item specifies a process function and the necessary process parameters. The order of the recipe items specifies the order of execution of the sequence logic. In another example (Bowen et al., 1975) the recipe is written in rather free format English (Fig. 6-11), where a key word specifies the type of function followed by the parameters. Comments are allowed freely between the function and the parameters to make the steps more readable. A custom compiler picks the key words and the parameters and sets them in a tabular format before execution. During execution these recipes essentially call appropriate subroutines and pass on the necessary parameters to carry out the process functions. A clear division also occurs between the process-oriented function specified in the recipe and the plant-oriented functions specified as sequence logic steps in the subroutines, which makes the logic simpler (Fig. 6-12).

For most batch applications, fixed-sequence recipes suffice. Variable-sequence recipes allow more flexibility and simplified specification of sequence logic steps at the cost of greater maintenance requirements, custom compilers, and increased system complexity. However, they provide elegant methods for controlling batch processes where large numbers of different grades of products are manufactured in one plant. Standard batch packages provide at least the simple recipe facilities, and some offer more elaborate functions.

The recipe functions complement the procedure functions described earlier (Fig. 6-7) to some extent. Where both functions are used, they provide powerful yet flexible ways of specifying the manufacture of batch products. These concepts are further elaborated in Part III.

Interfacing With Operating Staff and Plant Management

The hardware available for human-process interfacing in a typical computer-controlled batch system includes keyboards and printers, CRT monitors with and

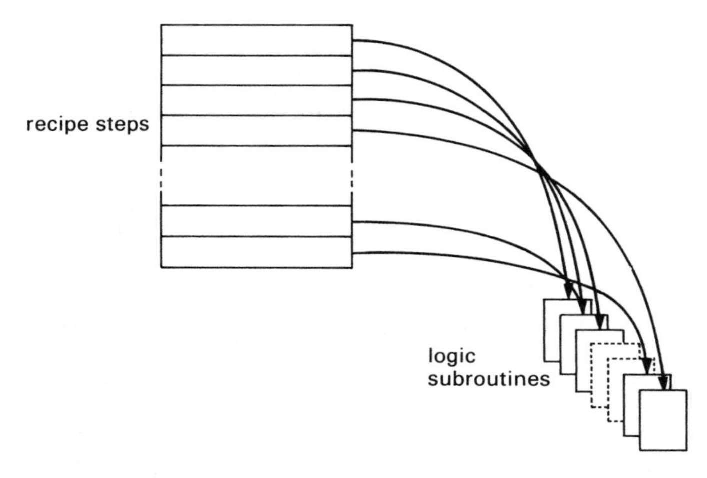

Figure 6-12. Variable-sequence recipe calling logic subroutines during execution.

without keyboards, dedicated panels with indicators and switches, graph plotters, pen recorders, and other devices.

Printers are generally used when hard copies are required (e.g., logs, reports, alarm notifications). A keyboard with a printer is generally available as a system terminal that links programmers, system engineers, and the computer for program developing and debugging. In the past when CRT monitors were not generally available, keyboards with printers represented the main interface for process plant operators and plant management. They are still used as backup devices for CRT monitors with keyboards. High-speed printers, such as line printers, are generally favored where long reports must be printed regularly. However, the line printers and the slower devices that print a character at a time have only limited capabilities for printing graphic data. For graphic plots such as trend displays, histograms, and so forth, where hard copies are required, digital plotters or pen recorders are often used. Digital plotters are slow, and pen recorders are generally suitable only for analog variables. An alternative method that is gaining acceptance is video copiers, where the required graphic display is first generated on a CRT monitor; the hard copy is then made by an electronic/photographic technique.

Because of their versatility in displaying alphanumeric and graphical data and their facilities for rapidly updating changes in plant and process conditions, CRT monitors and keyboards are increasingly becoming the primary modes of communication for operators and plant managers (Fig. 6-13). CRT monitors also provide color displays, which are used for emphasizing alarm conditions or specifying states such as motor on/off.

As previously stated, the human-process interface requirements in a batch control system are generally more extensive than those in a typical continuous-control system. In addition to displaying process variables, a batch control system must also display the mode of a unit, the current phase and step, and the current

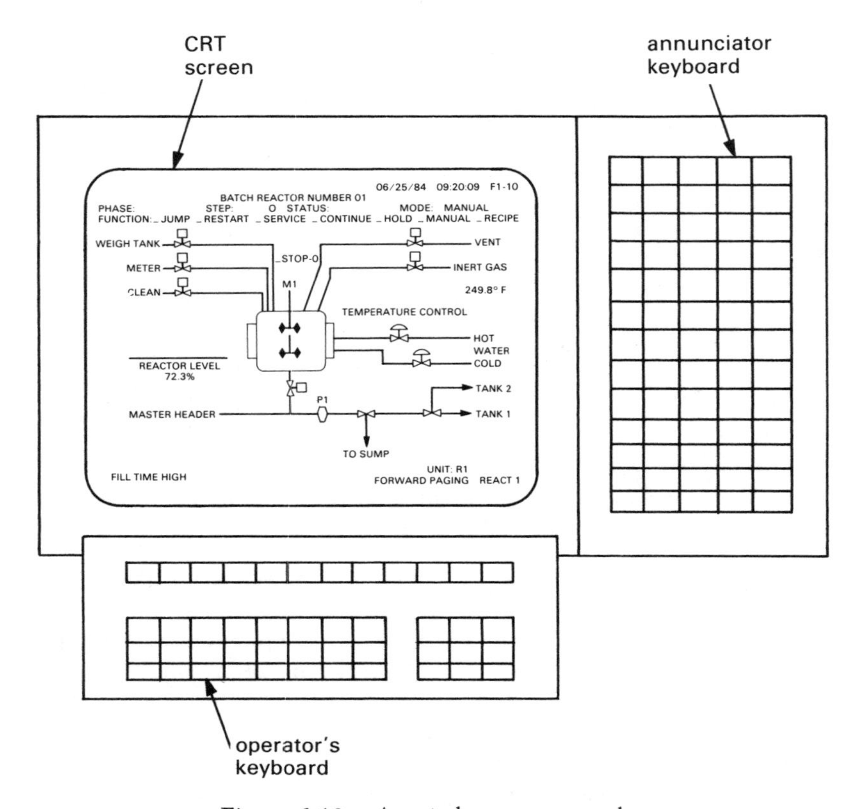

Figure 6-13. A typical operators console.

recipe, along with any existing alarm condition. The split-screen method of display, where certain areas of the screen are dedicated according to the type of information, is commonly used on CRT monitors (Fig. 6-14). In this example of a unit display the time and date are on the top line; the next three lines specify the display name and page number, phase name, step number, and status of the unit along with operator functions such as starting and placing the unit in hold or in manual state or displaying the recipe. The central part of the screen shows the flow diagram of the unit, the current values of selected analog variables, and the states of various plant items. The last two lines display the process and system alarm messages and the unit name. In this display, which is essentially split into three parts, the formats of the top and bottom parts are unchanged from one display to another. The middle part, representing the process unit, is user defined.

In the example in Figure 6-15 the state of the switch inputs and outputs for a unit are displayed in a tabular format where the point name, description, and

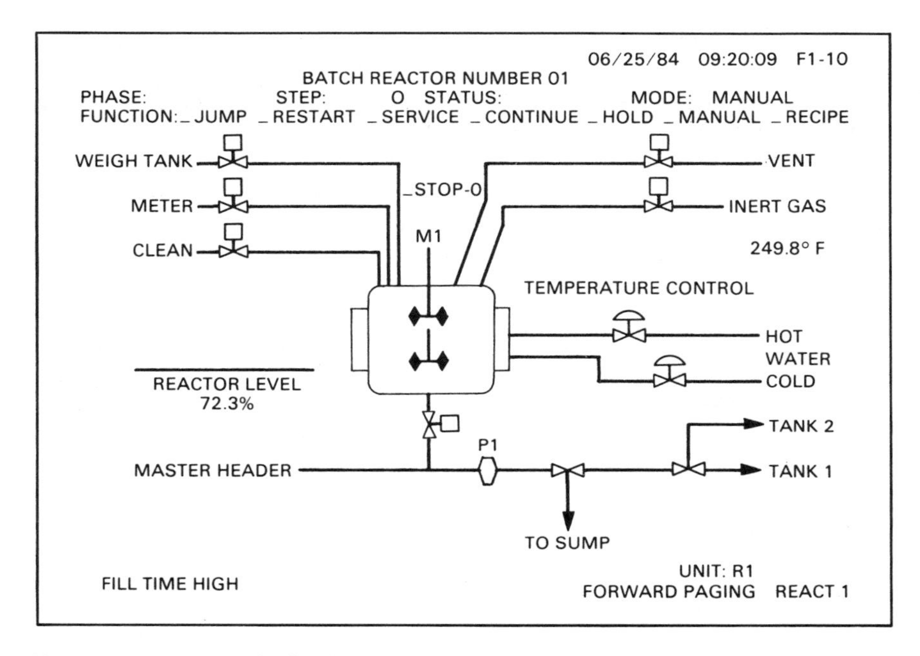

Figure 6-14. A unit display. *(Source: Fox 1/A IMPAC/BATCH Process Operator's Guide MI 854-184; copyright © 1984 by the Foxboro Company, reproduced by permission.)*

MEASUREMENT DISPLAY 08:33 07/24/82

REACTOR R1 POINT DISPLAY- PAGE1

POINT ID	TYPE	DESCRIPTION	STATE	ALARM
R1CS001	C/I	REACTOR FILL VALVE INPUT	OPEN	
R1CO001	C/O	REACTOR FILL VALVE OUTPUT	OPEN	
R1CS002	C/I	COMP. 1 FILL VALVE INPUT	OPEN	
R1CO002	C/O	COMP. 1 FILL VALVE OUTPUT	OPEN	
R1CS003	C/I	COMP. 2 FILL VALVE INPUT	CLOSE	
R1CO003	C/O	COMP. 2 FILL VALVE OUTPUT	CLOSE	
•				
•				
R1CS005	C/I	REACTOR PRESS RELEASE	CLOSE	
•				
•				
R1CS015	C/I	REACTOR DRAIN VALVE	OPEN	ALRM

Figure 6-15. A typical point display.

status and alarm conditions are set up in a predefined format. Here the user need only specify the order of the switches, and the standard display package will take care of actually displaying them in the predefined format. Other standard displays and reports include plant status display (Fig. 6-16), alarm summary report, historical report, real-time and historical trend reports, and bar chart displays of actual and anticipated occupation of plant items (Fig. 6-17). A control system vendor supplying a computerized batch control system usually provides the software packages to display and print these reports. The user usually must specify the display items and necessary parameters and also the flow diagrams for the graphical displays.

If vendor-supplied report formats for plant management reports dealing with inventory, energy usage, effluents, and so on, are inadequate, special formats may be generated. The system vendor will usually provide facilities to generate formats. The user may need to specify the formats and the algorithms for calculating the values to be outputted in the reports.

In conclusion, the inherent flexibility of the general-purpose digital computer allows reports of many different formats and sizes to be generated and displayed. A good reporting and display package lets the user generate formats in an easy and user-friendly way.

13:23 04/10/85

MILL LINE STATUS

LINE	UNIT	PHASE	MODE	STEP
L1	GRINDING	STARTUP	AUTO	3
L1	SCREENING	STARTUP	AUTO	17
L1	CONCENTRATOR		MANUAL	
L2	GRINDING	RUN	AUTO	32
L2	SCREENING	RUN	AUTO	29
L2	CONCENTRATOR	STARTUP	AUTO	18
L3	GRINDING		MANUAL	
L3	SCREENING	RUN	IN HOLD	45
L3	CONCENTRATOR	RUN	AUTO	36

Figure 6-16. A typical plant status display.

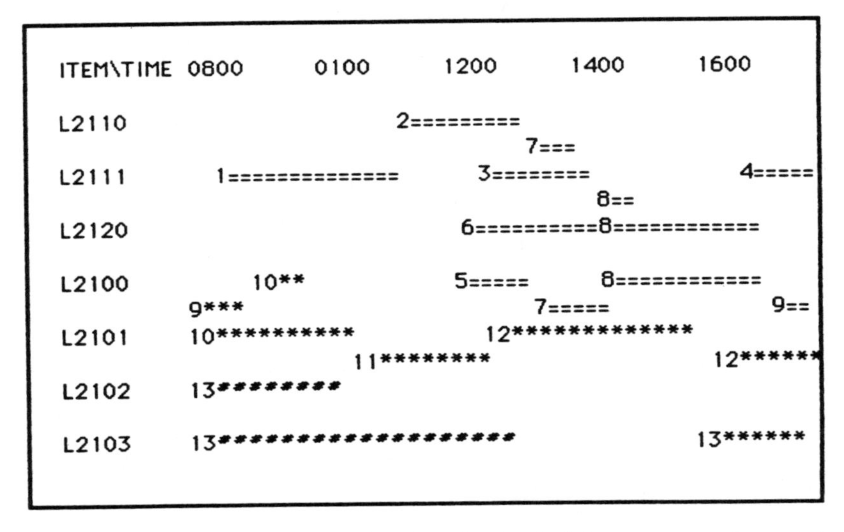

Figure 6-17. Example of a dynamic bar chart display. Each bar represents the period of occupation of a plant item, with the header indicating the phase number. Batches are distinguished by the characters or color in the bars.

DISTRIBUTED BATCH CONTROL SYSTEM

As stated in Chapter 3, a distributed batch controller (DBC) functions as a digital computer. When operated in a stand-alone mode (Fig. 6-18), the DBC is usually limited in the size of the process it can control (number of inputs and outputs and sequence logic steps) and in its flexibility for defining control strategies and operator's communications. These limitations are overcome when DBCs are configured in a network fashion with intelligent terminals and/or host computer(s) (Fig. 3-17). An *intelligent terminal* is an interface for the operating staff and the plant management, allowing them to communicate with many distributed controllers. It stores necessary process information and communicates with process operators and plant management in an intelligent and user-friendly way. An intelligent terminal usually allows the entry of sequence logic steps and data base information, which are then down-loaded to appropriate batch controllers. Host computers, where used, meet top-level process control needs, such as optimization, advanced control, scheduling of batches and process units, recipe handling, and management reporting.

A DBC also allows continuous control functions. Other than the sequence logic steps, generally very little programming is permitted in a DBC. All other permissible functions are specified as parameters to the data base(s).

Because DBCs are similar to digital computer systems for batch control, their

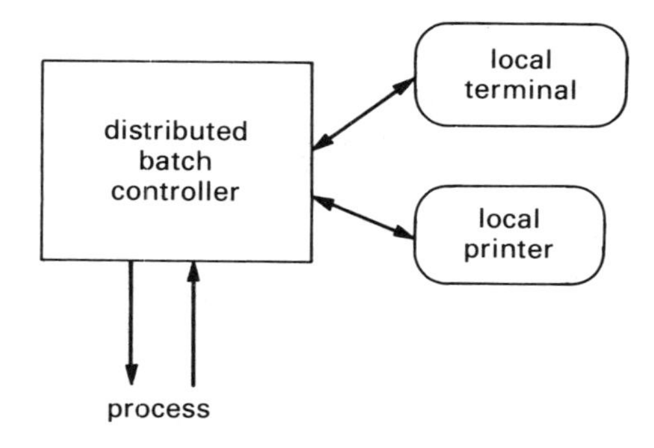

Figure 6-18. Distributed batch controller in stand-alone mode.

functions will not be discussed in detail. Discussions in the following sections are limited to those areas where the functions differ.

Handling Field Signals

A DBC generally provides its own process I/O interface. I/O cards and the multiplexer are mounted in the same rack or in adjacent cabinets. Thus, the DBC reads and manipulates the process inputs and outputs for its own area directly. When it is necessary to read and manipulate outside its process area, facilities for peer-to-peer communications are generally provided.

Because a DBC handles noticeably fewer process inputs and outputs than a digital computer does, it can usually scan them at a faster rate (usually two to five times a second). Reading contact input states, comparing them with the expected states, and taking necessary action to resolve discrepancies are functions generally carried out in the same way as in a digital computer. However, some DBCs do not perform contact input checking as a regular function. Instead it is user specified as interlocks or subroutines in sequence logic.

Alarm messages for drawing operator attention and the facilities for manually driving outputs are generally conveyed by an intelligent terminal remote to the controllers. Local operators' stations may also be provided for these functions.

Interlock Functions

Interlock functions in a DBC may be handled by Boolean equations and/or by sequence logic steps, as in computer-controlled batch systems (see *Digital Computer Systems*). Some controllers also allow logic blocks for this purpose. Logic blocks are analogous to Boolean operators, but they are more convenient for graphical representation of process displays (Fig. 6-19).

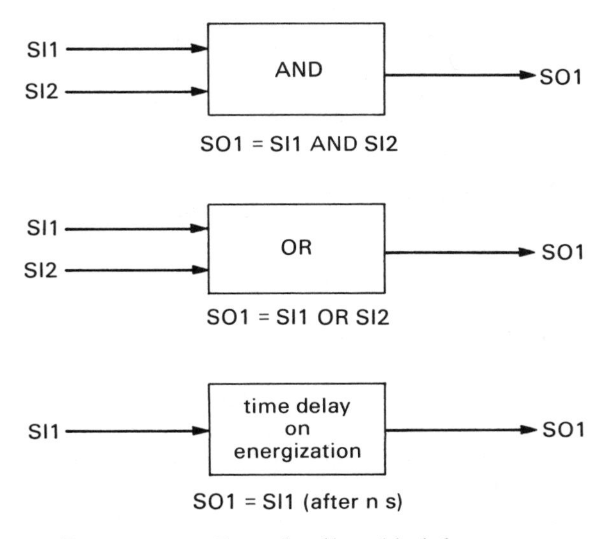

Figure 6-19. Example of logic block functions.

Common logic block types are AND, OR, XOR (exclusive OR), delay, pulse, and flip-flop. The inputs to these blocks may be switch inputs, internal flags, timers, counters, process alarms, or outputs of other logic blocks. The outputs may be switch outputs, internal flags, input to another logic block, or similar things.

The user usually specifies the logic block functions and the input and output identifiers when the data base is generated. Typically, the specification consists of four parameters: block type, input 1 identifier, input 2 identifier, and output identifier. A DBC usually provides a limited number of such blocks, and these are executed once every processor cycle.

Sequence Logic Functions

Executing sequence logic functions in a distributed batch control system is similar to that in a general-purpose digital computer. Multiple sets of sequence logic are usually allowed to run in parallel. The sequence logic is usually grouped into normal logic and service logic, as in digital computers, to take care of the normal and the alarm/failure conditions.

A vendor-supplied, high-level language allows the specification sequence logic, which is generally similar to that for a digital computer system (Appendix 2), although the number of instructions could be limited. General-purpose, high-level languages are generally not allowed for programming. In some systems the sequence logic steps and the data base information are entered directly by a terminal. In others they are entered via an intelligent terminal or a host computer, where they are processed before downloading to a DBC.

In a DBC phase linking—that is, specifying the order in which the phases are to run in a unit—may be stated explicitly in the phase logic. Thus, the last instruction in a phase may call the next phase to be started. The procedure function described in the section on sequence logic functions in computer-controlled batch systems for linking different phases for a unit is generally available through the host computer or the intelligent terminal.

Interfacing With Regulatory Control Functions

A DBC usually includes the regulatory control functions required for controlling a process area. In many cases batch and regulatory control functions are combined. As in a computer-controlled batch system, the sequence logic function lets regulatory control functions, such as set points, alarm limits, controller constants, and so on, be read and manipulated. Automatic changes of switch states on alarm limit violations are also generally provided as a standard feature.

Flexibility With Repeatability

Most distributed batch control systems permit a rudimentary recipe function. The basic recipe function, which allows the specification of sets of process variables for different grades of product, is usually available. Any enhanced recipe function, such as the use of logic locations or variable sequencing, generally requires a host computer.

Interfacing With Operating Staff and Plant Management

A local operator's station directly connected to a DBC, where allowed, can interface with the operating staff in a limited way. Live displays of process data are possible in an alphanumeric format, but a graphic display facility is usually not available. The displays are normally preformatted, and little flexibility is allowed from one application to another.

In a network environment, sophisticated human-machine communications are possible by using intelligent interfaces and host computers. An intelligent terminal is usually a CRT display unit with keyboard and alarm annunciators. As stated, an intelligent terminal has its own memory and processing unit. The display formats, including fixed graphics, are stored in their own memory, and the live data are updated at regular intervals from appropriate distributed controllers.

In addition to graphic and tabular displays, an intelligent terminal has some or all of the following functions:

Faceplate displays for analog variables
Live and historical trend displays
Historical data storage and reporting
Compiling and editing sequence logic steps and generating data bases for DBCs

Compared to a host computer, an intelligent terminal has limited memory and cannot store large amounts of historical data. It also generally has no programming capability in a high-level, general-purpose language. So for storing large amounts of historical data and generating sophisticated management reports, host computers are generally used.

PART III

THE PRACTICE OF AUTOMATION

CHAPTER
7

THE PLANNING AND IMPLEMENTATION CYCLE

INTRODUCTION TO DESIGN BY LEVELS*

Practitioners and users of batch control have not been completely successful at organizing and communicating what is known about the job a particular control system has to do and how it is to do it. We confuse specification with design, design with implementation, and functional requirements with constraints (which will be defined shortly). We are only now developing common languages for the definition of time-domain requirements, yet the time domain is what batch (and all sequential) processes are all about. There is no technique with the universal acceptance of, for example, the process and instrument diagram (P&ID), for documenting the time-dependent nature of a batch system. We will show that the diverse requirements of batch control really require several languages.

An additional problem is that system security factors are not always given appropriate prominence. System security includes much more than the reliability of individual electronic components. For example, it encompasses functional isolation of faults and alternative means of operator interface. Furthermore, many batch control systems are designed to make future changes difficult.

The following discussion—indeed the rest of the book—is organized around the

*This chapter is based on H. P. Rosenof, 1982, Successful batch control planning: A path to plantwide automation, *Control Engineering*, September, pp. 107-109.

control requirements of highly complex, multiple-sequence, multiple-product, chemical batching plants. The techniques have been used for food and pharmaceutical processes, batch wood pulping, and other applications.

BASIC PRINCIPLES

Specification and design are different but equally important. We can model the two as circles touching at one point (Fig. 7-1). When we finish the activities represented by this illustration, we have built nothing yet. Let us use the analogy of building a house. The specifications contain the requirements (i.e., three bedrooms). The design is a set of drawings intended to meet the intent of the specifications while reflecting the designer's detailed knowledge of building codes, material characteristics, industry practices, and so on. Only in the implementation stage is hammer taken to nail. Specification and design are the major subjects of Part III.

Considering specifications first, we note that they can be divided into two categories: *Functional specifications* are statements such as "there shall be 28 batch units," and "there shall be an alarm point for high level in the product storage tank." A good part of the rest of this discussion will consider how to express the relationships between such statements.

Other statements, however, do not directly address the immediate goal of the batch control system to control the batch process. They are statements such as "the central computer shall have at least 10% spare main memory," and "weighing of catalyst shall be by dedicated weighing system, separate from the control computer."

Again using the house-building analogy, we can say that nonfunctional specifications are like saying "all windows shall be double-glazed." Such specifications do not affect functional design, but they do affect implementation. Writers on systems analysis (DeMarco, 1981) call them *constraints*. The word has other meanings to people involved in process control; we prefer to call them *implementation requirements*.

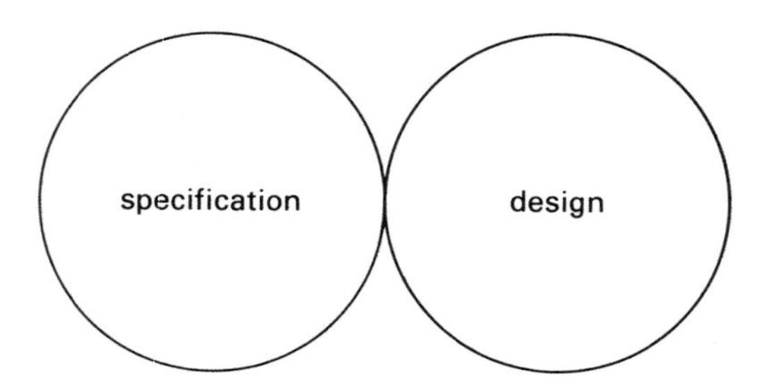

Figure 7-1. A graphic representation of the relationship between specification and design.

Our major interest is in the first kind of specification, the functional kind. A typical automated batch plant will have about eight *levels*. Each level may have a unique vocabulary, a unique method of communicating requirements. Each specification by level will be distinct but not really independent. They will have to be read in a specified order to be understood. For each specification level there will be a corresponding design level. Everything else stated for specification levels is true for design levels.

Figure 7-2 shows the completion/understanding sequence for specification and design. The specification should start at the general and proceed toward the specific. Then design should start at the specific and proceed toward the general. Although not shown, implementation usually follows the same path as design. There is no universal agreement that this sequence is correct. One often hears the phrase "top-down design," but this book recommends something quite different—bottom-up design.

Aside from the confusion and lack of communication that exists in this field, there are legitimate differences among people who are not confused. We have simply taken the approach that makes the most sense to us. If the design is bottom-up, there is a complete path from where the design is currently to the process. Readers interested in additional comments on the top-down/bottom-up decision are referred to the two articles by D'Angelo (1981).

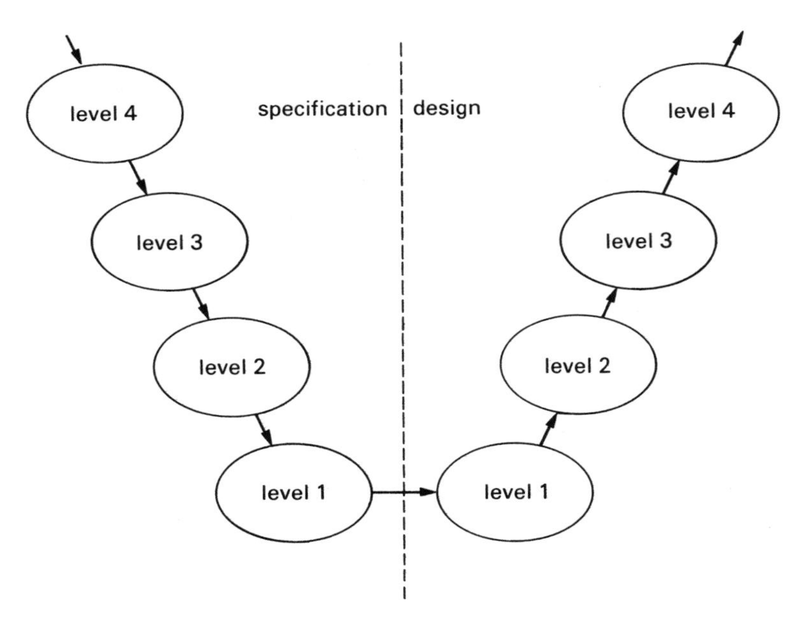

Figure 7-2. The sequence of completion and understanding for specification and design.

FORMAL DESIGN METHODS

Several highly formal design techniques are oriented chiefly toward commercial data processing users. Their intent is to guide the analyst in documenting some process (typically, order receipt) as a prerequisite to its automation. These methods have been used in real-time control applications; again see the articles by D'Angelo. Our method of batch specification can be seen as a *subset* of, for example, D'Angelo's structured analysis techniques (Demarco, 1978). Structured analysis will accept as valid any one of many different hierarchical breakdowns of the same process. The method that we detail will guide the engineer toward a model consistent with structured analysis but with the breakdown controlled by an additional set of rules.

DESIGN BY LEVELS: A FORMAL DEFINITION

A *level of control* is a part of a control system (meaning separately identifiable hardware, software, and/or firmware) characterized by a specific function, operator interface, computed data interface, command interface, and means of specification and design documentation. Levels communicate with each other according to a strict set of rules:

Instructions only go down (i.e., to a lower numbered level).
Each level must be able to operate safely without instructions from the level(s) above it.
Levels above a nonoperational level are irrelevant. It is better to leave them off.

The philosophy behind these rules is that the effects of a failure should go no closer to the process than the level at which the failure occurred. Good engineering and programming practice will serve to isolate the failure within the level.

Figure 7-3 details relationships between levels. (The ellipses are consistent with the standards of structured analysis. The drawing helps to illustrate that extensive training in formal design techniques is not required for understanding designs made by those techniques.)

Individual Levels Defined

The first level will be termed here "level 0." It is not part of the control system at all; it is the process—vessels and piping providing containment for the materials undergoing processing. The message is simply that the control system should not be required to compensate for problems in the process design.

The first level that is part of what would be considered the control system, "level 1," consists of hardwired safety devices, interlocks, and so on. These elements are

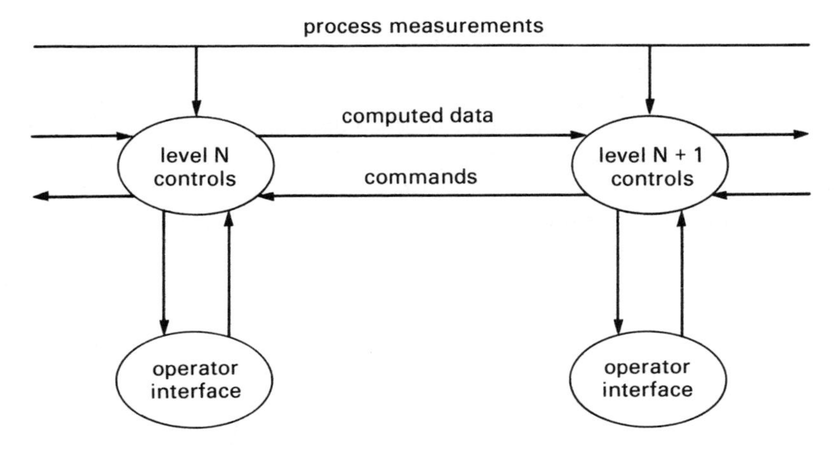

Figure 7-3. Routing of process measurement, control, and command data between levels. In practice, an item of data may bypass one or more levels.

the system's last chance to gain control over a process before we have to depend on the integrity of the process equipment itself to prevent damage and personnel hazards. The operator's interface to these is through traditional panel devices: indicator lamps, switches, and annunciators are typical. Specifications for controlling this level are by P&ID notation (the interlock symbol), logic diagram, and tabular lists such as for annunciator points. Design documentation is by elementary (schematic) electrical diagrams and similar mechanisms for other technologies.

Functional hardware found at this level is usually electromechanical relays, valve position (limit) switches, and the like. Typical functions are

Shutting off material flow to a vessel if vessel's net weight becomes excessive
Shutting off heat (and initiating cooling) to a reactor whose temperature exceeds a limit
Blocking a material charge to a vessel when the vessel's discharge valve is open

Clearly, this level exists not to make a product but to keep the plant safe.

Level 2 contains computational safety systems. They are almost the same as the safety systems of level 1, making use of interlocks, switches, and the like, and using similar forms of documentation. The difference is that in this level, we will accept some loss in simplicity in order to gain functionality. Typical functions here are

Differentiation of reactor temperature to determine whether an exothermic reaction is close to going out of control.
Multiplication of temperature differential across a reactor jacket, by the rate of

jacket heating medium flow, to calculate heat input to the reactor so that it may be limited.

Reduction of the charging rate of other materials being charged simultaneously, if one material cannot maintain its desired flow rate. (This "pacing" function is a standard feature of self-contained blending controllers.)

Finding the median value of a set of thermocouple measurements, to be used for control. (Sometimes the high value is taken. It is also possible to calculate deviation, from the lowest to the highest, for example, and initiate an alarm if the value becomes excessive.)

Once again, this is a safety level and will not make a product in any but the simplest batch plants.

At this level we will permit the use of some equipment that, for reasons of complexity and/or the possibility of a wide-scale fault, would not be used at level 1. This includes analog computing modules, programmable logic controllers, and emergency shutdown systems.

Levels 3, 4, and 5, the analog or continuous control levels, are functionally different, but since they are invariably embodied within the same hardware they will be considered together. Level 3 enables the operator to control outputs manually. It is the lowest level at which the typical batch process plant can make a product. Level 4 contains conventional analog control functions—PID, RATIO, and so on—operating as comparatively simple loops. At level 5, advanced (or coordinated) continuous control, loops operate together. Level 5 may be augmented by programmed control strategies using FORTRAN or another computer language.

For levels 3, 4, and 5 loop diagrams may function to document specification and design, since they are generally very detailed. The design is not complete, however, until documents like connection and wiring diagrams (and/or their software equivalents) are prepared.

Level 6 is sequential control. It may also be termed the level of basic batch control. A plant with automation to level 6 would have stand-alone solids batch weighing systems, stand-alone fluid metering systems, and perhaps some sequential control device, such as a drum controller or small programmable logic controller, operating a machine or individual process unit. This level is the ability to run simple, uncoordinated sequences in the plant, rather than all of the sequences required to produce a batch, which is the next level. At level 6 the operator can request the charging of a selected quantity of a selected raw material to a selected process vessel, for example. Operator interface is often through a unit control panel with thumbwheel switches (quantity), selector switches (material and vessel), and similar devices. Sometimes the interface is through CRT, typically with a graphic representation of the unit being controlled.

The traditional way of specifying sequences is by flowchart. For some good reasons flowcharts are falling into disuse, and various techniques are taking their place. One of the most generally useful is derived from structured analysis and is called *structured English*. Structured English descriptions can be considered as one-for-one replacements for flowcharts, but with some logistical features that make them much easier to use (Chapter 9).

Unfortunately, flowcharts and structured English share one important limitation: They show only one side of what is going on—the control side. If a vessel temperature increases when a particular valve is closed, this relationship is, at best, implicit and commonly completely hidden on a flowchart or structured English description. The graphical technique of process timing diagrams has been developed to show a control system as part of the process it controls—just as a P&ID does, but now in the time domain. (The analogy with continuous control is that when dynamic terms are put together around a closed loop, the source of each term is irrelevant to the resultant loop dynamics.) This technique has been used with success to show entire small batch processes on one piece of paper and to show particularly complex sequences within larger processes. Since the method requires extensive drafting support, its use to document all the sequences of a large process is not generally recommended.

Design documentation consists of commented code, annotated programmable controller ladder diagrams, and other methods similarly appropriate to the equipment being used.

At level 7, the level of complete batch control, the operator can give the control system commands like

```
MAKE A BATCH OF PRODUCT ABC
```

or

```
WASH MY CLOTHES
```

The control system achieves batch control by putting together individual sequences. (For a process like clothes washing there is only one sequence, and levels 6 and 7 are, in practice, identical.) Specification and documentation techniques are the same as for level 6. Methods of operator interface are also similar.

Level 8 is the level of production scheduling. Here the operator gives the control system a set of requirements such as

```
COMPLETE A BATCH OF PRODUCT ABC AT 8 A.M.
COMPLETE A BATCH OF PRODUCT DEF AT 10 A.M.
COMPLETE A BATCH OF PRODUCT GHI AT 12 NOON
COMPLETE A BATCH OF PRODUCT JKL AT 2 P.M.
```

This level represents a transition between what is generally thought of as real-time process control and a plant management data processing function. Documentation used for specification and design may be typical of the latter, including the data flow diagram (Demarco, 1978) similar to Figure 7-3 showing data flow between levels. The data dictionary identifies each variable and its functions. Implementations at this level may range from the quite simple to elaborate scheduling systems that take into account the timing of process steps, market demand, storage capacity, maintenance requirements, and so forth.

Level 9 is the level of plantwide automation. Questions are asked and answered about

Economic optimization of the production facility
Responses to changing feedstocks or material shortages
The maximum number of batches the control system must accommodate at one time
Overall plant energy management

In practice, most level 9 implementations are comparatively simple or nonexistent.

Higher levels may be defined to match the corporate organization—each plant reporting to the corporate production department, for example. These levels are outside contemporary control engineering practice and are not considered here.

Application Example

A typical specification-to-design progression will be used to illustrate the method.

Specifications

Level 9. The engineer responsible for the process determines that the weight of a particular material must be held to 0.05% of full scale when being charged to the reactor.

Level 6. Charging a material to such a high precision implies charging at a low flow rate. As a compromise, to prevent the overall charge time from becoming excessive, engineers decide to specify dual-rate charging.

Level 5. The switching mechanism for transition between flow rates.

Level 4. The mechanism to close valves when charging is complete.

Level 3. Controls for fast and slow feed rates.

Level 1. Interlock to turn off all feed when excessive charged weight is detected.

Designs

Level 1. Interlock to turn off all feed when excessive charged weight is detected.

Level 3. I/O (input/output) for valves and weight signal.

Level 4. Logic to open and close valves.
Level 5. The mechanism to switch between rates when required.
Level 6. The mechanism to read the recipe and calculate the switch point.
Level 9. Desired weight must be stored as a floating point variable.

This example points out that every level does not have to be implemented for every control function.

BAD EXAMPLES?

The design-by-levels terminology permits an analysis of previous or alternative control designs to identify potential weaknesses in comparison with designs adhering strictly to the design-by-levels method. Whether these designs are wrong or bad is up to the user, who should consider them specifically in the application context.

The first example concerns an exothermic reactor. We want to differentiate the reactor temperature to detect incipient loss of control. We decide to perform this differentiation in the computer. Only if differentiation is a "nice to have" feature rather than a "must have" is there the possibility of safe operation without the computer. If differentiation is a must have, then it is an interlock function, and the plant should not be operated unless all interlock functions are operational.

For the second example, differentiation is recognized as a must have interlock function and removed from the computer in favor of implementing it with simpler hardware. The rules of design by levels are not, however, fully met unless there is provision for an operator interface. The reason is that there might be a time when the monitoring function must be turned off—when charging with a hot material when the vessel itself is at ambient temperature, for example. The interface in this case could be a simple on/off switch. Otherwise, the upper level controls could turn off this alarm function and then fail, leaving the differentiation function out of operation.

The third example is a typical process control computer or distributed controller that detects certain failures in the field (e.g., an open circuit in signal wiring) or the equipment itself and makes them known to other functions within the computer. (A likely response to an input failure would be to control to the input's last valid value.) If a function within the equipment provides signals to both control and data logging functions, the user must carefully check the manufacturer's system security provisions and the user's own application of them to insure that a failure in the data logging function cannot propagate through the signal function and back into the control function. Although this is a theoretical consideration in computer control, the authors have never observed it. Figure 7-4 shows how the problem can become quite real in a modular control system. With a measurement going to both a recorder and controller, an electrical fault at the recorder can most definitely cause a control failure. Users in the nuclear power industry avoid this by dividing the plant into safety-related and non-safety-related areas and using isolation devices between the two. Another approach is simply to avoid sharing sensors and other facilities by control and display or recording functions.

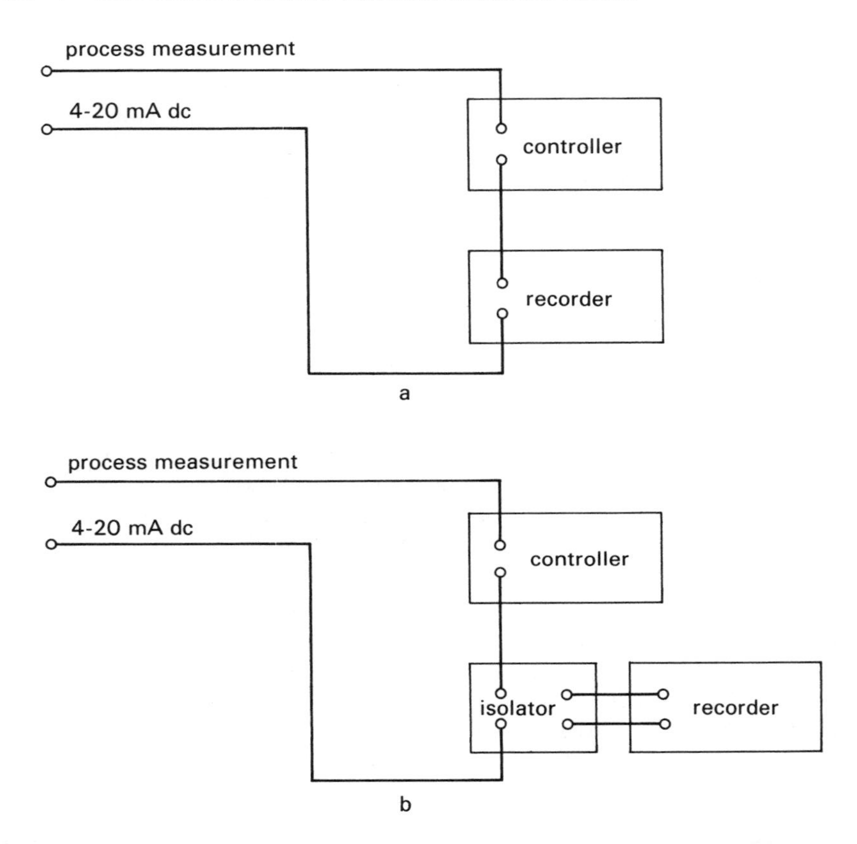

Figure 7-4. *a.* The controller in this loop performs a safety-related function; the recorder does not. An electrical fault at the recorder could disrupt the process measurement signal and so jeopardize plant safety. *b.* an isolator prevents such a fault from affecting the main loop.

HARDWARE IMPLEMENTATION

The design-by-levels technique permits a detailed description of batch process control requirements without reference to the selection of a particular hardware implementation. It actually demands such a description. Subsequent chapters discuss hardware considerations. Even here, however, the design-by-levels vocabulary is of great value because it provides orientation to the numerous options available.

CHAPTER

8

THE EXECUTIVE LEVELS

PRODUCTION SCHEDULING

Why Schedule?

There are several reasons for scheduling.

1. *To Meet Market Demand.* It may be as simple as arithmetic: dividing the production requirement over, say, a week by the batch size to find the number of batches needed. For a multiproduct plant the appropriate schedule can be used to maximize production.

2. *To Share Constrained Resources.* Some resources are simply not available to meet all potential demands. A shared material charging unit, for example, may have one flow transmitter for charging eight units. If two units are ready for their charges at the same time, one will have to wait. In many processes this waiting time has no process effect. In others it can ruin the batch, so the shared unit's allocation must be scheduled beforehand. In still other cases the incremental cost of providing a resource varies with demand (Fig. 8-1). An example is a pulp mill with cogeneration, in which excessive steam demand from the batch digesters, caused by starting several batches at once, can force the mill to buy expensive utility electric power.

3. *To Feed Downstream Units.* Where batch reaction is followed by continuous distillation, for example, the reactor(s) must provide a minimum feed at all times to the downstream unit. Better scheduling can minimize the need for intermediate storage.

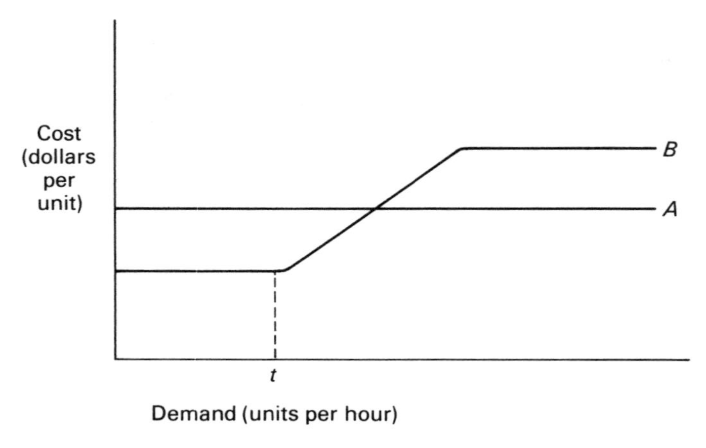

Figure 8-1. Curve A shows a resource with a constant cost over the range of demands considered. Curve B shows a resource that can be met, up to *t* units per hour, from within the plant but which must be purchased at higher cost to the extent that demand exceeds *t*.

4. *To Perform Maintenance.* A large plant may be scheduled to accommodate equipment downtime for maintenance.

5. *To Accommodate External Constraints with Limited Storage Capacity.* For example, a plant may have to be scheduled to accommodate the anticipated arrival of a tank car. Even when adequate storage capacity is available, such scheduling can improve plant economics by minimizing inventory.

Manual and Automated Scheduling

Three techniques commonly used to schedule batch plants are the intuitive method, the analytical method, and the simulation method. The *intuitive* method relies on the practitioner's ability to generate and solve a simple model of the plant. The model may be a mental model, or be supported by simple graphical representations of the process. An example of the intuitive approach is given in problem 1 in the next section.

Analytical methods model the process in precise mathematical terms, then solve the model by techniques such as linear programming. These techniques have proven their value in batch plant scheduling at a few centers, mostly overseas, but are becoming better known in the United States. Linear programming and related techniques have long been established, in the United States and elsewhere, for solving problems such as finding the refinery "slate" that gives maximum net revenue. Problems 2 and 4 give examples of these methods.

Simulation subjects a process model to influences that model the anticipated operating environment. In the simulation technique most often used for scheduling, the process model exists as one or more software modules, and the environment is represented by other modules that can act to change inventory levels, and so on. Some models are discrete, which means that the process interacts with its environment only through yes/no decisions. This kind of model would be used to study the effects of varying the number of highway toll booths servicing a given traffic pattern. Models that combine discrete with continuous interactions (such as for inventory drawdown) are generally of more interest for batch plant scheduling. The analyst generally sets up different scenarios and conducts experiments until the results are satisfactory. Problem 3 gives an example of this type of simulation.

This chapter gives an overview of automated batch plant scheduling. For more information, see the referenced articles and the rich operations research literature.

Typical Scheduling Problems

Problem 1: Find the Maximum Production Rate of a Plant

Although this is primarily a process design problem and not a control problem, it helps to establish background for more detailed scheduling problems, and its results can be used as a reasonability check for their solutions.

Some of these problems can be solved by common sense: If a clothes washer requires 30 min to wash a batch of clothes and a dryer requires 60 min to dry the same batch, then the "plant" can process one batch per hour (ignoring the time required to transfer clothes from one to the other). If the dryer can hold two loads from the washer, then, by intermediate storage, processing can be doubled to two batches per hour ("batch" is defined as a washer load). If washing and drying each take 1 h and the dryer has the capacity for two washer loads, then adding a second washer doubles capacity. These situations are illustrated in Figure 8-2. (For a more detailed discussion, see Barona and Bacher, 1983.)

General-purpose simulators are readily used to study more complex plants. Morris (1983) gives the details of a GPSS (general-purpose simulation software) model used for studying a chemical batch plant.

Problem 2: Schedule a Single-Product Plant for Minimum Cost

Batch plants are normally considered (at least by many control engineers) to operate according to strictly fixed recipes. One of the control system's responsibilities is to adhere to the recipe as closely as possible to achieve batch-to-batch uniformity. Some processes, however, are fairly tolerant of variations in times, temperatures, and even material amounts. These processing parameters do affect costs, so, with the required production rate fixed, the most economical schedule for operating the plant must be found.

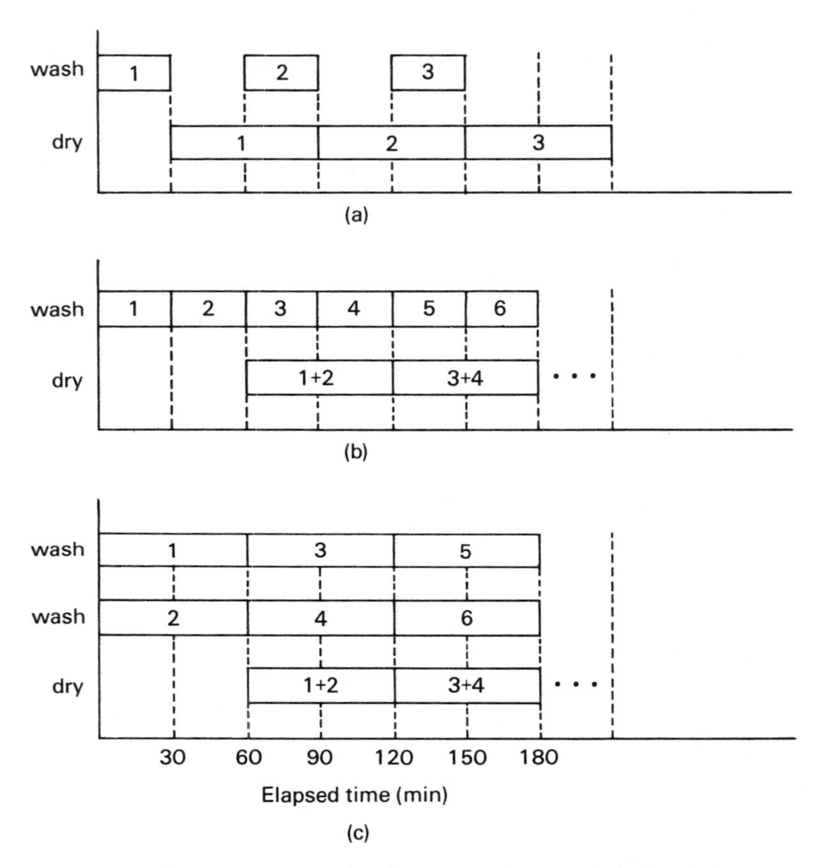

Figure 8-2. *a.* After 90 min, one washer-load is completed per hour. *b.* After 120 min, two washer-loads are completed per hour. *c.* With 1-h washing time, assuming the dryer has a capacity of two washer loads, a two-load-per-hour rate requires two washers.

Ellingsen (1985) describes the procedure he and associates used to solve this problem. A dynamic model of the process is developed, independent and dependent variables are identified, and limits are assigned to them. Operating costs per unit of steam, materials, and so on, are used to determine a cost function. Then a *nonlinear programming (NLP) algorithm*—a procedure to find values of variables that minimize (or maximize) the value of a function, where the equations describing constraints are nonlinear—is used to find the minimum cost based on a dynamic model of the process.

The NLP algorithm uses a trial-and-error technique and so is a substantial consumer of computer resources. The project did not justify the computer cost required for a conventional nonlinear programming solution. Initial results from the dynamic model were used to construct a simpler algebraic model that was solved by the NLP algorithm at a reasonable cost. Values of the independent variables found by this method were fed back to the dynamic model, and the model was run for one iteration. The approximate optimum thus found was fed to the regression model, and the regression model was solved by a less expensive algorithm.

A sensitivity analysis determined which independent variables required close control, and which could economically be subjected to less rigor. Finally, various local optima—conditions that do not lead to the minimum cost but do lead to reasonably economical operation—were investigated. For example, optimum plant operation was determined for the case of maximum-size batches. Maximum-size batches allow substantial downtime for maintenance. Although the "global optimum," the most economical schedule, did not use maximum-size batches, this local optimum solution allows the most economical operation consistent with the amount of downtime required.

A simpler situation is found in batch digesters used in the pulp and paper industry. Many paper mills are equipped for cogeneration. If the digester house uses so much steam that not enough steam is left to supply all of the mill's electricity requirements, then power must be purchased at a comparatively high price. Effectively, at this point, the steam is more expensive. Even a short-term purchase of power can lead to a large demand charge. In addition, rapid changes in digester steam demand can disrupt operations in the continuous process areas. Perron et al. (1975) use a combination of optimization and simulation to schedule a batch digester plant for minimum change of steam consumption. They developed a closed-loop control strategy that includes all six digesters of a paper mill. At each optimization step a plant simulation determines whether the optimization step can be accepted. The simulation also finds values for certain parameters not determined by optimization. The scheduling programs run in the process control computer and modify the cooking time controlled by lower levels.

Lemay (1979) describes a simpler solution in which a steam leveler program, operating in the regulatory levels, varies steam flow to the digesters to maintain a preset production schedule based upon the plant's required production rate. Fadum (1979) describes a similar system and includes a model for predicting required cooking times.

Problem 3: Schedule a Multiproduct Plant for Minimum Inventory

Sicignano, McKeand, and LeMasters (1984) describe the use of a simulation model running in a personal computer to schedule a solvent plant including both batch and continuous units. The operator begins a session by specifying the values of various independent variables, such as maximum unit charge rates and the anticipated shipping schedule. The computer calculates anticipated inventories over

12-h intervals. If plant operational parameters are not met (e.g., inventory of a material below mandated minimum), the operator has an opportunity to return to the previous simulated time (12 h before) and make a correction. The program operates over a 30-day planning horizon.

Problem 4: Schedule a Multiproduct Plant for Maximum Profitability

This problem is difficult, and much has been written about it in the past few years. The problem may be formulated in various ways: Production cost may be included or excluded. (Excluding production cost from the model implies, for example, that the cost per pound of a product is not significantly affected by the particular vessel chosen or the fraction of the vessel utilized. Obviously, the validity of this assumption must be established before using it.) The market may be assumed to be inelastic (it will absorb all of a plant's production at a fixed price per material), sharply limited (it will absorb a certain amount of a material at a fixed price), or elastic (as the product increases in availability, its value per pound decreases). We are aware of no simulations using elastic pricing; presumably in most cases a single plant is incapable of affecting the market enough to change a product's price.

Probably the most active author in this field is D. W. T. Rippin. Together with Egli, he described, in 1981, the computer program SRSBP (for short-range scheduling for batch plants). Using fixed batch sizes, this program accepts schedules of raw material arrivals and product shipment commitments, utility constraints and other information, and proposes the most economical production schedule—over a 60-day planning horizon, in the example given—considering the costs (and times) of product-to-product changeovers and interest for financing intermediate inventory.

The program uses a technique that is almost mandatory in solving problems of this type. In an effort to limit exhaustive evaluation of an excessive number of candidate solutions, these solutions are filtered through a set of rules, "heuristics," that, in this case, eliminate those that do not satisfy basic plant constraints. Then an "ideal" schedule representing minimum production cost is developed. Then additional constraints are applied (e.g., for maximum electricity demand), yielding a best practical schedule. When run on the appropriate computer, the program is fast enough to enable short-term recalculations to account for equipment failures and similar change.

The Batchman program (Mauderli and Rippin 1980) addresses the even more complex problem of multipurpose batch plants—plants in which a variety of equipment lineups are available to manufacture particular products. The Batchman program, faced with a greater computational load, similarly screens candidate solutions and operates in less detail (fewer influences recognized) than SRSBP.

King (1984) discusses a technique for scheduling shared units based upon the value of the units' services to the various "tasks" (usually individual batches, in our case, represented by process vessels) that are in contention for it. The tasks are assigned "demand units" by the user, and each shared unit is allocated to the task "bidding" the greatest number of demand units. The scheduling system functions within a simulation. The result is, among other things, a GANTT chart showing the process.

The user assigns demand units indirectly, using a set of rules. The rules are supported by a set of functions written in the artificial-intelligence language LISP. Demand units may increase with time, so if a task is denied service at one attempt, it may receive the same service later when it bids a greater number of demand units. (The shared units operate under rules directing them to accumulate the greatest number of demand units as "payment.")

This system was developed to schedule the resources required to support the space shuttle between flights. With contention rules reflecting a model of the marketplace, such a technique could be used in the process industries as well. Such a marketplace model might not use absolute completion dates as Rippin does, but instead it might assign "incentives" and "penalty clauses" for completions before or after the scheduled date.

RECIPE STRUCTURE*

When a batch control system operates according to a set of variables that may be changed for each batch by operating personnel, that set of variables is a recipe. *Recipe structure* is the arrangement of these variables. Using the case study approach with examples typical of chemical processing, this section develops a suitable recipe structure for each of these processes. The structures are of two general types: those requiring the recipe to fit within a predefined master sequence, and those without such constraints. Each type has significant advantages and disadvantages.

Simple structures

Example

A system consists of ingredient storage tanks, ingredient handling equipment, and one or more batching tanks. The system is used to make one product, and the formulation of that product changes so infrequently that the sequence may be

*The section on recipe structure has been adapted from H. P. Rosenof, 1982, Building batch control systems around recipes, *Chemical Engineering Progress,* September, pp. 59-62, by permission of the American Institute of Chemical Engineers. Copyright © 1982 by the American Institute of Chemical Engineers.

written directly into computer/programmable controller programs. All ingredients are metered as they flow into the batch tanks.

The recipe for this system consists of only one variable, the batch amount. Ingredient amounts are defined as fractions of the batch amount; since the formulation is essentially constant, these fractions are contained either within software or in tables accessible to programmers but without a recipe maintenance function. This process system is based on batching tanks, in which ingredients are assembled, calling shared resources for service. The sequence is arranged the same way. The step

```
CHARGE A PERCENT OF BATCH AMOUNT OF MATERIAL B
```

implicitly includes the specification

```
. . . TO THE BATCHING TANK FOR WHICH THIS LOGIC IS
BEING PROCESSED.
```

An instruction to

```
DUMP PRODUCT TO STORAGE TANK C
```

similarly would refer to the batching tank. We will refer to this characteristic (that of all material transfers being to or from one location per batch) as the *single-center* (SC) attribute and consider more complex recipes for these systems.

Example

A system is arranged the same as in the preceding example. Ingredients are charged in a constant order, but proportions may vary from batch to batch. The total batch amount is not necessary as a recipe variable; it is replaced by a separate specification for each material. Amounts are stored by order of charge in a table (Fig. 8-3) in which engineering units are interpreted by the control system's logic.

The operator may enter the appropriate values immediately before each batch, but in many applications the recipes are previously written, and the appropriate one is simply selected. Therefore a recipe maintenance function is needed that, in this case, should

1. Identify each recipe in the system
2. List the values of a selected recipe
3. Accept changes to a recipe
4. Delete and add recipes

amount	amount	amount
103.2	97.5	98.0
857.3	902.0	857.3
1520.	1508.	1617.
10.57	19.5	19.5
0.0	5.2	7.1
17.5	22.0	15.8
2.115	3.207	0.0
75.3	75.3	82.0
10.0	0.0	11.6
62.3	55.3	68.7

Figure 8-3. Recipes for a 10-ingredient batch process. *(After Rosenof, 1982b; reproduced by permission of the American Institute of Chemical Engineers.)*

The ability to duplicate a recipe is useful but not required.

If charge amounts of 0.0 are frequently encountered, the size of the recipe file may be reduced by storing charge ID tags along with the amounts. Instead of an amount of 0.0 in Figure 8-4, the absence of a charge at a particular point is indicated by the absence of its ID tag, called *logic location*.

Example

A system is arranged the same as in the first example. Out of many possible logic locations for charging, only a few are used for each product type. However, several ingredients may have to be charged at particular logic locations, and the ingredient types to be charged at each logic location vary with the product type being manufactured. For each ingredient to be charged, the recipe contains the logic location, ingredient type, and amount. For multiple charges at one logic location, the logic location number simply appears once per material (see Fig. 8-5). Note that at logic location 14, materials AQ, X5, and PK are to be charged.

The recipes of the preceding examples may include temperatures, ramp rates, sample points, and so on, in addition to ingredient specifications. Where changes are always made at the same points, it is convenient to store these variables in a table similar to that of the first example. Otherwise, all points where there might be changes should be defined as logic locations, and variables should be stored with corresponding logic location numbers (see Fig. 8-6). At four points in the batch the controls are required to ramp batch temperature to a new set point at a specified rate. At logic location 622 the temperature control is turned off.

logic location	amount	logic location	amount
01	403.7	01	516.0
02	1818.	03	3727.
18	2.3	16	185.5
35	27.8	42	660.2
51	1523.	44	48.3
52	3.2	60	3.8
53	50.1	72	19.1
74	320.8	81	175.
88	10.2		
90	12.6		

Figure 8-4. Recipes for a 10-charge (maximum), 99-ingredient batch process. *(After Rosenof, 1982b; reproduced by permission of the American Institute of Chemical Engineers.)*

logic location	material code	amount
02	VM	1058.
14	AQ	2.2
14	X5	15.6
14	PK	41.9
23	BH	800.7
50	J2	12.2
82	AQ	1.8
93	Z5	17.9

Figure 8-5. Recipe specifying logic location, material type, and amount for charges to the process. *(After Rosenof, 1982b; reproduced by permission of the American Institute of Chemical Engineers.)*

Example

In Figure 8-7 a system has some ingredients to be weighed in separate tanks and charged after weighing.

This example differs from the first three by the presence of additional tanks to be charged under recipe control. Systems of this type are referred to as *progressive-center* (PC) systems, since material transfer destinations progress through the system.

logic location	temperature	ramp rate
601	108.0	2.5
605	305.6	0.8
608	218.2	1.2
620	45.0	2.0
622	0.0	0.0

Figure 8-6. Logic locations used to specify control actions. *(After Rosenof, 1982b; reproduced by permission of the American Institute of Chemical Engineers.)*

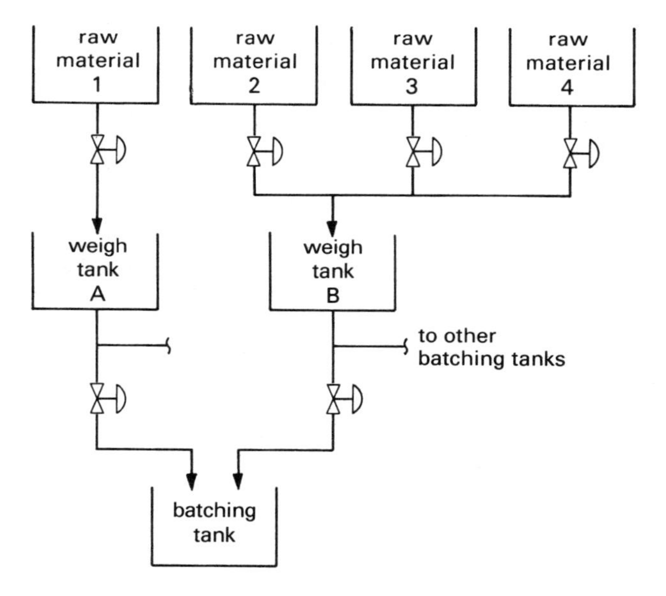

Figure 8-7. Arrangement of tanks for the weighing of raw materials. *(After Rosenof, 1982b; reproduced by permission of the American Institute of Chemical Engineers.)*

Charge Anticipation

The recipe structures of the preceding examples may be used directly as described, but this method is inefficient because the total idle time for charging is the sum of the individual preparation times. Efficiency is increased when charges are prepared before they are needed; to do this, charge anticipation must be added.

One way to do this involves no recipe changes. Instead of being designed to respond to requests from the logic of the unit to eventually be charged, the charging unit's logic is provided with the capability to read the recipe itself and prepare its charge independently. The charging unit synchronizes with the unit to be charged only when the charge is to take place, not during preparation. Multiple charging units may do their preparation simultaneously, and multiple batches can be in the system at one time, further increasing efficiency.

This method requires weighing tanks to be dedicated, each tank receiving a charge. Alternatively, when a weighing tank must be *shared* by tanks below it, the recipe may be modified to separate the preparation and charge functions. Responding to the prepare instruction, the logic of the receiving tank requests the charging unit to prepare the specified amount of the specified ingredient. When it reaches the charge instruction, the receiving tank waits for the charging unit to signal that it is ready. Then the charge takes place. This second method also enables partial charging of the ingredients prepared in the weighing tank (Fig. 8-8).

Figure 8-8 shows that every movement of material is controlled by an entry in the recipe. At logic location 110, 250.0 lb each of raw materials 2 and 3 are charged to weigh tank B. (The "A" indicates that the following value is to be treated as an amount.) At logic location 122, 55.3 lb of this mixture are charged to the batching tank. At logic location 127, 15.7 lb of raw material 4 are added to weigh tank B, and then all the contents of weigh tank B are charged to the batching tank at logic location 142. Logic locations 110 and 127 are processed by weigh tank B's logic, and logic locations 122 and 142 are processed by the batching tank's logic; material destinations are implied by the logic location numbers.

Since the requests to a shared weigh tank may come from more than one batching tank, weigh tank logic must have a way to prioritize requests. A description of these methods is contained later in this chapter. There are at least two general approaches:

Logic location	Source	Amount/Percent	Value
110	RM02	A	250.0
110	RM03	A	250.0
122	WTB	A	55.3
127	RM04	A	15.7
142	WTB	P	100.0

Figure 8-8. Separation of recipe into preparation and charge steps. *(After Rosenof, 1982b; reproduced by permission of the American Institute of Chemical Engineers.)*

1. By time, with simultaneous requests either prevented by system design, or serviced according to a preestablished order
2. By operator decision

Variable-Sequence Recipes

All of the recipe types discussed so far require a *master sequence*—one that includes all sequences that will actually be used. Ingredient types and amounts may vary, but if a charge is to occur at a particular point in the batch then that charge must be part of the master sequence (MS).

Recipes can also be written as a sequence of phases. Typical phase functions include CHARGE INGREDIENT, START TEMPERATURE RAMP, and WAIT FOR TEMPERATURE. The recipe (see Fig. 6-10) consists of a list of phase functions and, associated with each use of each function, a set of variables to be interpreted according to phase function logic. Recipes written this way make prior definition of all sequences unnecessary and are thereby more accommodating to the development of new processes. Implementation of these *variable-sequence* (VS) recipes is simple for SC systems. For PC systems efficient use of time requires charge anticipation for all separate weighing units, whether dedicated or shared. Charge anticipation is accomplished by adding PREPARE REQUEST functions as required.

Several varieties of sequence phases may be included in a VS system:

1. Instantaneous steps that change control system parameters, then exit. A CHANGE TEMPERATURE step may, for example, set final temperature and ramp rate values in the continuous control software and then exit.
2. Endpoint steps that wait until the process reaches specified values, or until the control system performs some action; then they exit.
3. Operator action steps that wait until the operator performs some action; then they exit.
4. Time steps that wait for a specified time (generally a recipe variable); then they exit.

Enhancements to VS recipes are required if the system is to control more than one batch in the plant at a time. Without enhancements the logic of only one unit in the system "reads" the recipe and directs other units' logic to prepare charges and perform other tasks. An additional unit with such capability is needed for each additional batch to be accommodated at one time.

When multiple batches are to be controlled with a VS recipe, units should be able to independently perform all process logic up to the point of synchronization with a unit not yet available. The most flexible arrangement is to provide recipe-reading capability to all tanks, minimizing idle time between batches. As soon as a tank has finished one batch, it is available to start the next one. With this capability available to only one tank in a multiple-tank unit, the tanks finish a batch, and start a new one, as a group. The group as a whole cannot begin work on a new batch until all tanks have completed the previous batch.

In a VS system, control of execution is generally transferred from master logic for a unit to the software modules that perform sequential process control. The unit master logic provides housekeeping functions to insure that the correct phase is called and that it accesses the correct data from the recipe.

To enable a phase to direct the execution sequence, we pass a parameter from it to the master logic. The usual value is +1, indicating that the sequence may progress according to the recipe. A value of −1 indicates that the previous phase should be repeated, +2 directs that the next phase should be skipped; other values have corresponding meanings. A value of 0 could be used to repeat the current phase after a predetermined or process-dependent time.

Example

A chemical process requires that batch temperature be raised to a certain value and held for a certain time. A test is performed. If the batch fails, the temperature is increased by a predetermined amount and the temperature soak is repeated. The test is performed again, and until the batch passes or some process limit is reached, the entire sequence may be repeated indefinitely.

The control system here must accommodate changes in sequence as well as allow a step-control software module to write into and perhaps read from the recipe values associated with another module. In a system with minimum sequence rules imposed on the process designer, the module may have to search backwards until it finds a change-temperature step, and adjust its displacement value accordingly.

Example

A chemical process requires that a test be performed at several points in the sequence. If the batch fails with a low result value, steps are repeated as described in the preceding example. If the test is failed with a high result value, a corrective step is to be started immediately, with the process then to resume at the step after the one that performed the test.

This sequence introduces two new requirements to the batch control system: the step module must be able to calculate a displacement to a fixed step number or pass the fixed step number rather than a displacement to the master logic. Second, the system must retain the "last step number" so that control can be returned to the appropriate point.

Variable-Sequence (VS) and Master-Sequence (MS) Recipes Compared

With the greater flexibility offered by VS recipes, one might question why an MS-type recipe would ever be used in a complex system. MS recipes, however, have advantages that in some applications may make them the preferred choice:

1. *Simpler Recipe Maintenance.* VS recipes require maintenance software that can create space to insert a new step or fill in the hole left after a step is deleted. The software must keep track of the engineering units associated with each variable used by each step type. It may have to perform complex, sequence-based checks before accepting a new or changed recipe. Software for MS recipe maintenance is comparatively simple.

2. *Greater Memory Efficiency.* MS recipes appear to require less memory space than comparable VS recipes. (This advantage is becoming less important as larger computer memories become available.)

3. *Implicit Synchronization.* Synchronization between units is generally implicit in MS recipes since they are written around preestablished synchronization points. VS recipes are written around units and may require the recipe designer to carefully specify each synchronization.

4. *Easier Process-Controlled Sequence Changes.* There are few, if any, practical batch processes that are completely specified in a recipe. Sequence reversals with VS recipes must be done carefully, and complex changes may indicate that an MS recipe structure should be used.

5. *Match to Process.* Some processes are structured so that MS recipes are clearly the logical choice. The flexibility of VS recipes is of no benefit.

6. *Emphasis on Parallelism.* Simple batch systems tend to proceed as a sequence of steps. Many complex batch systems are more accurately described as having parallel sequences. This characteristic is emphasized by system structures using MS recipes: points at which the logic for a batch unit must operate in synchronization with logic for other units are identified; otherwise, the units operate independently. A VS recipe, however, can look like one sequence of steps, and careful analysis might be required to find the points at which units operate together and those at which they do not. For complex systems, therefore, MS recipes may be easier for many to use than VS recipes.

7. *Easier Restarting.* Problems with process equipment can cause a control shutdown. Restoration of normal control is comparatively simple when the affected batch unit has been operating alone. At other times logic processing may be suspended for more than one unit. The use of logic locations enables multiple units to be restarted in synchronism, usually at a point immediately after that at which control has been shut down.

8. *Emphasis on Unit Operations.* The structure of a MS recipe emphasizes unit operations rather than the individual control actions composing the unit operations. Some chemical engineers have reported that this feature makes MS recipe designs much easier to understand.

Recipe Movement

The process recipe, as established in the control system by the engineer or supervisor, is in many systems not used directly for process control. Instead a *working copy*, made from this *master copy*, is made available to the batch logic and to the operator. The master copy is never overwritten by the working copy. Thus a high level of security is maintained for the master copy, but the authorized operator is able to change the recipe for a particular batch.

This arrangement also allows the recipe to be modified by the logic. The process may require, for example, that if the batch fails a particular test, a subsequent charge must increase. The logic would write the new value to the working copy, but the master copy, and therefore later batches, would not be changed. Because this subsequent charge may take place in a downstream vessel, recipes generally "move" through a plant along with their corresponding batches. At the end of the batch the working copy can be compared to the master copy to highlight recipe changes since the beginning of the batch.

The recipe may also move in a direction opposite that of the material. This movement occurs when the recipe is transmitted to a material charging unit. This copy of the working recipe typically contains a reference to the *specific* charge that is to take place so that the charging unit can perform the charge(s) specifically required at that point in the batch. This working recipe copy is usually not copied out of the unit but simply overwritten for the next charging operation.

BATCH INITIATION

In a single-product plant the starting of a batch can be as simple as pushing a button. In plants capable of producing a limited range of products, the operator may be presented with several pushbutton switches, each to start the manufacture of a different product. Where ingredient amounts, processing times, and other variables can change with each batch, the operator uses thumbwheel switches or other means of numerical data entry, usually combined with an "enter" switch to prevent inadvertent changes (Fig 8-9).

CRT terminals provide similar functions in a compact space. They are also used for purposes not usually associated with hardwired panels such as the entry of an alphanumeric batch identification code, which is later used for sample labels, the batch report, and possibly the product itself. The CRT is also better at reporting problems found by the system as it attempts to start the batch, such as VESSEL IN USE, RAW MATERIAL LOW, and VALUE OUT OF RANGE.

In some plants operator action is not required to start each batch. These may be continuous-batch plants, in which the batch part of the process is responsible to

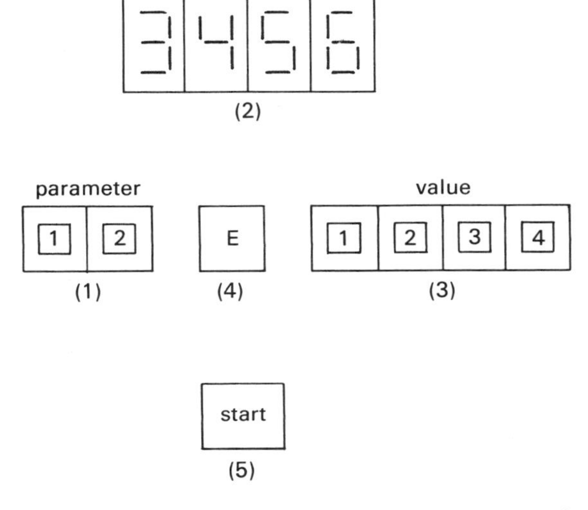

Figure 8-9. The operator specifies a parameter of interest by setting the corresponding number at the parameter thumbwheel (1). The present value of that parameter appears at the display (2). To change the value, the operator sets a new value at thumbwheel switch (3) and presses button (4). Once all parameters have been set, the operator starts the batch with button (5).

maintain a desired flow rate into the continuous part. Batch-by-batch scheduling is accomplished within the control system based upon the desired flow rate, the quantity of product produced per batch, batch equipment out of service, and other factors.

CONTINUOUS-BATCH CONTROL

Viewed from the outside, the continuous-batch plant uses raw materials (typically in batches) to produce a continuous stream of product (Fig. 8-10). Viewed from the inside, the plant is equipped similarly to other batch plants, but by the use of more than one process train and/or intermediate material storage it provides continuous product flow. The need for these plants is usually dictated by downstream, continuous-processing steps, such as distillation and extrusion. A common continuous-batch plant is the reciprocating engine, in which combustion takes place in batches to produce force in batches. Intermediate product storage is provided by the inertia of a flywheel, so variations in speed are minimized by the time the output is used.

Many continuous-batch plants operate roughly the same way—by using several

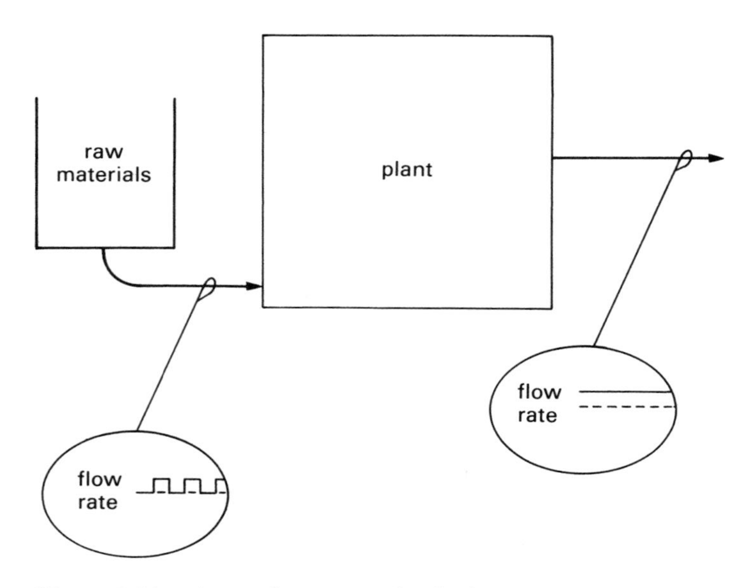

Figure 8-10. A typical continuous-batch plant viewed from outside.

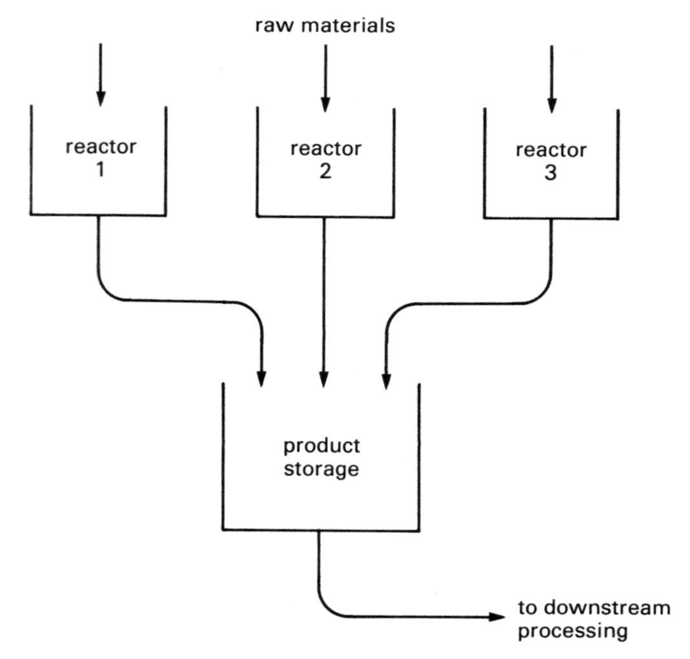

Figure 8-11. A continuous-batch plant viewed from the inside.

batch reactors and converting discontinuous flow to continuous flow just upstream of the plant outlet (Fig. 8-11). In this case, automation can help to reduce plant capital costs by minimizing the need for storage capacity, and operating costs by minimizing the need to finance the stored material.

Conversion from discontinuous to continuous flow can also take place upstream of a continuous reactor: Best (1975*a*) describes batch-by-batch material weighing into a continuous reactor, made necessary by the unavailability of suitable flowmeters (Figs. 8-12 and 8-13). This application not only provides a continuous flow but allows for each batch to be weighed twice—an advantage not available with the more obvious parallel arrangement (Fig. 8-14). This double weighing permits detection of a failure within the weighing equipment.

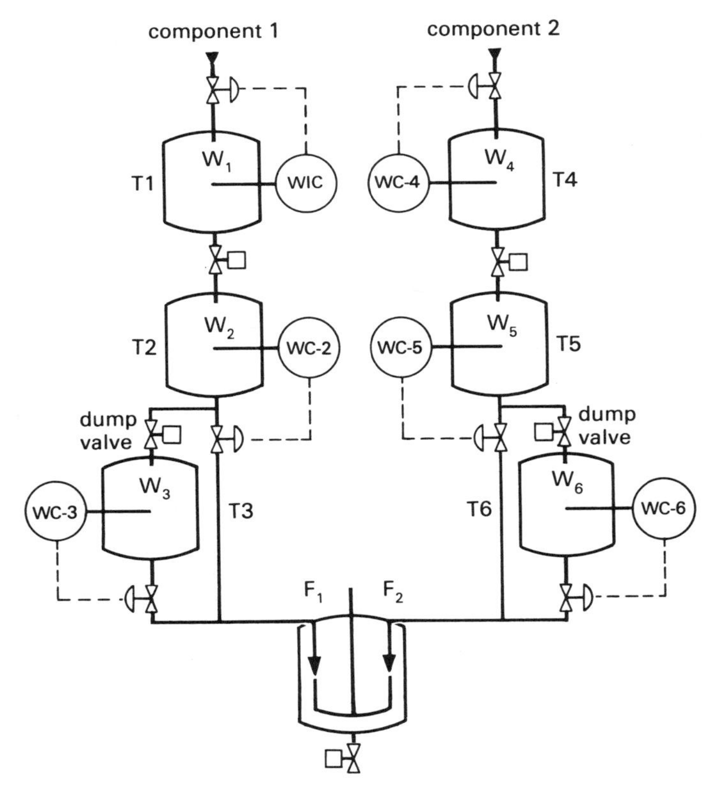

Figure 8-12. Conventional control with flowmeters in this dyestuff process would not provide the required accuracy in ratioing components of vessels 1, 2, 4, and 5 with load cells. *(After Best, 1975a; reproduced by permission of the Instrument Society of America.)*

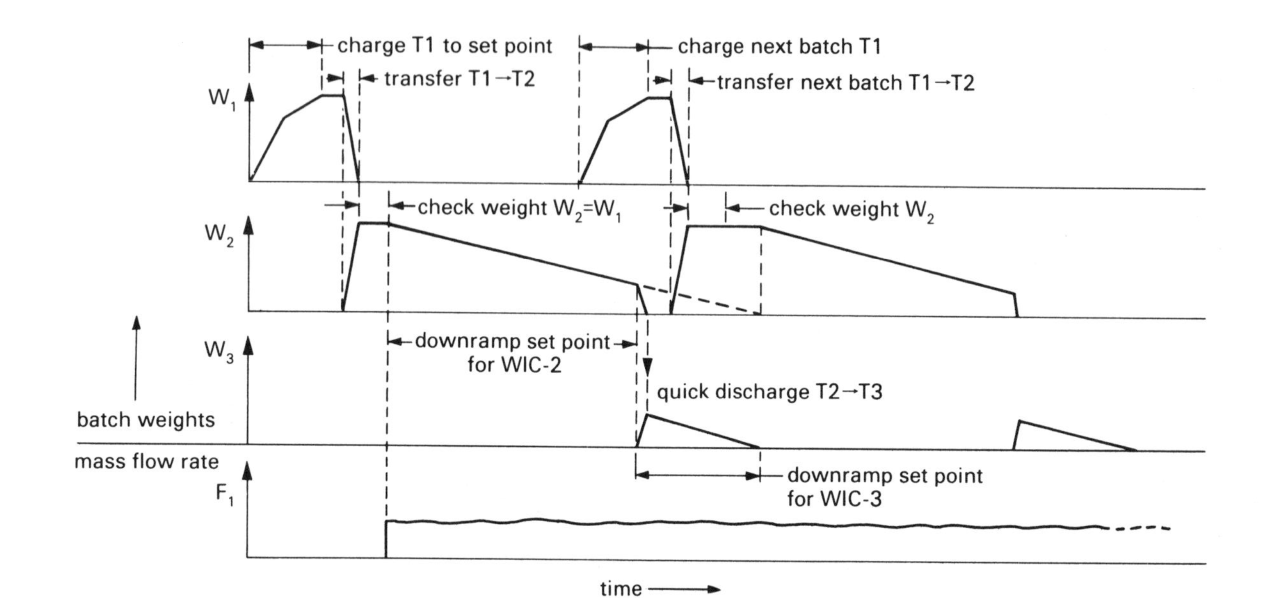

Figure 8-13. Timing diagram illustrating the operation of the system in Figure 8-12. Weighing vessel T2 cannot accept the contents of T1 and simultaneously check this weight while dispensing liquid to the synthesis reactor. Therefore, auxiliary vessel T3 is added to the train. Toward the end of the discharge ramp, at a preset weight value, the remaining contents of T2 are quickly discharged or dumped into T3. The auxiliary vessels T3 and T6 thus feed product to the synthesis reactor during the last part of the discharge cycle. These "dribble" flows are also controlled by downramping the weight set point. *(After Best, 1975a; reproduced by permission of the Instrument Society of America.)*

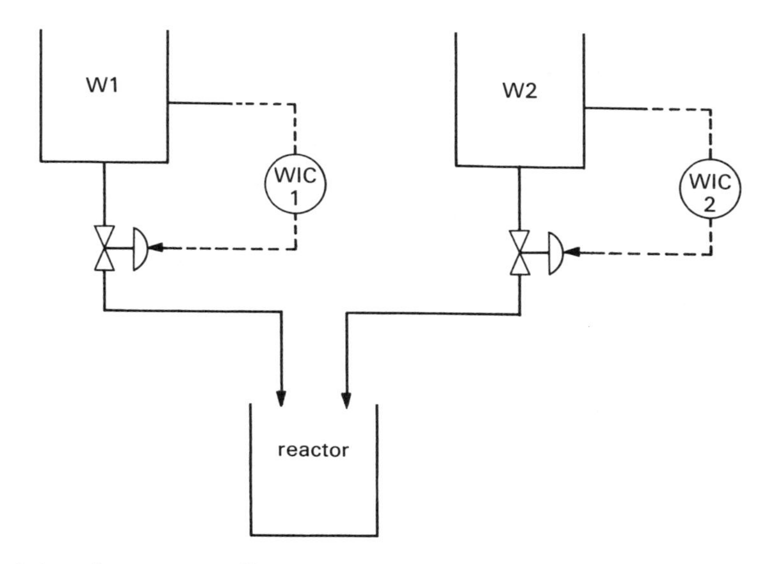

Figure 8-14. In a more typical lost-weight charging system, each component is weighed as it goes into its charging tank, W1 or W2, and again, by the same load cells, as it is slowly discharged into the reactor. Charging must stop when W1 and W2 are being recharged.

DEFINING BATCH UNITS*

As discussed under *Recipe Structure* and to be discussed further under *Interunit Synchronization*, all but the most trivial batch-process plants are divided into process units. Each unit is capable of being controlled by a separate stream of sequence logic. Coordination with other units takes place as required. Otherwise, one sequence stream must control the entire plant, and in many cases will be a serious bottleneck.

The following criteria help us define the units:

1. The larger the number of units, the greater the number of independent actions that can be controlled concurrently.
2. The larger the number of units, the more interunit communication will be required.
3. Units should be chosen to minimize the amount of interunit communication.
4. The larger the number of units, the greater will be the overhead requirements for module scheduling, the burden on the operator, and so on.

---

*The sections on defining batch units and interunit synchronization (except for merged synchronization have been adapted from H. P. Rosenof, 1982, Interunit synchronization in batch systems, *Advances in Instrumentation* **37**(part 2):863-877, by permission of the Instrument Society of America. Copyright © 1982 by the Instrument Society of America.

5. Material transfers should only take place when all units involved are ready. Confirmation is generally done with software, but where a material transfer between tanks requires two valves to be open, each valve should usually be a part of the same unit as the tank closest to it.

Commonly, units are defined to match process vessels. On occasion, two or more vessels may operate so closely in conjunction with each other that it is practical to combine them into one unit. Individual pieces of equipment like pipelines and valves are usually associated with process vessels, but sometimes a plant's flexibility is enhanced if they are kept separate. Some systems are benefited by the use of units without corresponding process equipment, such as for an operator data entry function. In most cases the computer's operating system provides all the contention logic required for system resources.

INTERUNIT SYNCHRONIZATION

Charging Systems

Batch processes require the movement of material and coordination of resources among items of plant equipment. Rarely is all control performed by one hardware-software module; usually the plant is controlled by multiple modules of software, and possibly hardware, that must communicate with each other in the performance of these functions.

Why Synchronization?

Figure 8-15 depicts a common situation in batch control. Batching tanks 1-3 are piped to receive charges from tank 4. Plant cost is minimized by having material to

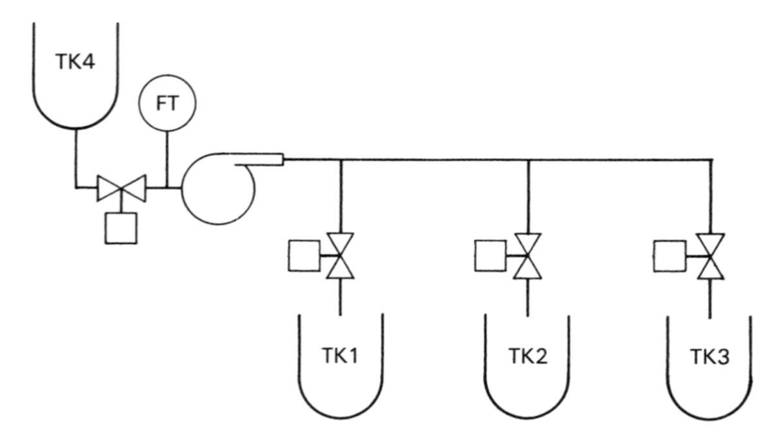

Figure 8-15. A shared charging unit serving three process vessels. *(After Rosenof, 1982c; reproduced by permission of the Instrument Society of America.)*

each batching tank flow through one flow transmitter. With this arrangement we must control the system so that only one of tanks 1-3 is being charged at any time. It is most convenient to allow each charge to run to completion before beginning the next one.

Our first attempt to control this part of the process might be as shown in Figure 8-16. One computer program is written to control the entire process for all three batch tanks. The charge from tank 4 is assumed to be at step 10. If any tank is at step 10, it receives its charge from tank 4 at this point.

This structure is unlikely to be found in current systems or in many previous ones. Its most obvious drawback is inefficient use of time—while one tank is being serviced, no control actions are performed on the other two. In addition, a program

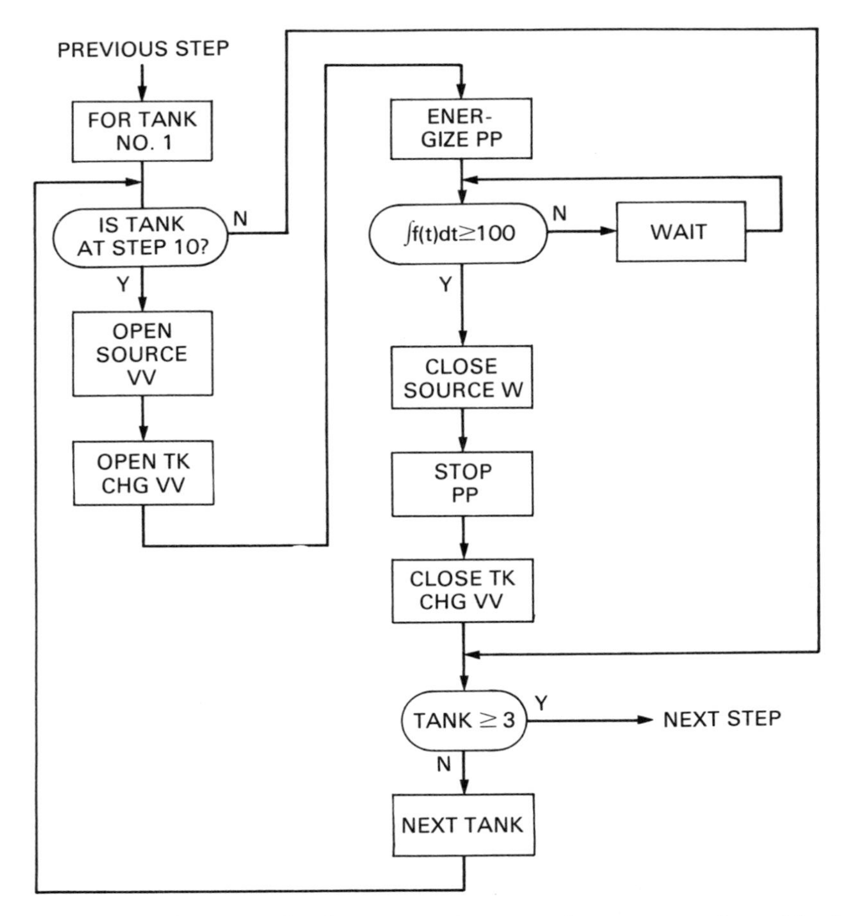

Figure 8-16. While one tank is being serviced, no control actions are performed for the other two. *(After Rosenof, 1982c; reproduced by permission of the Instrument Society of America.)*

this large will be difficult to write and debug and will subject all batch tanks to loss of control if any error in computer-programmable controller (PC) processing is encountered. Note, however, that there is an implicit arbitration of contention for the shared resource. Even if all three batch tanks reach step 10 at the same time, the computer will not attempt to charge more than one at a time.

Real-time multitasking operating systems in computers and PCs help solve these problems and allow the control software to be divided into modules and the modules to be sequentially executed for various functions (Fig. 8-17). They also provide communication between modules.

Figure 8-18 shows a module for step 10. The module is called by the "executive" up to three times per "scan," once for each batching tank. If tank 1 is at step 10, its processing waits for the completion of the charge. Other tanks at other steps are not affected. If tanks 1 and 2 are both at step 10, however, there is a possibility that tank 4 would attempt to charge them both at the same time. To prevent this from happening, we must introduce synchronization between the units.

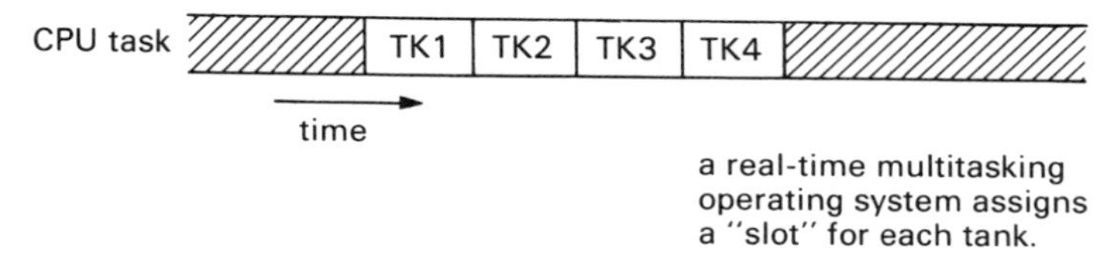

Figure 8-17. Division of software into modules. *(After Rosenof, 1982c; reproduced by permission of the Instrument Society of America.)*

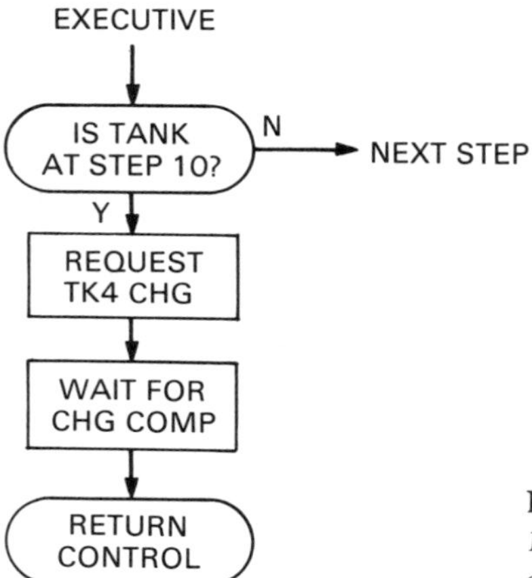

Figure 8-18. Module for step 10. *(After Rosenof, 1982c; reproduced by permission of the Instrument Society of America.)*

Centralized Synchronization

A *supervisory unit* can be added to the system to arbitrate requests from the logic of the three tanks. The function of this unit is depicted in Figure 8-19. When a tank's logic reaches step 10, it sets a bit (or changes a hardware output state). These signals are labeled Qn (n = tank number, 1-3) for queue. The supervisory unit checks all Q signals in sequence, at intervals between sequences typically of 1 s. When it finds a request for a charge, it directs the tank 4 logic to perform the charge via the T (tank) signal, which has the value 1, 2, or 3 (or 0 for no action required). When the charge is completed, the tank 4 logic sends the C (completion) signal to the supervisory logic, from which a completion signal Cn is sent to the appropriate tank. The V signal represents the value of material to be charged and is used in applications where the amount is not fixed.

Throughout this discussion we assume that each signal is controlled by only one unit and is readable by all. This characteristic is typical of PCs, distributed control systems, and some computer systems. Even where, as in some computers, these signals can be true globals (controlled by any unit), adhering to this principle should lead to increased system security.

The arrangement of Figure 8-19 is practical but limited. It would require that all valves needed for charging be part of the tank 4 unit, preventing compliance with criterion No. 5 for the definition of units.

Of course, the logic functions discussed here are not the only ones that can be

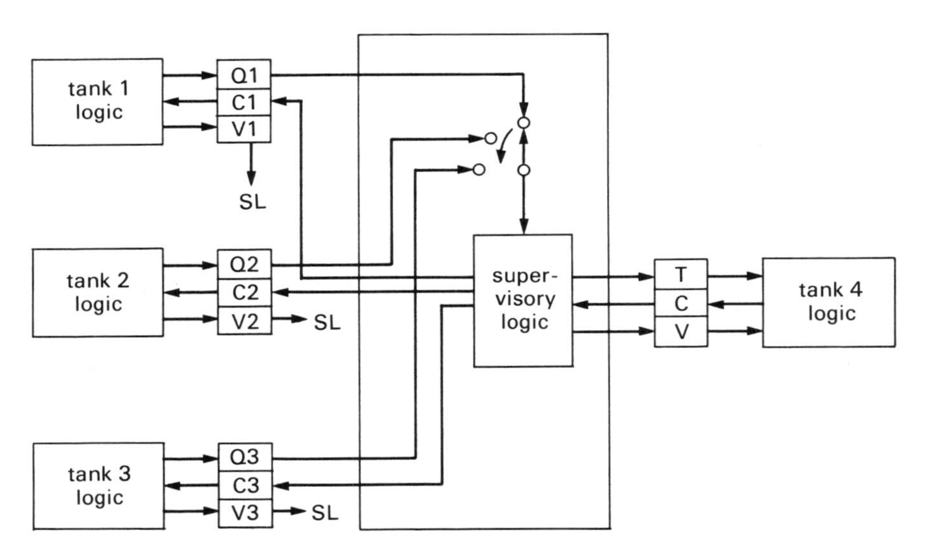

Figure 8-19. Centralized synchronization using a supervisory unit. *(After Rosenof, 1982c; reproduced by permission of the Instrument Society of America.)*

implemented by the supervisory unit. The supervisory unit can set priorities by criteria other than scan position; it can communicate with the operator, record charge data in computer bulk storage or on hard copy, monitor raw materials, stop a charge in progress if plant conditions require, and provide numerous other capabilities.

Are there any weaknesses to this approach? Situations are described in the *Distributed Synchronization* section that would produce very complicated supervisory units but are dealt with simply when the supervisory unit is eliminated. Since it is a part of every unit-to-unit interaction, the supervisory unit can cause a shutdown of an entire plant if it experiences a hardware or software failure. In certain kinds of computer systems, memory can be conserved by eliminating the supervisory unit. First, though, consider a synchronization method that has proven useful when the process does not demand the features that the more complex methods provide.

*Merged Synchronization**

Figure 8-20 shows a process in which compressed air is used to agitate the contents of the batching tanks. The plant air supply has only enough capacity to agitate one tank at a time; thus agitation is a shared resource. Unlike most material charging systems, the compressed air system operates asynchronously with the tanks once a tank is at an appropriate step. The only control action required is the opening and closing of the solenoid valves, and the only synchronization is an interlock to ensure that two or more valves are not open at the same time.

Merged synchronization refers to the merging of all the shared unit's logic into the

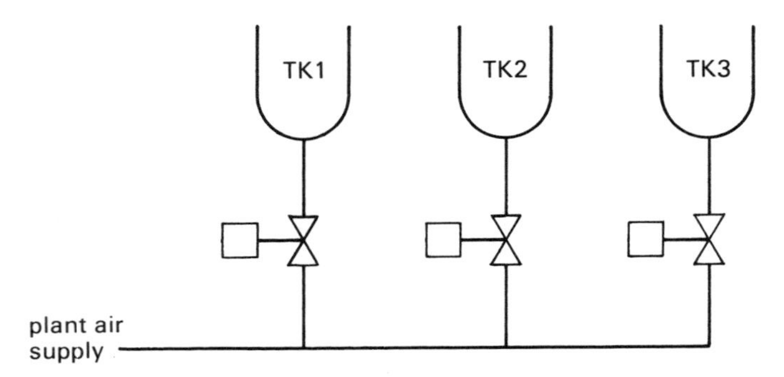

Figure 8-20. The plant air supply is able to service only one tank at a time. *(After Rosenof, 1982c; reproduced by permission of the Instrument Society of America.)*

---

*The text of the section on merged synchronization has been adapted from H. P. Rosenof, 1980, Multiple PC's share scans, *Instruments and Control Systems*, March, pp. 118-119, by permission of the Chilton Co. Copyright © 1980 by the Chilton Company.

logic of the receiving units. Signals are passed between units along a "ring." To expand the ring, we need only open up the wiring (or perform the software equivalent) between any two units and insert another unit. Bypassing a failed unit is not much more difficult. This approach is similar to the "token passing" used in some contemporary communications systems.

In Figure 8-21, when the scan-active signal is on, it means that one PC is currently scanning, and no other PC may begin a scan. The other PCs must remain at step 0 of their scan sequences, performing logic functions that do not need air

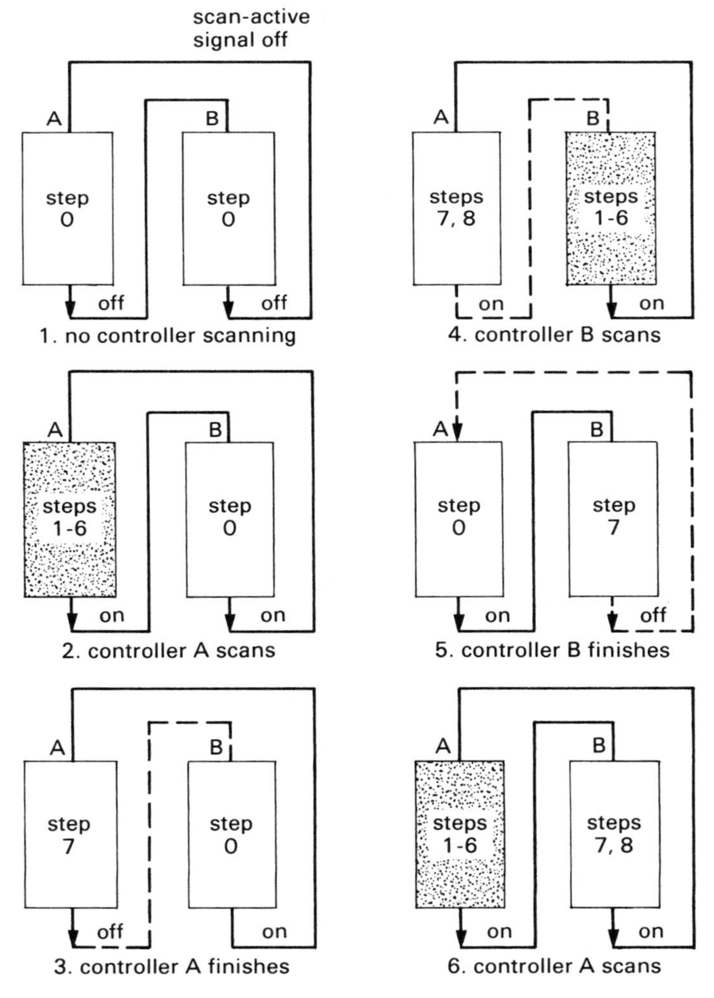

Figure 8-21. Merged synchronization. The technique is shown, for simplicity, applied to two programmable controllers. *(After Rosenof, 1980; reproduced by permission of the Chilton Co.)*

(such as housekeeping, retransmitting the scan-active signal, etc). When a PC finishes its scan, it sets the scan-active signal to off, and the next PC begins its scan. Thus, the PCs scan in sequence: No PC can hog all the air, and no PC can miss a turn.

Failure modes for this method can be set up in two ways, if we assume the signal will be lost if a failure occurs. If the scan-active signal is on to indicate scanning, then all PCs can operate independently in case one PC fails. If off indicates scanning, all PCs will stop if one fails.

Some form of logic must be provided to begin scanning at system start-up. In most cases scanning will be inhibited at start-up, which allows the operator to initiate control when all other parts of the system are ready to start. For this system an external switch on a PC can be used to temporarily bypass the scan-active signal. Thus one PC can begin scanning. When it completes its first scan and turns off the scan-active signal, the other PCs will start scanning in sequence. This approach combines merged synchronization *between* the controllers with centralized synchronization *within* each controller to distribute the compressed air to the process units it controls.

Distributed Synchronization

In Figure 8-22 tanks 4 and 5 are a long distance from the rest of the process equipment. Piping costs are minimized by using pipe to carry charges from tanks 6 and 7 to tanks 4 and 5. Centralized synchronization as described will resolve contentions for tanks 6 and 7, but would allow, for example, tank 4 to be charged by tank 6, and tank 5 to be charged by tank 7 at the same time. Each tank would then receive a mixture of the two materials.

Again, in Figure 8-22, assume that one or more process tanks is to be charged simultaneously by tanks 6 and 7. Or imagine a situation in which a batch tank must be charged by a shared unit with material at the same temperature as that already in the batch tank. After the charge an acceptable leak test must be made before processing.

The logic of the supervisory unit can be designed to accommodate any of these situations. As the number of these special cases increases, the supervisory logic becomes more complex to the point where it may be among the most intricate parts of a large and complicated system. It can even lead to a project management problem during system design, with several people changing the supervisory logic at the same time and thus interfering with one another. In addition, eliminating the central supervisory logic would remove one source of failure that could affect the entire plant.

Figure 8-23 depicts a simple *distributed synchronization* arrangement in which two signals are used per pair of units being synchronized. The first problem affecting the example in the figure is solved by defining the common pipe as a

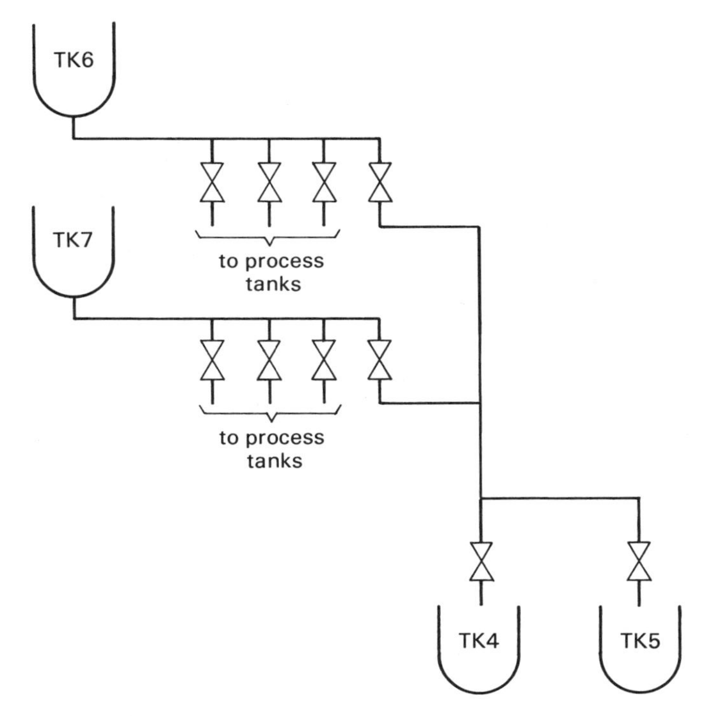

Figure 8-22. Tanks 4 and 5 are charged through a common length of pipe by tanks 6 and 7. *(After Rosenof, 1982c; reproduced by permission of the Instrument Society of America.)*

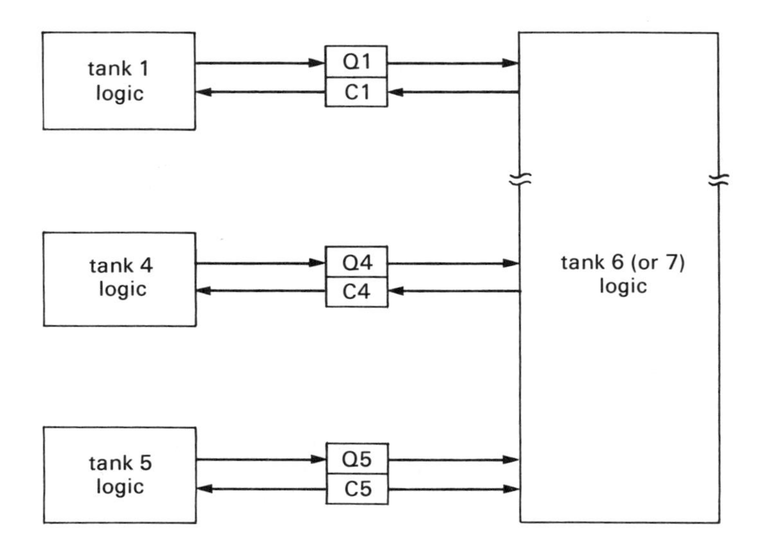

Figure 8-23. A simple example of distributed synchronization. *(After Rosenof, 1982c; reproduced by permission of the Instrument Society of America.)*

batch unit and by *embedding* synchronizations between the pipe and tanks 6 and 7 and between those tanks and tanks 4 and 5. A process tank (4 or 5) synchronizes with a charge tank (6 or 7); then the charge tank synchronizes with the pipe.

The second problem associated with Figure 8-23 is solved by placing synchronization in series. Once both charging units confirm assignment, they are simultaneously requested to begin their charges. In a computer system that allows synchronization logic to be written as separate modules (a minimum of four: two to establish the handshake; two to break it—for initiator and responder), there may be no changes required in the modules written to accommodate the solutions to the first problem. Tanks 6 and 7 are equipped to be responders to all process tanks and initiators to the shared length of pipe. The initiator's handshake establishment logic may be divided into smaller modules or into different modules provided for this special case. In this case "special" modules affect only the functions that require them—there are no interactions between these modules and the standard logic.

The third problem, like the second, requires that some logic be divided into smaller modules or that different modules be written. The number of signals required need not increase if the signals change their states for each part of the handshake.

To solve the first and second problems *simultaneously*, make the pipe length batch unit responsible for calling for charges from tanks 6 and 7. Then tanks 4 and 5 synchronize only with the pipe lengths to obtain charges from tanks 6 and 7, whether individually or together. Here the pipe length-charge tank synchronization is embedded within the process tank-pipe length synchronization.

General Considerations

Matching Synchronization Points. A complex system could have several points at which the same combination(s) of units might have to be synchronized. Logic must be carefully designed so that control errors do not occur due to a mismatch between synchronization points. Several means are available:

1. As part of the synchronization handshake, the calling unit provides to the other unit(s) a complete definition of the actions the other unit(s) is (are) to perform.
2. Different sets of signals are used for the various synchronizations. Unless all units are at the same synchronization point, the handshake will not take place. (This method is most applicable to distributed synchronization.)
3. Each unit makes available to other units an "index," whose value represents its current point in the logic (most applicable to centralized synchronization).

Error Handling. The synchronization plan should include provisions for handling software and hardware failures with minimum effect on those parts of the process and control system not immediately involved. Specific provisions may be made for specific failures, but generally there are many possible problems—some identified, and others not identified.

A general approach is to establish a status signal for each unit and to make it available to all other units. This signal is on when the unit is in normal operation and off when the unit is in an off-normal state (Hold for suspensions of sequential control in favor of predefined output values, manual to enable control by the operator in typical systems). During a handshake (only), an off state of any participating signal causes the other units to go to hold. For centralized synchronization no other unit's off-normal state should cause the supervisory unit to stop processing, but if it does all units with handshakes active should go to hold.

Testing. Several features should be available to expedite control system checkout:

1. The ability to determine the status or value of every interunit signal.
2. The ability to manipulate interunit signals so that any side of any synchronization may be manually controlled.
3. The ability to determine and change any interunit variables that affect any synchronization.
4. The ability to "trace" each unit's progress through the synchronization. (The trace function yields a record of the statements processed by each unit.)
5. The ability to use "dummy logic" as a temporary substitute for working logic. (Dummy logic allows the engineer to isolate synchronization logic from other functions. The dummy logic substitutes for the software not under immediate test.)

Avoiding Deadlock. Deadlock occurs when there is no time in the future at which a synchronization can proceed without reinitialization. This is different than a simple delay in synchronization while some other process action runs to completion. Deadlock conditions occurring as the result of failed shared equipment, for example, can be resolved with proper error handling. Error handling logic cannot, however, detect the situation in Figure 8-24, in which a tank in the tank farm has synchronized with pipeline A and is waiting indefinitely to synchronize with pipeline B, while a process vessel has synchronized with pipeline B and is waiting indefinitely to synchronize with pipeline A. These conditions can be prevented only with careful design.

Batch Movement

Movement of a batch from one vessel to another is, in many ways, similar to charging operations except that, in this case, material movement is away from the controlling vessel rather than toward it (Fig. 8-25). In this simplest case of a series connection between two tanks, the only precondition for the receiving tank to be

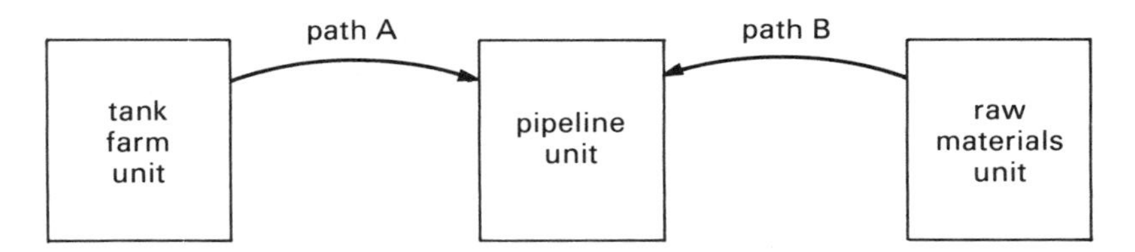

Figure 8-24. A "deadlock" situation—neither path A, from the tank farm to the process vessels, nor path B, from the process vessels to the tank farm, can be completed.

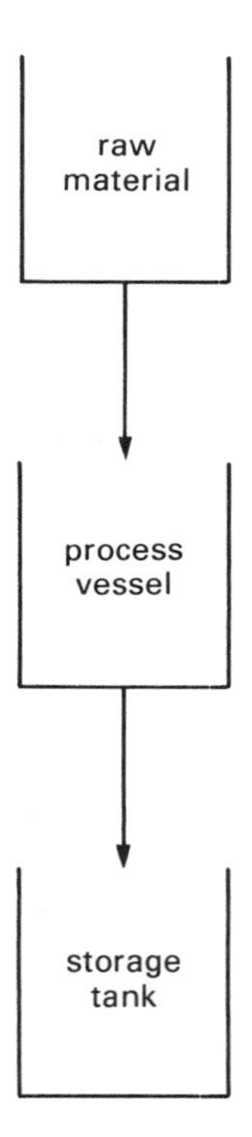

Figure 8-25. Movement of a raw material into a process vessel is not a batch movement. Subsequent transfers are batch movements, since the materials have received some processing.

able to accept a new batch is that it be empty of the previous batch. Even this is not a requirement if the receiving tank serves to collect numerous batches of the same product. Then the only requirement is that there is enough volume available in the receiving tank.

Along with the material, a lot of information has to be transferred. The receiving tank must receive instructions defining subsequent processing of the batch. This is another requirement that distinguishes batch movement from material charging; in the latter, it is only necessary to specify materials, amounts, and so on, to be charged. In practice, the entire recipe is usually transferred with the batch (Fig. 8-26). Typically, these simple transfers are allowed to take place under full automation, although a requirement for operator approval can be built in if needed.

As plant configurations increase in complexity, so do the controls needed to operate the batch transfers. Figure 8-27, for example, shows a reactor able to feed three different product storage tanks. These tanks might be used, for example, to store each of the three products the reactor is capable of manufacturing. In this case the positions of the valves are dictated by the product identification, which

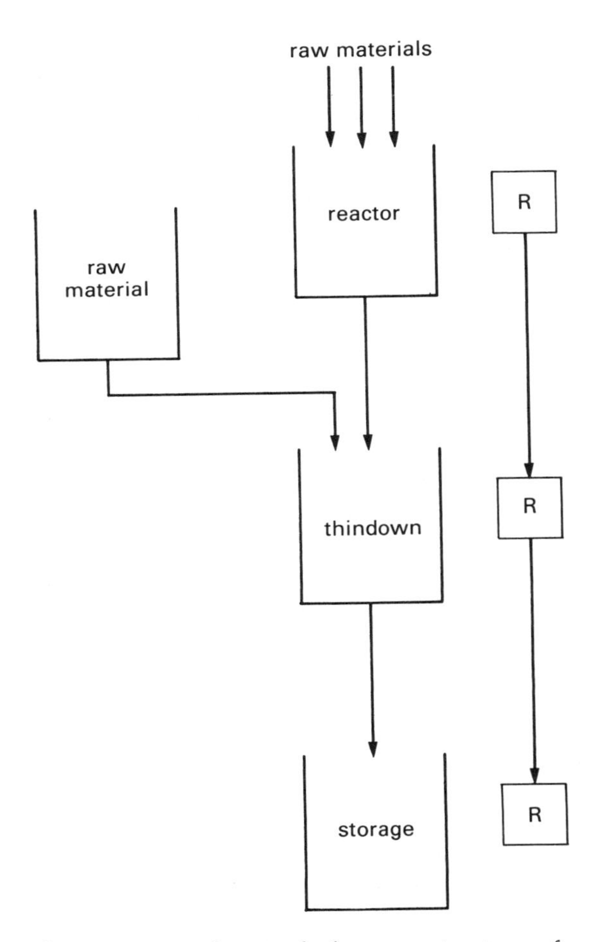

Figure 8-26. The entire recipe, directing further processing, is transferred among units along with the batch.

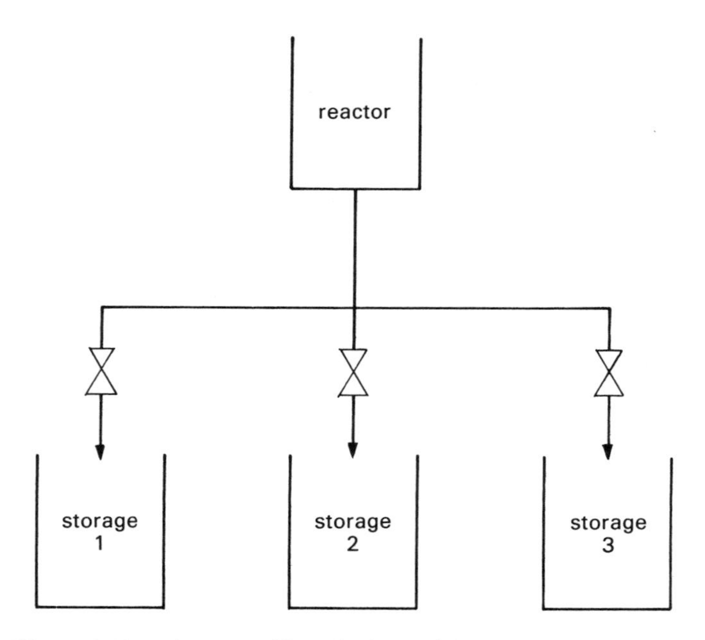

Figure 8-27. A reactor able to feed any of three product storage tanks.

is readable by the logic from the recipe or elsewhere depending on the system being used.

When two reactors discharge into the same tank (as in a continuous-batch system) and for some reason only one may do so at a time (e.g., rate control with a sensor at the receiving tank), the receiving tank may be treated as a shared unit. Note the difference between production schedules and actual plant conditions: The two reactors may have their batch start times set so that they normally would never complete batches, and require the receiving tank, at the same time. If production is delayed at one reactor, however, the contention may still occur, and the automatic system should be designed to handle it.

The possibilities are not yet exhausted: Figure 8-28 shows a situation in which each of three reactors can discharge into any of three receiving tanks. It may be that tanks 4, 5 and 6 store different products, and that the reactor above each tank is the one that normally produces the product it holds. Whether required by equipment unavailability or market conditions, the plant must, on occasion, use a reactor to make a product usually made by one of the others.

The reactor units have logic that reads the product identification. Most of the time the identification matches the product stored in the tank below, and the only valves needed open are the reactor discharge and storage tank inlet valves. The horizontal valves remain closed. When the product is to flow horizontally, as from

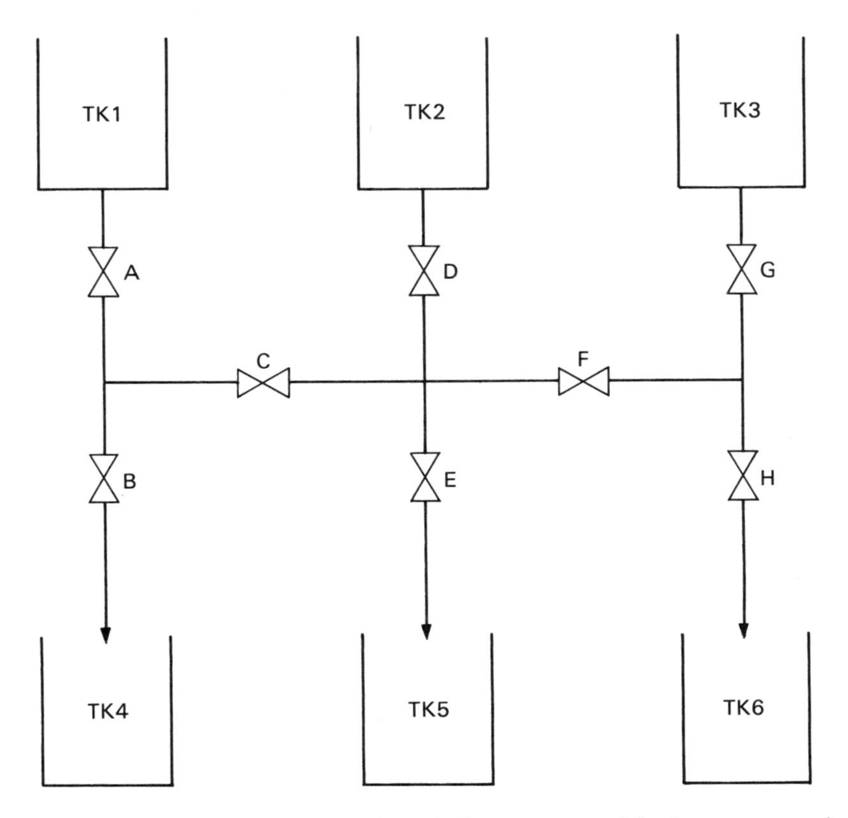

Figure 8-28. Three reactors, each able to discharge into any of the three storage tanks.

reactor 1 to reactor 5, then not only must one or both horizontal valves be open (in this case, valve C), but one or more *vertical* valves must be *closed* (in this case, valve B and horizontal valve F).

To begin analyzing this problem, we prepare Table 8-1 to show the valve lineups associated with each possible transfer. First, confirming common sense, the discharge valve beneath each reactor must be open when its reactor is transferring material out. At all other times the valve may be closed. Since a closed valve is obviously a prerequisite for the reactor's operation, except at discharge, we simply define each such valve to be part of its reactor's batch unit, to be opened by reactor logic only for discharge.

Similarly, the inlet valve to each storage tank *must* be open when material is being transferred in, and it must be or may be closed when material is not being transferred in. We define each inlet valve to be part of the same batch unit as its associated tank. When the tank is to receive a product, the valve is open; at all other times it is closed.

Table 8-1. Valve Lineup for Figure 8-28

ID	Transfer from	to	Valves A	B	C	D	E	F	G	H
a	1	4	O	O	C	X	X	X	X	X
b	1	5	O	C	O	C	O	C	X	X
c	1	6	O	C	O	C	C	O	C	O
d	2	4	C	O	O	O	C	C	X	X
e	2	5	X	X	C	O	O	C	X	X
f	2	6	X	X	C	O	C	O	C	O
g	3	4	C	O	O	C	C	O	O	C
h	3	5	X	X	C	C	O	O	O	C
i	3	6	X	X	X	X	X	C	O	O

O = valve open, C = valve closed, X = don't care

We could define valves C and F to be one shared unit and require each reactor to synchronize with the shared unit before discharging. Upon synchronization, valves C and F would move to positions required by the particular transfer (a through i). This setup would "work" because it would not violate Table 8-1, but it would be inefficient.

Examination of Table 8-1 or the configuration itself shows that mechanically the plant can at one time handle any of seven pairs of simultaneous charges as well as one triplet: e, ai, ei, af, ah, bi, di, and aei. A control system artificially preventing these might not be acceptable in some plants.

We get closer to a solution when we define each of valves C and F to be a separate batch unit. Then when a particular transfer requires one or the other or both valves closed, the reactor logic can synchronize with the appropriate unit or units to drive the valve(s) as required and then proceed with the transfer. Pair ai is the only one that can really take advantage of this, however, because all of the others, and the triplet, place *two* constraints on one or more valves. If a shared unit can only synchronize with one user at a time, then that shared unit can only respond to one user's requirements at a time. The second user might find its conditions satisfied, but it cannot enforce them (Fig. 8-29). A scheme enabling one shared unit to synchronize with more than one user at a time would likely be nonstandard and highly complex.

A standardized and structurally simple way is to use the *virtual valve:* Treat the plant as though it were arranged as shown in Figure 8-30. This gives us the valve position chart shown in Table 8-2. In fact, there are still only two horizontal valves C and F controlled by simple logic, as shown in Table 8-3. This implies that either the L or R member of each pair might be eliminated from the system. They cannot be eliminated, simply because to do so would force the remaining shared valves to

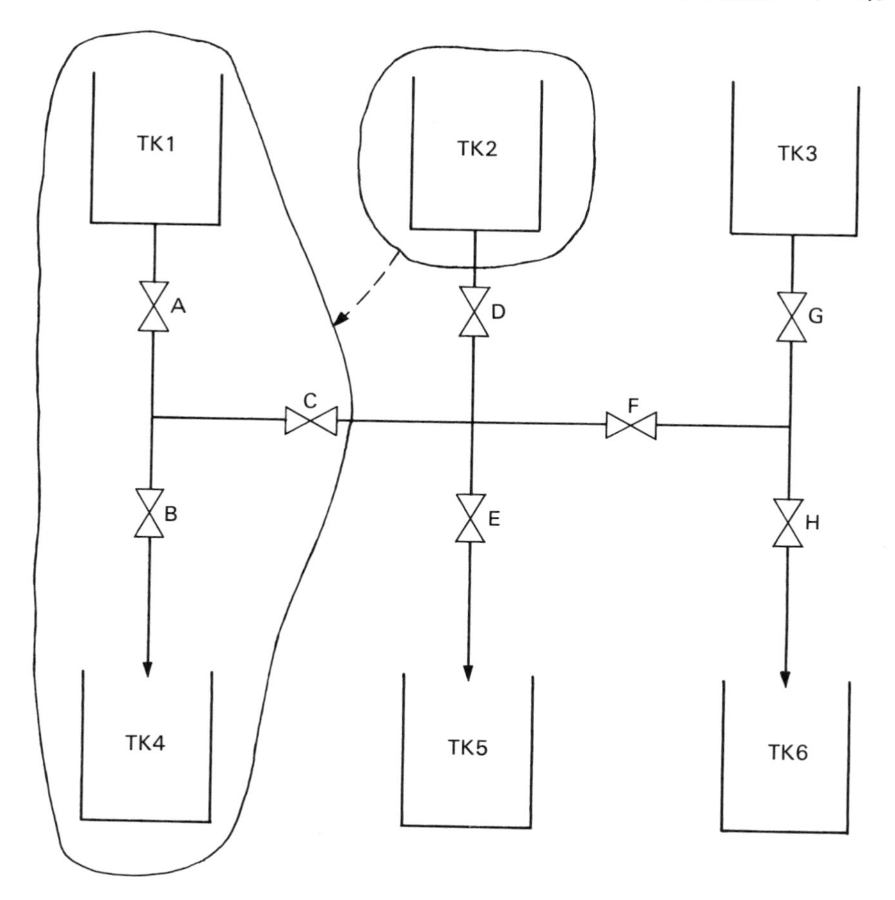

Figure 8-29. Reactor 1, to discharge to storage tank 4, synchronizes with valve C to hold it closed. Tank 2 is ready to discharge to tank 5, but it cannot do so yet because valve C is not available for synchronization.

have to synchronize with two users simultaneously. Also note that the process is safer when both virtual valves are used to open the real valve.

Another solution is to look even more carefully at plant geometry: Valve C is required to be open only when reactors 2 or 3 discharge to tank 4 or when reactor 1 discharges to tanks 5 or 6. We can design reactor logic so that this synchronization only takes place when needed. When valve C is not needed, it is ignored. In our example of Figure 8-30, charges from reactor 1 to tank 4 and from reactor 2 to tank 5 can take place simultaneously because neither requires valve C.

Suppose that reactors 1 and 2 are both ready to discharge to tank 4: Since tank 4 can only synchronize with one reactor at a time, we have no real problem. We only

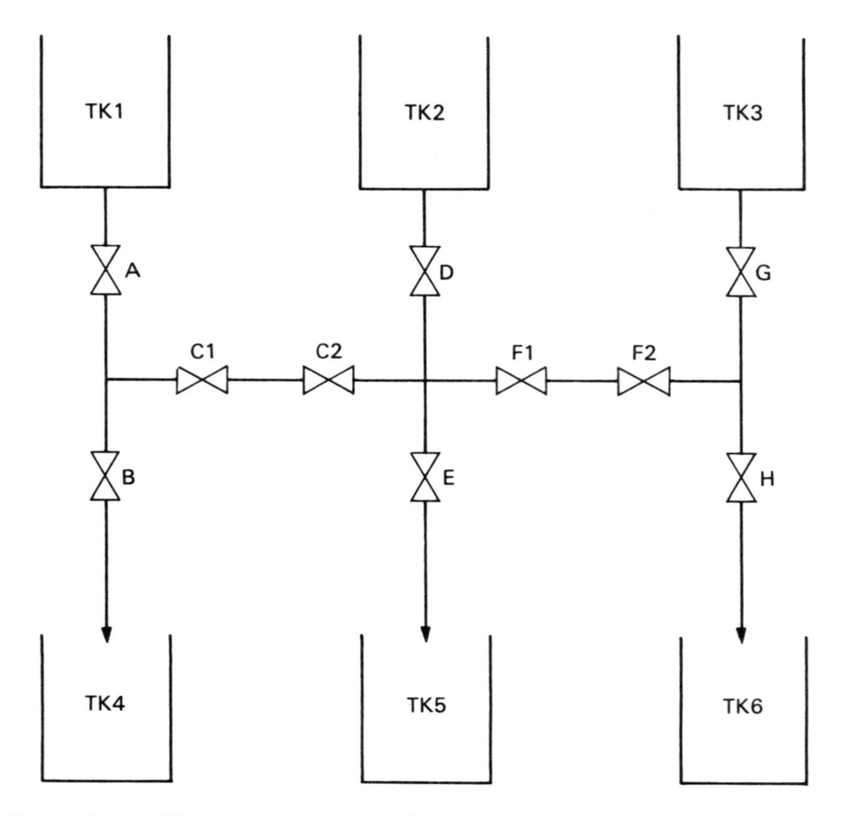

Figure 8-30. The plant is modeled as if it had two valves in each horizontal position.

Table 8-2. Valve Lineup for Figure 8-30

ID	*Transfer* from	to	*Valves* A	B	C1	C2	D	E	F1	F2	G	H
a	1	4	O	O	C	X	X	X	X	X	X	X
b	1	5	O	C	O	O	C	O	C	X	X	X
c	1	6	O	C	O	O	C	C	O	O	C	O
d	2	4	C	O	O	O	O	C	C	X	X	X
e	2	5	X	X	C	X	O	O	C	X	X	X
f	2	6	X	X	X	C	O	C	C	X	X	X
g	3	4	C	O	O	O	C	C	O	O	O	C
h	3	5	X	X	X	C	C	O	O	O	O	C
i	3	6	X	X	X	X	X	X	X	C	O	O

O = valve open, C = valve closed, X = don't care

Table 8-3.

C1	*C2*	C
O	O	O
O	C	Not valid
C	O	Not valid
C	C	C
O	X	Not valid
X	O	Not valid
X	C	C
C	X	C
X	X	C

Because many of the input combinations are not valid, either AND or OR logic may be used to generate the Open signal to the valve. AND logic should exhibit less vulnerability to a possible extraneous signal.

have to make sure that no valve opens until all synchronizations are complete. This is good practice in any case.

In use, the arrangement will not be quite this simple. Reactor 1, though not synchronizing with valve C to discharge to tank 4, will still have to confirm that the valve is closed. It might have been opened by operator command. Administrative controls or a nonstandard logic design will possibly be needed to prevent valve C from being opened by operator action at the wrong time, since at times it will not be in synchronization with any other unit but still be required to be closed. Valve F would be controlled similarly.

These tankage arrangements become even more complex when multiple paths are possible (Fig. 8-31). As the number of valves increases, it may become impractical to give each one the status of a batch unit, let alone two or more batch units.

Solutions for these more complex arrangements are found in using one or a very small number of batch units to represent all of the relevant valves. Synchronization is not with the valves but with a software mechanism that determines which valves are needed to be open or closed for a particular transfer, checks them to determine whether they are available, and lines them up properly if they are. At the end of the transfer, open valves are closed. Synchronization with this mechanism, then, gives a message to the calling unit advising it whether or not it can proceed.

Techniques like these are found in refinery oil movement and storage (OM & S) systems, which handle crude, intermediate products and final products and, according to Czech (1982), can typically handle between 50 and 75 operations at a time. Czech lists several approaches to automated path selection:

A plant model with penalty values. The computer uses the available path that has the lowest overall penalty.
Table lookup of predefined paths.
Leaving path selection up to the operator.

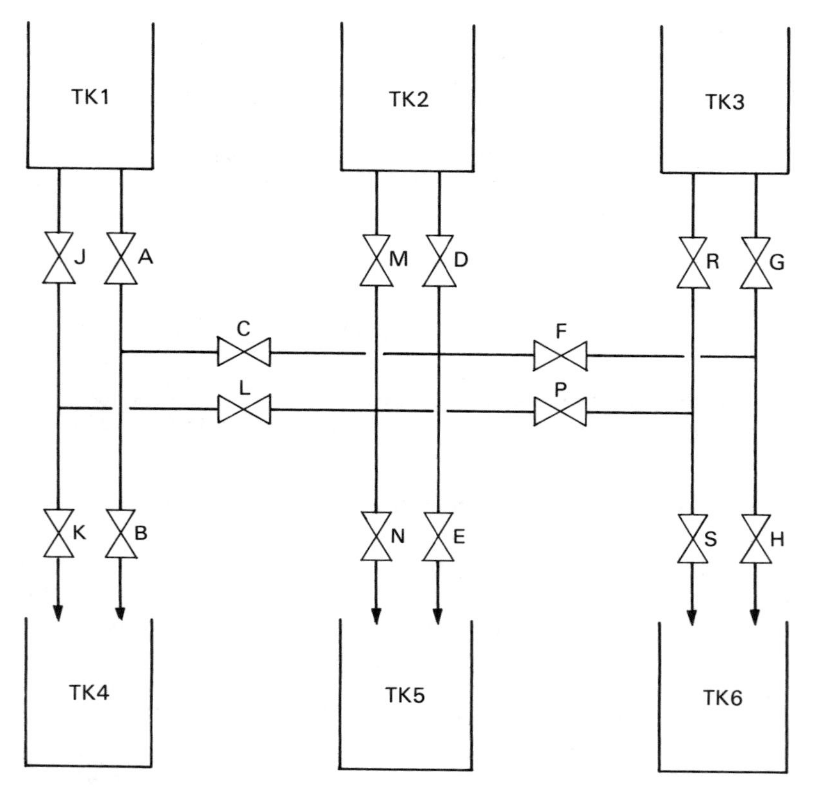

Figure 8-31. Two paths exist for each charge from a reactor to a storage tank. Any pair of charges is mechanically feasible.

Following path selection, the sequence generation function specifies the order in which field operations are to be performed. The sequence control function is responsible for implementing the sequence, starting the transfer, stopping it and terminating the transfer (i.e., releasing equipment to other transfers).

Alternative Strategies for Batch Movement

Batch movement as considered so far assumes that the destination of a batch is determined by recipe (i.e., is predetermined) or by the operator. A third, less common, approach is to have the control system make the decision according to a preestablished set of rules. Typical considerations might be

Compatibility with existing or previous batches
Storage space availability

Feed requirements of downstream processes
Minimization of resource use (i.e., use of the smallest downstream vessel able to accommodate the batch)
Process operations required that are only available with certain routings.

Once the destination is chosen, the transfer operation takes place, as described for preprogrammed and operator-directed transfers.

Plant Resource Allocation: Feedback and Feedforward Control

Feedback control techniques for continuous processes are well understood and commonly used. Even the best feedback control loop is imperfect, however, because an error must occur before it can be corrected. There is no before-the-fact control action to compensate for a disturbance before it causes a process upset. Feedforward control attempts to detect these disturbances, evaluate their anticipated effects according to a process model, and act to minimize those effects. "Feedback trim" acts on the remaining errors, those not accounted for by the process model, for which disturbances are not measured or which result from errors in the model.

Similarly, production scheduling in advance may be regarded as analogous to feedforward control: A process model is used to predict behavior, and that prediction is used as a basis for control actions (i.e., the determination of batch startup sequences and process vessel allocation). The equivalent of feedback trim is exercised at the sequence levels, which can make adjustments required by model errors, oversights, and contingencies.

BATCH LOGGING AND REPORTING*

Report Types

Batch processes make use of logging and reporting techniques common for continuous processes, as well as some techniques that are unique to production in discrete quantities. The *current data report,* or "snapshot," is borrowed directly from the continuous-control world. This report consists of measured and computed variables as recorded at one particular time, which may be hourly, at shift end, at some other predetermined time, by operator demand, or upon an event occurring. Measured variables are temperatures, pressures, flows, and such like. Computed variables are instantaneous heat transfer, time rate of change of measured variable, and mass flow computed from volume flow, for example. Operator-entered data

---

*The section on batch logging and reporting has been adapted from H. P. Rosenof, 1985, Data logging and reporting for effective batch control, *Instruments & Control Systems,* July, pp. 29-32, by permission of the Chilton Co.

may also be reported back. As in a continuous plant, a snapshot report describes the operation of the plant at a selected time, allowing determination of instantaneous production costs, detection of certain equipment performance problems (heat transfer surface fouling, for example), and comparison of equipment utilization with nominal ratings. Unlike continuous plants in steady-state operation, the snapshot report for a batch plant does not characterize the batch. The batch will reflect the processing it received over time, from the beginning of the batch to its completion, and thus the data must be measured, computed, and entered by the operator over this time.

The *historical report* is also borrowed from continuous-process control systems. It is based upon a historical record of measured, computed, and operator-entered variables. The record may retain values for months. Typically, the most recent data are recorded with the greatest sample frequency (i.e., once per 15 s). As time proceeds, the data are progressively characterized so that, for example, data in the system for a week may be represented by hour-by-hour arithmetic averages (Fig. 8-32). Some systems use data compression techniques to further reduce the amount of memory required. In addition to averages, the data may be represented by maximum values, minimum values, and totalizations (e.g., totalization of electric power load to represent energy consumed). When applied to a batch plant, the historical report can be used to track plantwide utilities consumption and equipment utilization.

The *alarm log* provides a record of off-normal conditions and their resolutions. Its value in a batch plant is the same as its value in a continuous plant. It provides the means to identify recurring equipment problems, for example.

None of these logs and reports is very useful for tracking and recording batch

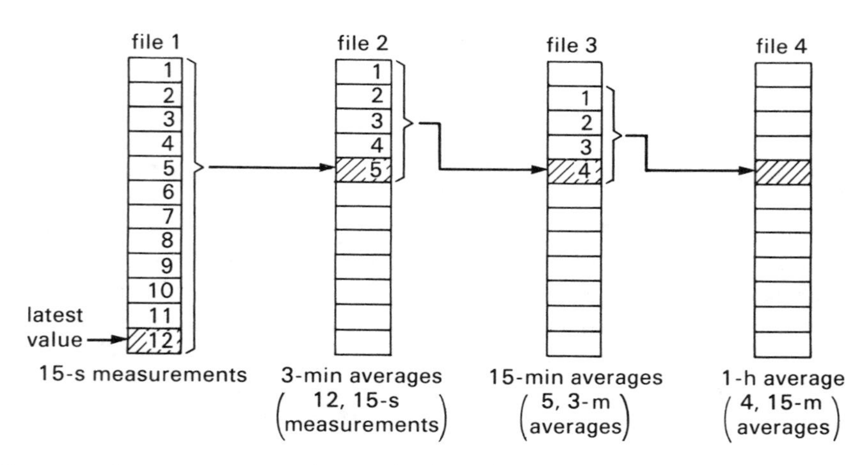

Figure 8-32. Compression of historical values. *(After Rosenof, 1985; reproduced by permission of the Chilton Co.)*

production. From a design-by-levels perspective all of these derive data almost exclusively from the lowest control levels. A satisfactory batch report would freely receive data from all process control levels, up through the batch levels.

Specifically, data for the report should be available from these sources:

1. *The Recipe Image.* This is simply a copy of the recipe that was used to make the batch. It may or may not be identical to the master recipe for the product, since, in some systems, the working copy of the recipe may have been changed by the operator or by the logic. Be sure to choose the correct working copy, the one that reflects all of the changes made up to the end of the batch.

2. *Batch Identification Data.* This is generally entered by the operator as part of the batch start-up sequence, and it may be entered automatically when automated production scheduling is in use. These data serve to distinguish multiple batches made with the same recipe, and they are important for quality control, especially where record keeping is needed to conform to governmental or other external regulations.

3. *Recipe-Matched Data.* Some of the data accumulated by the system correspond directly to recipe data. For example, the system should record the amount of each material charged to check conformance to the recipe.

4. *Recipe-Implicit data.* Some data do not directly correspond to the recipe but are generated in response to recipe-directed operation. For example, the recipe may specify a sample to be taken and the temperature at which the request to the operator is to be made. The system may record the time at which the sample is taken, which does not directly correspond to the recipe, as well as the temperature, which does. If the sample is analyzed and the results recorded within the control system, these results also become recipe-implicit data.

5. *Summary Batch Data.* These data summarize utilities consumption, equipment run times, temperature maxima, and so forth, for the entire batch or predefined portions.

6. *Operator Scratchpad.* Literally, free-format text storage into which the operator can enter, typically via CRT, remarks such as equipment problems, instructions to the next shift, and operator employee numbers.

7. *Event Log.* A record of occurrence times of such "events" as alarms, operator instructions followed or ignored, and equipment status changes. Where sample times and similar pieces of information are not recorded as recipe-implicit data, they may be recorded here.

8. *Custody Transfer Data.* An end-of-batch accounting record containing storage tank number, shipping method, and so on.

Log Files

In the simplest case of a single-vessel, single-product batch plant, one computer file, or one set, is usually sufficient to meet all reporting needs. (If the system can start a subsequent batch before printing the result of the current batch, the file(s) must be duplicated.) In practice, different types of data are accumulated by different subsystems, so the data organization will probably be like that shown in Figure 8-33.

Where a system of logic locations is used, the sizing calculation for logging is similar to that for the recipe. For example, if a recipe will accommodate up to 99 material charges selected from among 50 logic locations and 200 materials, the log file must have 99 slots. Each slot must be able to record the actual amount of material charged according to the requirements of the corresponding recipe entry. It is not strictly necessary to use log files to record the recipe data (logic location, material identification and nominal amount) associated with each slot because

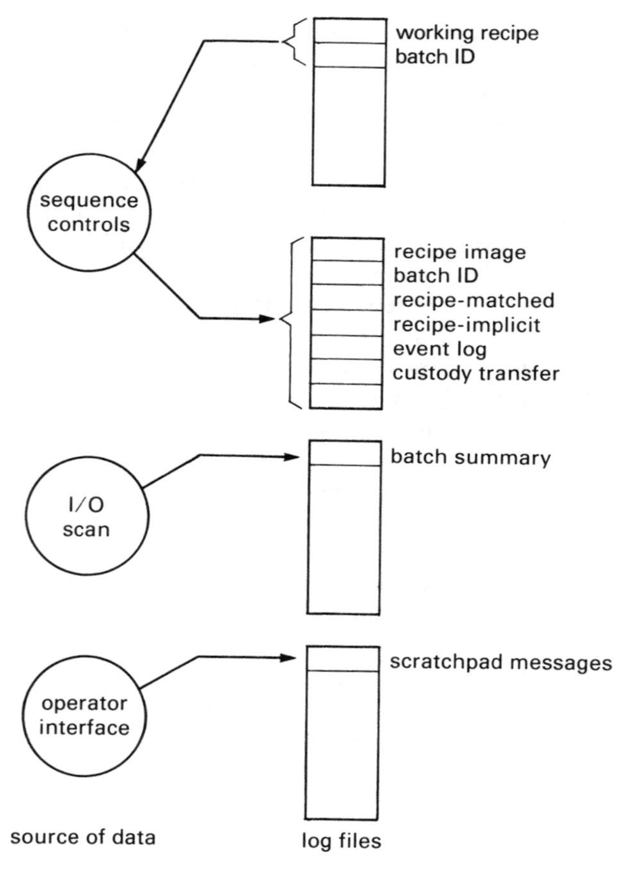

Figure 8-33. Simplified arrangement of log files in a typical application. *(After Rosenof, 1985; reproduced by permission of the Chilton Co.)*

they are available from the recipe itself. Doing so, however, provides the user with a record of the recipe as executed, securing the report against subsequent changes to the working copy. These logic locations, and so on, become recipe-matched data; the recipe image data is available separately, as is the master copy of the recipe.

For a process with more than one vessel, capable of processing more than one batch at a time, separate log files are maintained for each batch. In effect, the log moves with the batch (Fig. 8-34). In practice, a "pointer" is associated with each recipe, and the batch logic writes to the file identified by the pointer. Where data do not come from or through the batch logic (batch summary data is most likely), some means must be provided to associate the data with the appropriate batch. Once the contents of the log files are reported, transmitted to a central computer, and/or archived, as required, they are generally cleared and made available to a subsequent batch (Fig. 8-35).

As long as the batch plant manufactures only one product, or different grades of the same product with variations in materials, and so forth, within the same sequence, one configuration of log files is required: As long as some amount of material A is always charged at the same point in the batch, then the same slot in the log file may be used to record the actual amount charged, even if that amount varies widely between grades. Where sequences can change, however, log file configurations must be allowed to match the sequences.

For MS recipes, sequence variations are encoded as null data or by use of logic locations. In either case sequence variations are accommodated simply by sizing

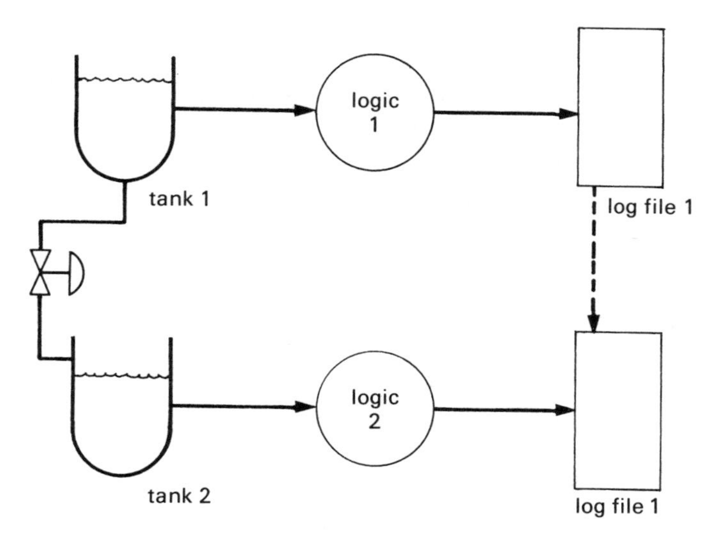

Figure 8-34. The log files may be pictured as moving with the batches. If the batch resides in more than one tank, however, then each such tank's logic uses the same log files. *(After Rosenof, 1985; reproduced by permission of the Chilton Co.)*

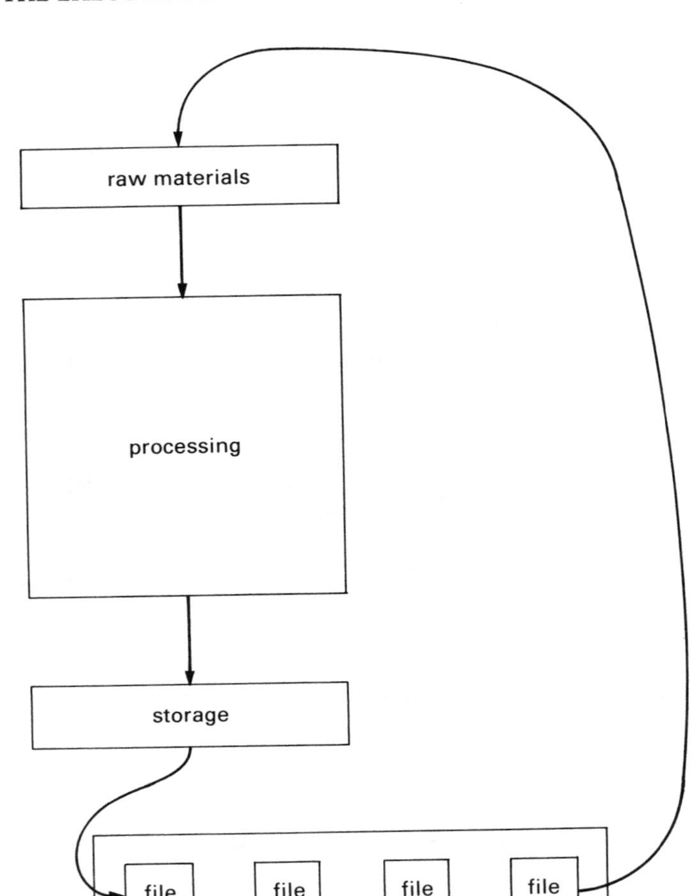

Figure 8-35. Log files in the "inactive" area are available to be printed or transmitted to another computer. Once these processes are complete, the file is marked "available," cleared, and made available for a new batch. If no file is available, the inactive file with the oldest data may be used.

and configuring the log files for the worst case. A slot is allocated for the charge of material A, for example, whether or not that material is used in a particular recipe. Where the material is not used, the slot is not written to and remains filled with null data.

For VS recipes log configurations are determined by the phase sequence: A material-transfer phase will use log files differently than a temperature-change phase. These individual phases are limited, however, and each phase will generally use log files the same way each time it is used. The engineer is presented with two possibilities: Configure log files separately for each sequence, or install the means to automatically configure log files to match each sequence. In either case these uniquely configured log files would store batch identification data, recipe-matched data, recipe-implicit data, and possibly summary batch data and custody transfer data. The remaining information would not require unique configuration (Fig. 8-36).

Batch logging facilities are usually designed to receive each item of data once. This may lead to a problem if a phase is repeated. For example, consider a case in which a process endpoint can only be determined by laboratory analysis of a sample. If the endpoint has not been reached, the previous phase must be repeated. If the duration of processing for the previous phase is recorded simply by writing to the appropriate slot in the log file, that slot will be written to twice, once each time the phase is run. The second entry will overwrite the first, so that the log file value will represent only the second running of the phase. The solution is to read the value in the slot, add the new value to it, then store this sum back in the slot. This way the log files can be made correct for any reasonable number of executions of a phase.

For a recording of maximum temperature, the reading of the file would be followed by a comparison: Only if the current temperature exceeds the value read,

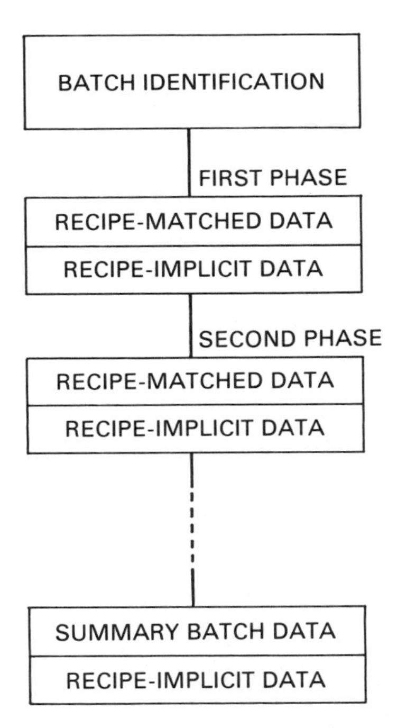

Figure 8-36. Log files specifically configured to match a variable-sequence recipe.

would the slot be overwritten. The temperature at which a sample was taken would generally overwrite the previous value, without the need for reading the previous value first.

If it is necessary to record the time of each running of the same phase, simply define start-of-phase as an event: If the phase is run five times, the system records five events in chronological order as long as the capacity of the event file is not exceeded. If the batch system is required to record each datum separately, without totalization, comparison, overwriting, or other slot-conserving techniques, then the log files must be configured in real time. As each phase is run, whether for the first, second, third, or later time, the proper space is assigned, and proper entries are made to allow the reporting facility to reconstruct the history of the batch.

Event data may be used to generate a tabular display as shown in Table 8-4. Nominal values either come from engineering analysis of how the process should function or statistical analysis of previous batch records. They may also be displayed graphically, as shown in Figure 8-37. The horizontal and vertical axes are both time with identical scales. Horizontal lines, one for each milestone event, are drawn with displacements from the origin representing their nominal times of occurrence. Actual occurrence of the event is plotted along the corresponding horizontal line. Since the horizontal and vertical scales are the same, a perfect batch—one that perfectly matches nominal event times—will be represented by a straight line with a 45° negative slope. Deviation from the ideal is represented by deviation from this straight line and is readily observed. This technique was first proposed (Morrison, 1984) for manual use, but it is readily computerized, especially where the computer has a facility with a plotter or CRT (with videocopier) for point-to-point line segment definition.

Reporting

Where each slot in the log files is used for the same purpose each time, then clearly only one report needs to be implemented. For systems using MS recipes without

Table 8-4. Event Data in Tabular Form

	Actual Time		*Schedule Time*	
Event	*Clock*	*Process*	*Clock*	*Process*
Batch start	08:00	00:00	08:00	00:00
Reactant A charged	08:30	00:30	08:25	00:25
Sample 1 taken	09:10	01:10	09:05	01:05
Heatup 1 complete	10:40	02:40	10:35	02:35

Source: From H. Rosenof, 1985, Data logging and reporting for effective batch control, *Instruments and Control Systems*, July; copyright © 1986 by the Chilton Co.

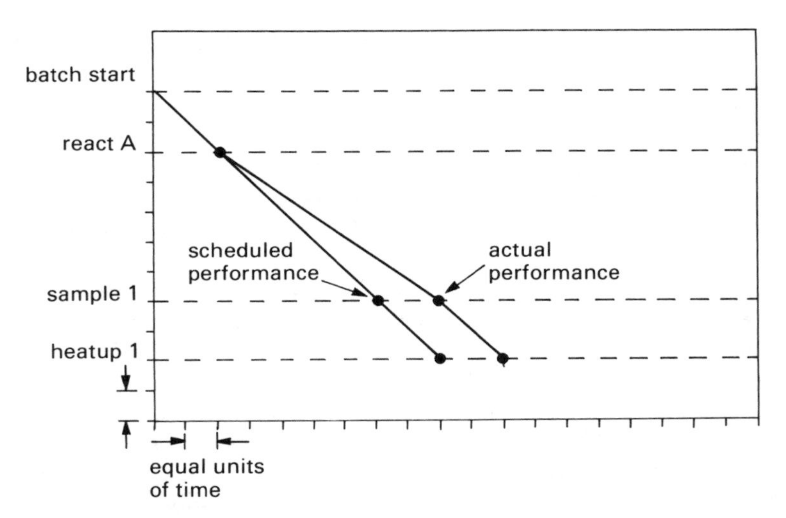

Figure 8-37. The values of Table 8-3, plotted as discussed in the text. *(After Rosenof, 1985; reproduced by permission of the Chilton Co.)*

real-time reconfiguration, reporting remains comparatively simple, especially if it is permissible to print null data as zeroes. It is not even very difficult to implement a report writer that recognizes and then skips over the null data.

For VS recipes the reporting format must uniquely match the sequence. The engineer may choose to design a separate report for each sequence, but it is also possible to use a program that configures the report as the report is being printed. Earlier we discussed a program that would automatically configure log files to match the recipe sequence; this is the equivalent for the report itself. The program would use a predefined report section for each phase, repeating each section as necessary.

The report pages themselves may be written by means of an explicit program using PRINT or similar statements or by a general-purpose report configurator/writer, which allows the user to specify header text, column headings, and so on, and the variables whose values are to be printed. These specifications are stored in memory and used by the report writer to print specified pages as required.

Most batch reporting utilizes text, but the computer can be used to present data in the form of trend lines (temperature profiles), bar graphs (lab test results), and similar displays.

Logged data are available for plantwide reporting of the total number of batches run per process train, the total amount of product manufactured per day, and other totals. In some applications logged data from various plants are transmitted over communication links to a central computer for consolidated reporting.

THE ROVING OPERATOR IN AN AUTOMATED BATCH PLANT

Roving operators are responsible for charging materials from bags and drums, taking and labeling samples, operating manual valves, hose stations, and other equipment, and investigating problems reported by the control system. Since most of these activities must occur at specific events or times within the process, there is an opportunity to use the process automation system to help schedule and prioritize them.

The simplest method is to use "messages" originating from sequence logic and displayed to the control room operator, who is alerted to a new message by an indication on a CRT or by illuminating an annunciator point. The CRT indication is used when messages are centralized on the same CRT display page; the control room operator simply calls up that display. Illuminating an annunciator point, with a corresponding pushbutton, is used when messages appear on a display page associated with the unit originating the message. One annunciator point and one pushbutton are assigned to each unit. By pressing the pushbutton, the operator accesses the display assigned to the unit and can read the message. In either case the roving operator is requested, through the plant public address-intercom system, to peform the required action.

The control room operator in these cases is required to prioritize and schedule these roving operator requests typically on an ad hoc basis. By simply establishing a shared batch unit to represent the roving operator, we can use the process control computer to do much of this work and take advantage of many techniques already established for general shared units.

Finally, the control system can communicate directly with the operator through discrete devices like pushbutton switches and indicator lamps or, more recently, environmentally protected CRT terminals.

The textile industry for years has used remote lamps to indicate to the roving operator the correct time to perform chemical additions. The operator charges the chemical and then reports completion by actuating a pushbutton. CRTs may be used to communicate more complex and lengthy messages in both directions, and they can enable the roving operator to access measurement and other information contained within the computer's data base.

CHAPTER
9

THE SEQUENCE LEVELS

All sequential processes, including batch processes, are governed by basic sequences. These sequences define the time-dependent progression of the process and, in practice, represent the highest level of automation implemented in many of today's plants.

The *start-up sequence* takes a continuous plant from its idle state into full production. A start-up sequence must handle various equipment constraints, such as for thermal stress. A sequence may have to bring a plant through or near an unstable operation region or, conversely, impose some variation upon an otherwise unchanging plant state. For example, "Dither" for a turbogenerator prevents its resonant coupling with other plant equipment. The continuous plant operates indefinitely, unlike a batch plant where new batches must be frequently started.

The *shutdown sequence* takes a plant from operation to a safe shutdown state. There can be several reasons why plant shutdown cannot be accomplished simply by stopping pumps and closing valves: Plant equipment that generates heat—whether an exothermic batch reaction or an overhead projector—may need to have that heat removed after shutdown to prevent dangerous temperature excursions or to prolong the life of components.

Some conveyor belts should not be stopped while fully loaded (Fig. 5.3). In general, it is most convenient to empty plant vessels and piping, where appropriate, before shutdown. (This convenience may not preclude a quick shutdown in the event of a genuine emergency.) Shutdown sequences apply to both sequential and continuous processes.

The *batch sequence* takes a plant from an idle state, causes it to produce a batch of output or product, then returns it to the idle state. Multiple-batch sequences may be in simultaneous operation in a plant that can accommodate batches in several stages of completion. Most batch sequences, therefore, are divided into unit sequences that enable various sections of the plant to operate independently, where appropriate.

The *continuous sequence* serves to maintain desired operating conditions in an otherwise continuous process. An example is filter backwashing, which does not produce a "batch" of anything because the filter can be left in the backwash condition for an arbitrary length of time. Another example is the traffic light sequence, resolving contentions between opposing traffic flows and enabling any flow to prevail for an arbitrary duration.

SERVICE AND HOLD LOGIC

Transfer to service or hold logic can be initiated by lower control levels due to plant conditions, from the same level due to operator actions or a logical decision, or from higher levels due to supervisory actions (Fig. 9-1).

It is preferable, and sometimes possible, to resolve plant equipment problems without recourse to off-sequence control. If a piece of shared equipment is not available, for example, and the process is stable, the control system can simply notify the operator and then make the process wait. The operator can then, if practical, initiate an alternative means of continuing the batch. If a piece of equipment is backed up, the control system may act, at its lowest levels, to switch over in case the primary equipment fails. Certainly a genuine emergency should be communicated to process equipment as directly as possible — for example, by using a relay interlock to shut off the flow of flammable material in case of a fire alarm, with off-sequence logic left to try to minimize damage by conducting an orderly shutdown.

Clearly problems originating at the plant equipment level should be resolved at the lowest control levels. In contrast, there is another class of plant problem that should be dealt with at higher levels, with lower levels serving as a last resort to minimize safety and equipment risks in case the higher-level measures fail. For example, we can postulate an operator's failure to perform a manual operation within a predetermined time. This failure is detected as a time-out within the normal sequence. A typical response would include a second message, a notation in an event log, and the cessation of the batch sequence until the process was restarted.

The service/hold branches may, of course, be contained within the main logic, but in medium-to-large systems it is generally more convenient to keep them separate. Separation also helps make the main logic sequence more readable.

As a more subtle example, consider a "semibatch" process involving the simultaneous addition of reactant and catalyst in preset proportion to a reactor. An exothermic reaction is taking place, and heat is being removed by a flow of cold water around the reactor jacket. At the interlock levels a cooling problem can be

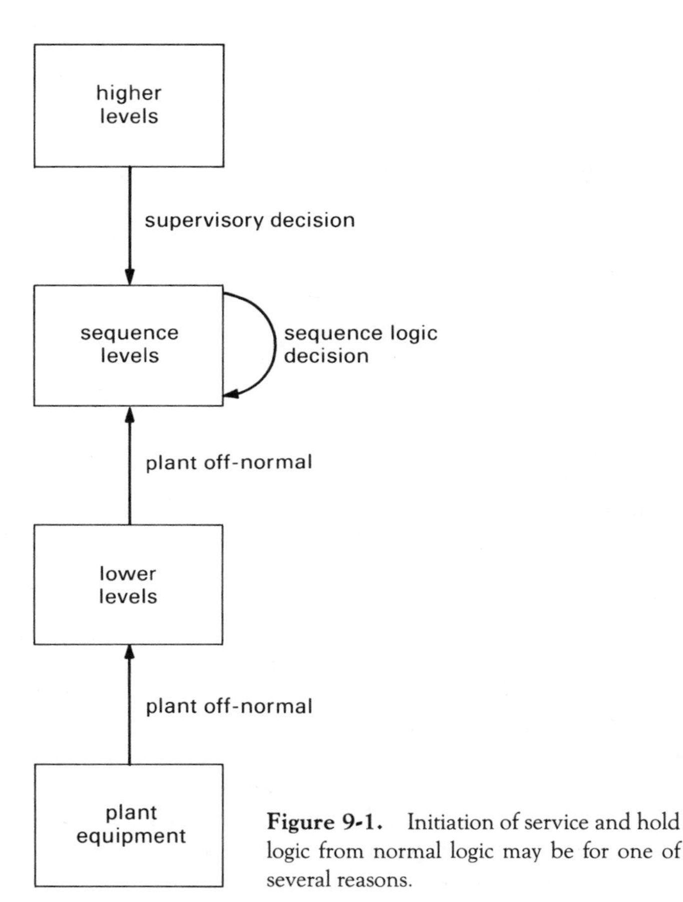

Figure 9-1. Initiation of service and hold logic from normal logic may be for one of several reasons.

identified by monitoring the reactor temperature, the rate of change of reactor temperature with time, and the output signal to the cooling water flow control valve. (An output of, say, 90% or more indicates that the temperature loop is nearing an inability to remove heat as it is being generated.) The response at the interlock levels might include venting air from the reactant and catalyst control valves, thereby stopping the transfer and adding a quench material to stop the reaction and ruin the batch.

At the sequence level a time profile of the anticipated material transfer can be preestablished. If transfer exceeds the profile by a preset amount, the reactant and catalyst control valves can be closed and the unit brought to a hold condition, enabling the operator to take control. The plant has not been placed in an emergency condition, and the batch has been preserved. The operator takes control, but the interlock levels are still in operation, protecting the plant in case an erroneous restart is attempted.

The ability of the logic to branch out of sequence also simplifies certain control strategies. For example, consider the specification

```
REQUEST OPERATOR TO TAKE SAMPLE 45 MINUTES
AFTER TEMPERATURE EXCEEDS 160 DEGF
```

The implementation is simple if the logic has nothing to do during the intervening 45 min. However, the implementation may not be simple if numerous actions are required over those 45 minutes, preventing any reliable prediction of the logic's state at sample time. Some systems contain provisions for separate timers that run for the preset interval—in this case, 45 min—then interrupt the main sequence to request the sample and, if necessary, wait for acknowledgment and/or entry of test results.

Similarly, service/hold logic can be used to collect preprogrammed responses to unavailability of shared equipment. Usually the process is designed so that it can wait indefinitely for service. Sometimes the process may be in an unstable condition, precluding an indefinite wait, so that a service/hold action is required.

Service/hold logic also serves to assist in unit shutdown. Shutdown may result at any time from a computer processing error, an operator request, failure of critical equipment, or by a preplanned action at the end of the batch. As stated, shutdown may not be as simple as turning everything off. Simply suspending the normal sequence is, similarly, not a complete solution because material transfers may be taking place that, if allowed to run to completion, could result in a vessel overflow, a batch that cannot be brought back into spec, or some other serious consequence. (Proper lower-level design minimizes these possibilities.) Even if these material transfers are simply terminated at an arbitrary shutdown time, it may be extremely difficult, without record keeping, for an operator or the control system to resume control later by running the remainder of the transfer. Finally, in extreme cases it may be necessary to quench a reaction to avoid endangering plant and personnel, add material to prevent a batch from solidifying, or perform some other positive action as part of a shutdown sequence.

Providing for an orderly shutdown is the responsibility of the hold logic. The hold logic can perform any function that normal logic can, as well as a few—such as driving outputs to predetermined hold states—that normal logic cannot. Typical hold functions are

Suspending a set-point ramp, allowing regulatory controls to maintain the controlled value at its prehold set point.

Capturing charged amounts, either in memory or as a message to the operator, or both, so that the remainder of the material may be transferred once the plant is back in normal operation.

Driving all or selected outputs to predetermined hold states. It should be possible for normal logic to change these desired states so that hold states can be made to match the condition of the process. Some of these outputs may be used to drive quenching or similar equipment.

Initiating a report to give the operator a detailed picture of plant status at the time of the shutdown.

Notifying other, so far unaffected, units of the shutdown so that they may take any appropriate action.

The hold logic should permit the engineer either to allow for the resumption of normal processing without further interruption or to require the operator to take additional action such as restarting at a phase transition.

It is mandatory that transfer to the hold path be capable of initiation by input conditions that are not being used by the normal logic at the time. A list of inputs with associated desired values should accompany each unit. The values should be changeable with normal logic, and normal logic should be able to enable/disable the branch to hold logic in the event of a discrepancy. For example, consider a situation in which the temperature of a batch is being changed. This activity is not directly related to the status of a programmable controller being used for material charging. However, the engineer must be able to configure the system so that the temperature change can be suspended if the PC fails.

SIMPLIFYING SEQUENCES*

A batch control system design normally depends on structuring of functions in an organized, modular fashion. Even though the unit control functions may be well defined within the context of the total process, not enough attention is ordinarily given to structure. As long as the logic performs as expected, the user will be satisfied and the designer will have little incentive to improve the situation. However, great penalties may result in terms of

Increased complexity in the logic, making the design documents and source code difficult to follow

More effort required for debugging and modification

Increased effort required in field support

The need to use detailed flowcharts and other specification/design documents in the checkout and modification of the logic

---

*The section on simplifying sequences has been adapted from A. Ghosh, 1982, Modular structuring of batch control logic, *Advances in Instrumentation* **37**(part 2):783-796, by permission from the Instrument Society of America. Copyright © 1982 by the Instrument Society of America.

The more complex a sequential logic system is, the greater the payoff from structuring it properly. Several levels of hierarchy may support the main logic. Here the term "hierarchy" is used not in the sense of design by levels but in terms of fineness of addressing. In other words, street addresses in the United States can be expressed in a hierarchical form like this:

State
City
Postal Zone
Street
Number
Apartment

The first three are expressed as a five-digit code. A hierarchical structure for sequence logic may be

Phase
Step
Statement

The *phase* is a process-oriented function. In dealing with a phase the operator, supervisor, or engineer understands the effect that the phase has on the process but not necessarily the details by which the effect is achieved. A typical phase is

```
HEAT REACTOR
```

with recipe specifications for final temperature, temperature ramp rate, and such alarm limits as maximum time. Heating the reactor may involve opening steam blocking valves and drain valves, activating safety interlocks, and switching states (e.g., manual to automatic) in the regulatory controls. The individual who chooses to invoke this phase does so only in consideration of its effect on the process, not of the way in which it opens the drain valves.

Similarly, other common phases are

```
START BATCH
CHARGE MATERIAL
COOK BATCH
TAKE SAMPLE
FINISH BATCH
```

In systems using variable-sequence (VS) recipes the phases are chained to define batches, much as letters are used to form words:

`LAP` and `PAL`

are combinations of the same letters that form different meanings. Similarly, phases may be used several times:

```
COMMITTEE
```

uses only six letters to form a nine-letter word. For example, one product may require heatup before a particular material is charged, whereas for another product the opposite sequence may be needed. A particular sequence of phases is sometimes referred to as a *procedure*.

In systems using master-sequence (MS) recipes either the phase order is fixed, with recipe-controlled branches taking place within the phases or transfer to the next phase is controlled by recipe instructions processed at the end of a phase. In practice, phases normally represent points at which a batch may be safely restarted.

The *step* is a plant-oriented function. In dealing with a step the engineer or programmer understands the effect the step has on the plant but not necessarily the details by which the condition of the plant is changed.

A typical step is

```
OPEN STEAM BLOCKING VALVE, WAIT FOR CONFIRMATION
```

Opening the blocking valve may involve a series of actions, including direct manipulation of the output, waiting for confirmation through inputs from limit switches, changing the valve's desired hold state, issuing an operator message, and other functions. The engineer who specifies this step as part of the heat reactor phase does so in consideration of the need for this plant change to be performed, not of the way in which the change is to take place.

It is often useful to establish the rule that each step corresponds to one suspension of execution of the process logic, such as for valve travel time, data entry by the operator, or the achievement of a process endpoint. The reasons are as follows:

1. The step number, if displayed to the operator, uniquely indicates what the logic is waiting for.
2. Computer/process debugging is facilitated for computers that offer single-step operation.
3. Operator-directed return to automatic control is simplified for computers that offer selectable-step reentry.

It is generally advisable to group all necessary setup actions (calculations, recipe data references, etc.) at the beginning of a step, follow them by output actions, and keep the suspension of execution at the end. Thus, with selectable-step reentry, preliminary actions must be performed before the process is affected, and the process actions must be taken before the logic waits for confirmation of the results.

Although we can define phases and steps with minimal knowledge of the control system, the *statement* level consisting of executable code does require knowledge of such system resources as the programming language, I/O addressing conventions, and layouts of recipe, logging, and other files. A batch control system is comparatively easy to program if its programming language makes generous use of process-oriented commands such as

```
HILIM(STMFLW,7.5)
```

to set the high alarm limit of the measurement input referenced by STMFLW (presumably, a steam flow) to 7.5 in the measurement's engineering units, or

```
WAIT 60
```

to cause the logic to suspend execution for 60 s.

General-purpose statements, as distinct from process-oriented statements, perform such functions as recipe data retrieval and calculations. For example,

```
LET GOAL=CONCN*NOMAMT
```

multiplies NOMAMT, the nominal amount of material to be charged, by a factor CONCN, the fraction of nominal concentration, based upon sensor measurement or lab analysis, to compensate for varying strength. This result is tagged GOAL and is used for further processing. Operator interface can similarly be simplified by statements such as

```
CRTMSG(11)
```

to display message number 11 on the unit CRT.

FLOWCHARTING

The *flowchart* (Fig. 9-2) is the tool most commonly used today for sequence specifications. It uses graphic symbols to represent control system actions and provides a common language for the process engineer, control engineer, and programmer. Numerous batch control systems have been implemented successfully by this technique. However, flowcharts do have some drawbacks:

They are undisciplined. Unless comprehensive standards exist, different individuals will generally choose varying levels of detail for their representations.

In our experience they are inefficient in their use of paper. Comparatively little information is contained on even fairly large drawings.

Many find them difficult to read. The processing can turn corners, make U-turns,

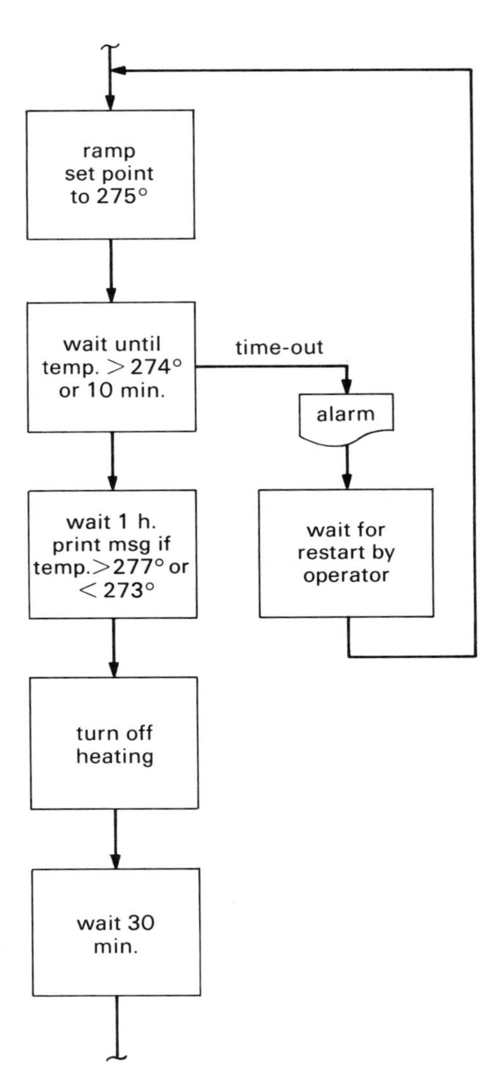

Figure 9-2. A typical flowchart *(After Rosenof, 1981; reproduced by permission of Technical Publishing, a company of the Dun & Bradstreet Corp.)*

travel from one side of the sheet to the other, and perform various other acrobatics. The problem gets worse as the flowchartist tries to make more efficient use of space. The problem gets much worse as the flowchart is revised, and as page-to-page connectors are used.

They are inefficient to use as checkout tools.

They must be carefully maintained over time to retain any real value.

Concurrency is difficult to show; a construction such as that of Figure 9-3 is not permissible. For batch sequences, concurrency is a necessary part of shared-unit and other combined operations. Variations such as *Petri nets* (Fig. 9-4) and *graphic batch processing language* (Fig. 9-5) offer a partial solution, but they are not widely known and are not consistent with the division of a plant into units.

Similar to Petri nets and graphic batch processing language diagrams are the *sequential function charts,* which are becoming popular in Europe. In France they are controlled by Standard NF C03-190 and are called Grafcets. They show steps as rectangles, with (in the French standard) each rectangle connected to the square containing the step number. Steps are drawn from top to bottom in the order in which they are processed. Simultaneous sequences of steps are shown by drawing multiple vertical paths connected by horizontal lines that show starting and stopping points.

Sequential function charts show fixed sequences fairly well but do not seem flexible enough for use with recipe-driven chemical, pharmaceutical, and similar batch processes. In contrast the hierarchical method, which the authors advocate, readily accommodates changes in the sequence of execution brought about by operator request, process conditions (anticipated or contingency), or recipe. We anticipate that the use of this method will be limited in the process industries to relatively simple applications. An example of a sequential function chart is shown in Fig. 9-6. For more information see Lloyd (1985).

Flowcharts require a lot of drafting effort for original preparation and for changes. Computer-aided design packages help but do not eliminate the problem.

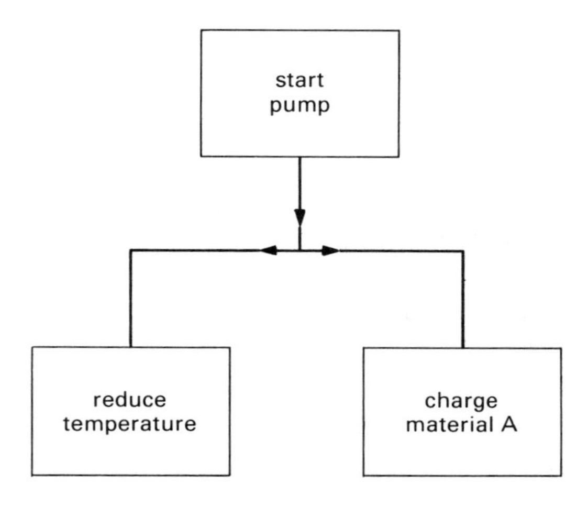

Figure 9-3. The standard flowchart does not show processing in two directions at once.

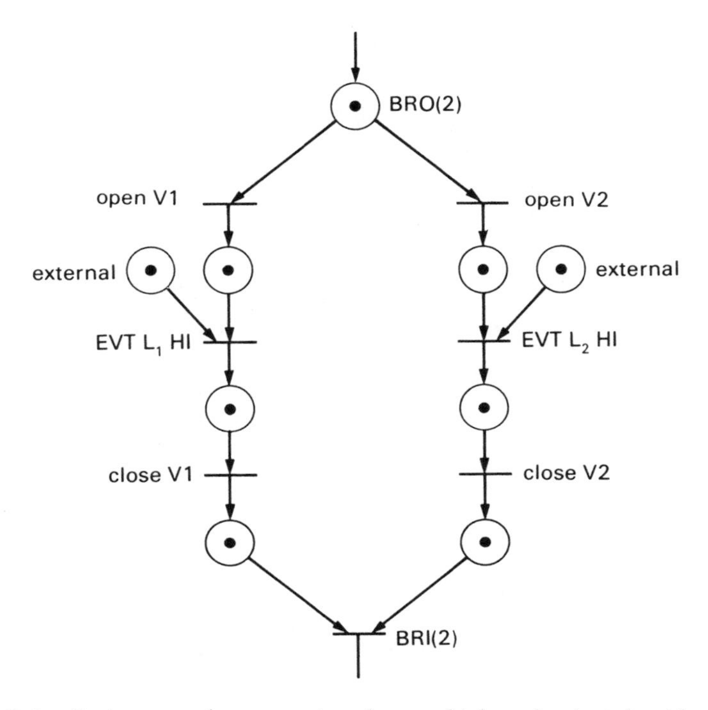

Figure 9-4. Petri net can show execution along multiple paths. A circle with a dot is a "place." A horizontal line is a "transition." When all of the places connected to a transition have "tokens" (dots), the transition is ready to "fire." When the transition fires, processing proceeds to the next transition. External places wait for external events *(After DaGraca, 1975; reproduced by permission of the Chilton Co.)*

For best results with a flowcharted specification, we recommend

1. Setting standards: symbol types, left-right and up-down progression, maximum percent fill on first release, and so on
2. Using a hierarchical breakdown (Fig. 9-7)
3. Revising flowcharts, on completion of coding, with specific coding information (Fig. 9-8)

STRUCTURED ENGLISH

Structured English or *pseudocode* is another documentation method not as widely used as it should be. In structured English, as the name implies, the logic is

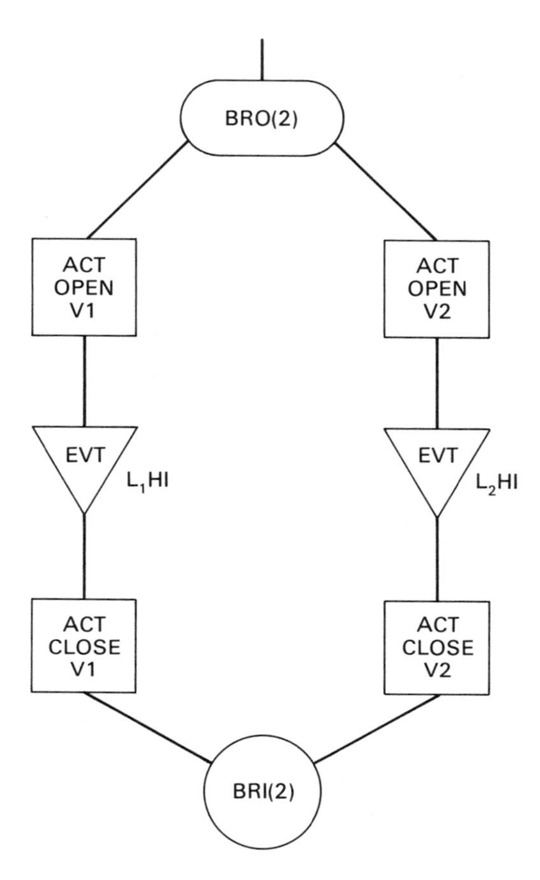

Figure 9-5. The graphic batch processing language looks more like a conventional flowchart, but, like Petri nets, also shows multiple execution paths. The intent of the logic—to fill two tanks simultaneously—is much clearer here. *(After DaGraca, 1975; reproduced by permission of the Chilton Co.)*

represented by English narratives written within a set of rules of structure and style for keeping the logic clear, unambiguous, and, at the same time, detailed and uniform. The following paragraphs outline some of these rules, which, however, may vary somewhat depending on the application.

In structured English a section of logic function is divided into statements, which are of two broad types: nonexecutable and executable. When a section of logic function is coded, each executable statement (but not each nonexecutable

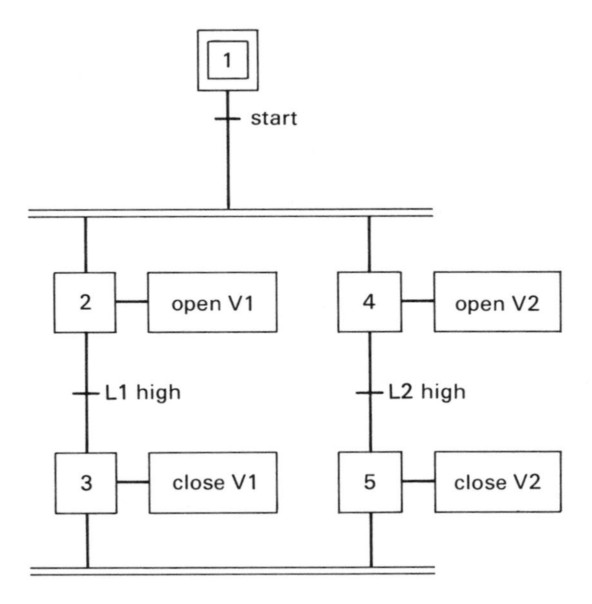

Figure 9-6. The parallel sequences of Figure 9-4 redrawn by the Grafcet method.

statement) has a corresponding line(s) of code. Nonexecutable statements are headers, comments, and labels. Executable statements are of the following types: imperative, decision, and repetitive. They are functionally analogous to the corresponding constructs in flowcharts.

Whatever the type, each statement must start on a new line. A continuation line may be indicated by a preceding hyphen. A header or comment needs to be distinguished from a normal executable statement. This may be done by using a set of preceding hyphens. For example:

```
-----SUBROUTINE FOR SETTING THROUGHPUT AND
     - CHECKING WASTEWATER FACILITY
```

Labels are needed at the entry point of a subroutine and also at any point where a jump is made. Two leading asterisks may be used to distinguish a label from a normal logic function. For example:

```
**CHECK-CONFIGURATION
```

An imperative statement describes a logic function to be performed. It should state clearly and concisely its function. Any plant hardware should be clearly

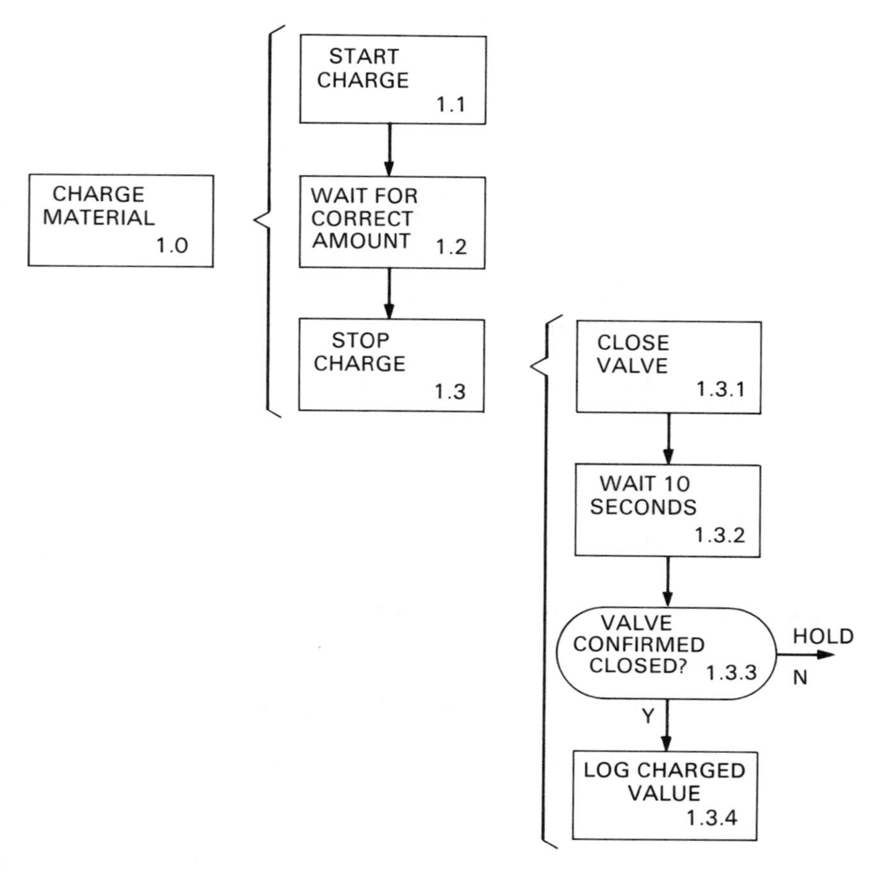

Figure 9-7. A hierarchical set of flowcharts shows the entire control sequence at an overview level, with each overview function broken down into successively finer levels of detail.

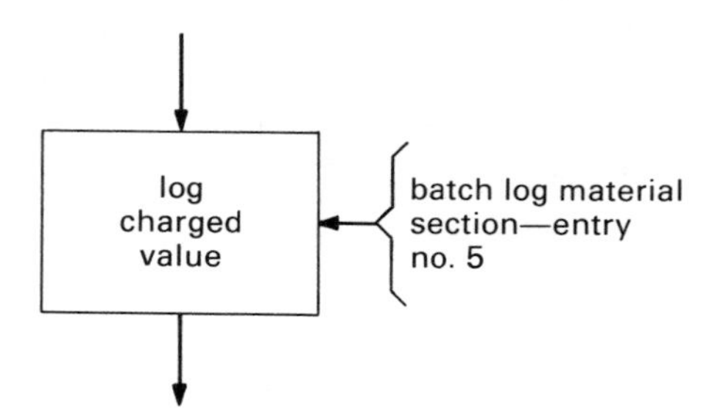

Figure 9-8. Coding details added to flowchart symbol.

identified, preferably with its tag number. A variable must have an explicit value or address. For example:

```
DRIVE THE SETPOINT OF THROUGHPUT CONTROL
- BLOCK (S1SC01) TO SPECIFIED VALUE (RECIPE ITEM 72)
```

In a decision statement IF, THEN, and ELSE functions are used where each branching starts on a new line, with the branches properly indented. For example:

```
IF COMPRESSOR (S1C051) IS NOT RUNNING
   THEN ALARM AND PRINT "COMPRESSOR DOWN"
   |
   ELSE OPEN COMPRESSOR DISCHARGE VALVE (S1C052)
```

Multiple levels of branching may be used in a decision statement, but we recommend no more than three or four levels, because this reduces the line size and makes the logic less readable. For example:

```
IF BOTH REACTORS ARE SCHEDULED
   THEN IF ALL 6 TANKS ARE SCHEDULED
   | THEN PRINT "VALID CONFIGURATION"
   | |
   | ELSE PRINT "INVALID CONFIGURATION"
   |
   ELSE IF ONE REACTOR IS SCHEDULED
     THEN. . .
```

As these decisions are nested to greater and greater levels, the description becomes progressively harder to read. An alternative description would be

```
IF BOTH REACTORS ARE SCHEDULED AND IF ALL 6
-TANKS ARE SCHEDULED
   THEN PRINT-"VALID CONFIGURATION"
   ELSE CONTINUE (This line optional)
IF BOTH REACTORS ARE SCHEDULED AND IF ONE OR MORE
-TANKS IS UNSCHEDULED
   THEN PRINT-"INVALID CONFIGURATION"
IF ONE REACTOR IS SCHEDULED. . .
```

Avoid operations like NAND and NOR because they are not standard English terms, and they make the logic obscure.

Each decision statement normally has an IF, THEN, and ELSE term; however, when the ELSE is only for continuation, it may be eliminated. For example:

```
WAIT FOR WEIGH TANK HIGH LEVEL ALARM (S1C023)
IF NO ALARM IN 180 SEC
  THEN DISPLAY "WG TK TAKING TOO LONG TO CHARGE"
SHUT WEIGH TANK INLET (S10020)
```

For multiple branching, several IF, THEN, and ELSE terms may be used or a CASE statement may be used. For example:

```
-----SERVICE LOGIC ENTRY DUE TO EITHER
    - 1. LOOPING
    - 2. OPERATOR HOLD REQUEST
    - 3. OPERATOR SERVICE REQUEST
    - 4. INPUT DISCREPANCY, CRITICAL
    - 5. INPUT DISCREPANCY, NONCRITICAL
    FOR CASE 1 GO TO: **LOOP
    FOR CASE 2 GO TO: **OPHOLD
    FOR CASE 3 GO TO: **OPSERV
    |
    |
    |
```

A Repetition statement specifies a set of logic that may be repeated until a certain condition is satisfied. You can use a Decision statement with a Jump statement. For example:

```
**OPEN DISTR VALVE
  IF DISTRIBUTION VALVE FLAG (S1C071) NOT ENABLED
    THEN PRINT "DIST VALVE NOT ENABLED"
    | WAIT 1 MIN
    | GOTO: **OPEN DISTR VALVE
    |
    ELSE OPEN DISTRIBUTION VALVE (S10071)
```

Another way to specify repetition is to use a REPEAT-UNTIL term. For example:

```
**OPEN DISTR VALVE
  REPEAT-UNTIL DISTRIBUTION VALVE FLAG (S1C071)ENABLED
  | PRINT "DIST VALVE NOT ENABLED"
  | WAIT 1 MIN
END REPEAT-UNTIL
OPEN DISTRIBUTION VALVE (S10071)
```

The examples given for structured English are for illustration only; the details of the rules and styles may vary from one project to another. However, the rules and conventions should be agreed on and clearly spelled out at the beginning of a project so that parts of the logic design documents generated by different individuals for the same project have the same standard and style. The design documents will then be clear and unambiguous to users.

Here are the general guidelines for writing in structured English:

Use simple statements.
Clearly define the required function in a statement.
Identify the plant hardware address wherever possible.
Indent where necessary.
Avoid negative logic.
Avoid excessive nested logic.

After you complete the design document in structured English, insert the necessary coding for each statement at appropriate places in the document to establish a one-to-one relationship between the design intent and the actual coding and produce a single document (Fig. 9-9). A word processing facility in the control computer or in a dedicated word processor may be used to strip the structured English statements or the coding when only the coding or the design document is needed.

The proper structuring of the control problem *must* be done before you write the structured English narrative. You cannot simplify a complicated batch control logic merely by writing it in structured English. First, the control requirements need to be analyzed. Long and complicated pieces of logic should be partitioned into modules (subroutines) in a hierarchical fashion (see previous section on simplifying sequences). The optimum size of each module should be such that the structured English narrative for each does not exceed one sheet of standard 8½-by-11-in. paper.

The main advantages of structured English over other methods of design documentation are that it is easy to generate and modify. Structured English requires a word processing facility, which is available in larger control computers, personal computers, or dedicated word processors. Other methods, where graphical or tabular representations are needed, require laborious drafting time for generation and modification. Another great advantage is in the single design and coding document, which reduces the possibility of mismatch between the two at initial coding and subsequent updates, thus reducing housekeeping efforts. Structured English is easier than other methods to understand and modify for persons not knowledgeable in computers, programmable controllers, or relay logic because it more closely resembles the normal English language.

The drawbacks of structured English are in the lack of generally accepted guidelines or rules and styles. It is up to an individual project team to agree on a set of standards. If not properly modularized, structured Engish can be obscure for a long and complicated piece of logic. Also, like flowcharts, parallel operations are not always easy to document (see *Process Timing Diagrams*).

```
<*-----COMPONENT CHARGE PHASE MAIN LOGIC                                    >
<*     -THIS PHASE CHARGES COMPONENTS 1 TO 5 TO THE REACTOR.                >
<*     -THE AMOUNT OF EACH COMPONENT TO BE CHARGED                          >
<*     -IS A RECIPE VARIABLE.                                               >
<*                                                                          >
<*      INITIALIZE THE COMPONENT COUNTER TO ONE                             >
C10010                STORE[IV001,1]
<*                                                                          >
<*    **CHARGE-COMPONENT                                                    >
<*      IF THE AMOUNT TO BE CHARGED FOR THE APPROPRIATE                     >
<*      COMPONENT IS NOT ZERO                                               >
C10020                STORE[TV001,0,FR001,IV001]
*                     IF[TV001,0.0,C10070,C10070,C10030]
<*                                                                          >
<*         THEN PRINT AND DISPLAY-"CHARGING COMPONENT NO--",                >
<*         COMPONENT NUMBER                                                 >
C10030                PRINT[1,IV001]
*                     CRT[1,IV001]
<*                                                                          >
<*         IF THE COMPONENT NO IS LESS THAN THREE                           >
*                     IF[IV001,3,C10040,C10050,C10050]
<*                                                                          >
<*            THEN CALL SUBROUTINE: **CHARGE-BY-FLOWMETER                   >
C10040                DO[C10100,40]
C10045                GOTO[C10060,45]
<*                                                                          >
<*            ELSE CALL SUBROUTINE: **CHARGE-BY-WGH-TANK                    >
C10050                DO[C10200,50]
<*                                                                          >
<*         PRINT AND DISPLAY-"CHARGING COMP FOR COMPONENT--",               >
<*         COMP. NO.                                                        >
C10060                PRINT[2,IV001,I]
*                     CRT[2,IV001,I]
<*                                                                          >
<*      INCREMENT COMPONENT COUNTER BY ONE                                  >
C10070                ADD[IV001,1,IV001]
<*                                                                          >
<*      IF COMPONENT COUNTER IS LESS THAN SIX                               >
<*         THEN GO TO: **CHARGE-COMPONENT                                   >
*                     IF[IV001,6,C10020,C10080,C10080]
<*                                                                          >
<*                                                                          >
<*         ELSE PRINT-"REACTOR CHARGING COMPLETE"                           >
C10080                PRINT[3]
<*                                                                          >
<*      GO TO THE NEXT PHASE                                                >
*                     PHASE[2,0,C20020,INIT,80]
<*                                                                          >
<*                                                                          >
<*                                                                          >
<*                                                                          >
```

Figure 9-9. Logic in structured English with coding.

SIMPLIFYING CODING

The preceding observations are addressed to those responsible for translating process needs into a specification for batch control logic. They are intended to be applied to a wide variety of computer-based systems and, to a lesser extent, systems using PCs. The discussion which follows is addressed primarily to those responsible for programming computer-based control systems. These techniques were developed for use on specific computer systems, so that equivalent techniques to achieve these same ends may be appropriate for other systems.

Subroutines

A *subroutine* is a program module that can be called, or transferred to by another program module. The difference between a subroutine and a part of the main program that may be accessible with a GOTO is that returns from a subroutine to the main code (usually at a point just past that at which control was transferred to the subroutine) are automatically handled by the computer. Subroutines reduce computer memory requirements and facilitate checkout. Batch control systems generally permit the placement of sequence logic in subroutines, so frequently used process actions or calculations need not be duplicated throughout the code.

If the engineer specifying the logic is familiar with the system's subroutine structure, he or she may specify subroutines within the structured English description. If the batch control system has already been chosen at the time of logic specification, the permissible structured English vocabulary and syntax options may be adjusted to maximum advantage.

Contemporary programming practice discourages the use of GOTO statements in favor of subroutines. The term "spaghetti code" is used to describe unstructured code, made difficult to follow by extensive use of GOTO statements (Fig. 9-10). Subroutines, by contrast, encourage the writing of code that proceeds in a straight line (Fig. 9-11). Readers can get a general understanding of the code's intent by reading the main code, and they can then read the subroutine's listing for functional details.

Conditional subroutine calls such as

```
IF TEST1='FAIL'
  THEN GOSUB REPEAT
```

replace conditional branches such as

```
IF TEST1='FAIL'
  THEN GOTO 25
```

```
    STATEMENT
    STATEMENT
    IF A THEN GOTO B
G   STATEMENT
    IF C THEN GOTO D
    STATEMENT
B   STATEMENT
D   STATEMENT
    GOTO E
    IF F THEN GOTO G
    STATEMENT
    STATEMENT
```

Figure 9-10. Spaghetti code.

```
STATEMENT
STATEMENT
IF A THEN DO SUBROUTINE B
STATEMENT
STATEMENT
DO SUBROUTINE C
STATEMENT
IF D THEN DO SUBROUTINE E
```

Figure 9-11. Straight-line code made possible by subroutines.

Branches and subroutine calls can both be minimized if the language in use can support additional conditional statements:

```
IF TANK1='NOT READY'
   THEN LET TANK2='BACKUP'
```

replaces

```
    IF TANK1='READY'
       THEN GOTO 17
    LET TANK2='BACKUP'
17 . . . . . . .
```

and

```
IF A=1 .OR. B=1
   THEN LET C=A+5
```

replaces

```
    IF A=1
       THEN GOTO 62
    IF B=1
       THEN GOTO 62
    GOTO 63
62 LET C=A+5
63 . . . . . . .
```

Systems featuring subroutine nesting allow one subroutine to call another, so individual subroutines may be kept smaller (Fig. 9-12). Systems featuring subroutine reentrancy allow one subroutine to be independently called by many users.

Storing Program Details in Tables

Consider a plant with three identical batch units. If I/O must be addressed directly, then three similar but different phases must be written, debugged, stored, and so on. This is necessary because stopping an agitator motor, for example, requires that each phase address a different hardware output to deenergize it. However, if the logic is separable into sequences and I/O references, then the sequences can be written in terms of data base reference instead of output tag. This technique is usable for interfaces to regulatory controls as well. If, for example, 10 valves must be opened simultaneously, the valve output reference numbers may be stored as a

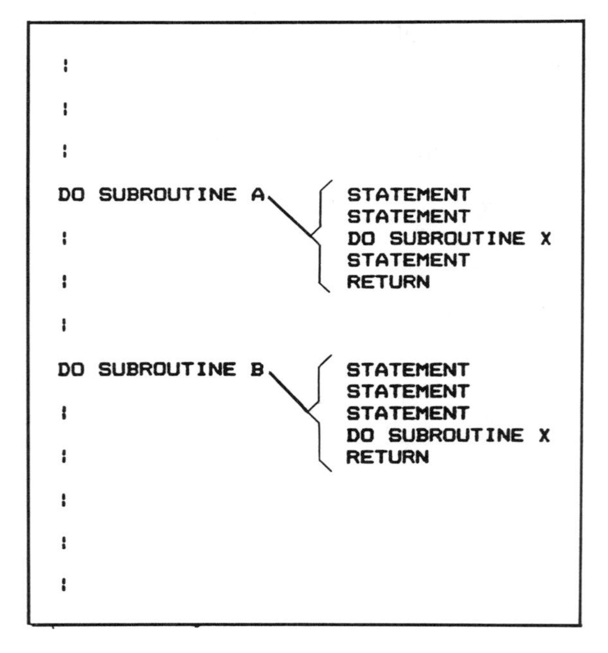

Figure 9-12. Nested subroutines. One subroutine (A or B) can call another (X).

table of values so that these actions can be directed by a simple FOR...NEXT loop rather than by 10 separate statements.

Now consider a plant with three similar process units that, except for size, are identical. Each unit may be assigned to run the same mix of products. At one point in the process the agitator must be turned on at a recipe-specified weight, expressed as a percentage of maximum weight. Since for each vessel this occurs at a different level, a different recipe would seem appropriate for each vessel. However, the use of unit-related data can make this unnecessary. The unit's maximum weight is represented by MAXWGT (different for each unit); and the weight at which the agitator is to start is AGITON percent, from the product's recipe. Then if the weight for agitator starting is determined from

```
LET AGITWT=MAXWGT*AGITON
```

the calculation is valid for each unit, and one recipe may be used for all units.

As another example, consider a plant of 20 main process units in which units 1-10 receive a material charge from unit 21 and units 11-20 receive the same material from unit 22. The number of the charging unit, 21 or 22, may be stored in the unit data table so that the request for service may be issued to the correct unit. Using a technique like this rather than hard coding makes changes much easier.

Unit data tables are also useful where process operations differ slightly between units. They allow multiple phases to be replaced by only one phase. For example, consider a situation in which two material charging units are identical except that the contents of one have to be at or above a specified temperature before the transfer can take place. The logic to check the temperature, initiate heating if necessary, and wait for the temperature condition to be satisfied is contained within the phase used by both batch units, but it is executed only if the designated entry in the unit data table has the required value.

PROCESS TIMING DIAGRAMS

For clarity and convenience structured English has replaced flowcharts in many organizations. However, structured English and flowcharts do share one limitation: They show only the control side of the system. Perhaps the first lesson the student control engineer learns is that once a loop is closed, the controls really become part of the process. The behavior of a process under closed-loop control depends on all of the constant and dynamic terms around the loop. Their source, whether process transport delay, valve, or controller, is of little importance. Thus, continuous control systems are often first defined on process and instrument diagrams (P & IDs), which show both the loops and their relationship to the process.

The situation for sequential control systems is more complex. The P&ID does little to specify or otherwise communicate the time-variant nature of the sequential process and its controls. The individual reading the specification for a sequential control system, whether in flowchart, structured English, narrative, or other form, must make assumptions about the behavior of the controlled process in order to make sense of the specification. Sometimes these assumptions are made easily with small risk of error, as for a shutdown sequence. Here the specification is likely to consist of little more than a list of pumps to be stopped and valves closed, along with possible alarms and messages for failure of equipment to respond properly. However, process behavior is not always readily predictable from a reading of the control specification. In batch boiling, for example, transition from a nonboiling condition to boiling is likely to be accompanied by subtle changes in the control system's operating mode. These changes may not even be taking place at the sequence level, but, as in this case, at the regulatory levels, so they may not even be covered by the sequence specification.

One of the authors proposed the use of "process timing diagrams" to achieve simultaneous definition of control actions and the process behaviors that are to accompany them (Rosenof, 1981). The technique extends the use of simple time-temperature curves to cover a wide variety of conditions. Since publication, process timing diagrams have been used for documenting comparatively simple processes (many can be fully shown on one sheet of paper) and for selected portions of larger processes, such as transition to boiling. Because a lot of drawing is involved, the technique is not recommended for documenting large processes. When used appropriately, process timing diagrams have been shown to be effective tools for communication among engineers and between engineers and operating personnel. The authors' general recommendation, then, is for structured English logic specifications to be used as controlling documents for all sequence logic to be used for a particular process. Where appropriate, these specifications should be preceded by process timing diagrams representing agreement between process and control engineers covering the sequence/process interaction.

The basic process timing diagram is shown in Figure 9-13. All parameters of interest are shown at all times, so they are not likely to "fall through the cracks." Concurrency is easy to show because timing relationships between subprocesses are now explicit. In comparison, concurrency is difficult to show with flowcharts or structured English. Progress is linear and horizontal, in the manner of reading. If

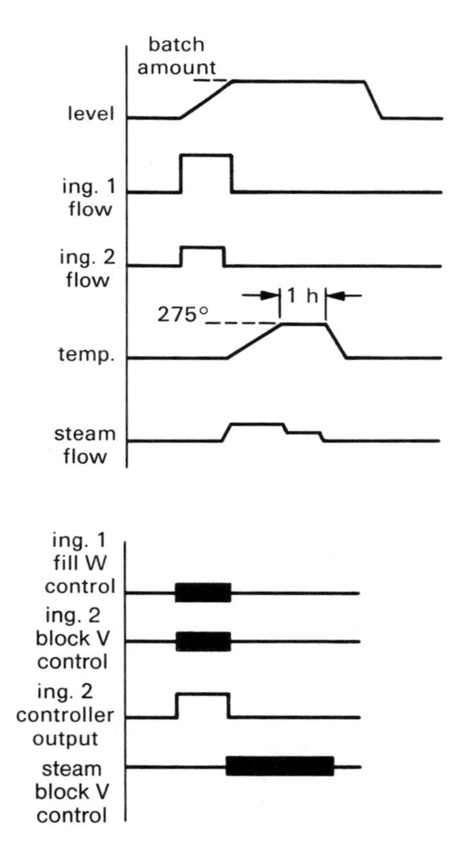

Figure 9-13. Basic process timing diagram. *(After Rosenof, 1981; reproduced by permission of Technical Publishing, a company of the Dun & Bradstreet Corp.)*

several sheets are needed to describe a particularly complex process, the user is not distracted by having to search all around the next sheet for a connector, as would be the case for flowcharts. Changes are accommodated with comparative ease, and standard drawing sizes are convenient. (Plain, 80-column computer coding sheets are excellent for sketching.)

Steps and Phases

Phase and step transitions may be shown as vertical lines; the user may use different thicknesses or types of lines to indicate the transition type. Steps and phases may be identified along horizontal lines above the process parameters (Fig. 9-14).

Internal Signals

The minimum information required to document process-control interactions is control outputs and process parameters. The system may be made easier to understand by showing signals internal to the control system as well: for example, set points, alarm limits, deviations, and signal selector outputs. Internal signals may be grouped to form a field, or they may be dispersed among other signals (Fig. 9-15).

Primarily and Secondarily Controlled Parameters

Controllers are frequently cascaded to improve control over one parameter that is influenced by another. The output of a temperature controller, for example, may become the set point of a steam flow controller, and that controller's output may become the set point to a valve positioner, which is another controller. The temperature controller is part of the primary loop, and the steam flow controller is part of the secondary loop. The valve positioner could be termed part of a tertiary

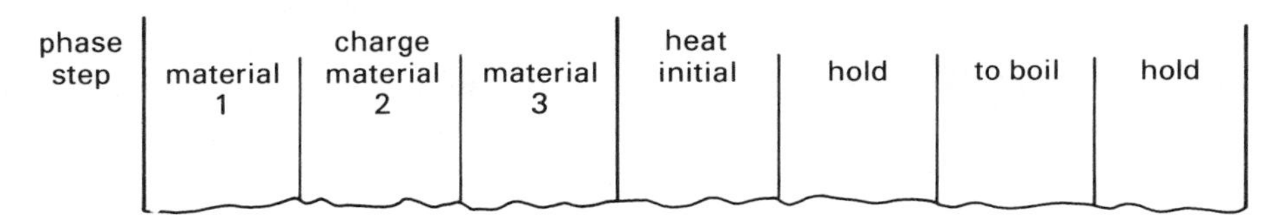

Figure 9-14. Phase and step transitions. *(After Rosenof, 1981; reproduced by permission of Technical Publishing, a company of the Dun & Bradstreet Corp.)*

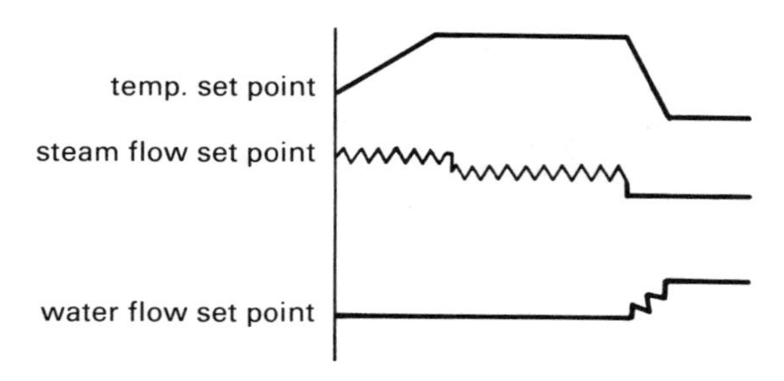

Figure 9-15. Use of internal signals, such as set points, in process timing diagrams. The steam flow set point is "secondarily controlled" over part of the time shown; this is explained in the text. *(After Rosenof, 1981; reproduced by permission of Technical Publishing, a company of the Dun & Bradstreet Corp.)*

loop, but for our purposes a *primarily controlled parameter* is of more interest, a parameter that a complete loop is intended to control. Parameters controlled within a loop, to influence a primarily controlled parameter are termed *secondarily controlled.*

Primarily and secondarily controlled parameters differ in their behavior over time. Presumably the control system is correctly tuned and holds the primarily controlled parameter close to its constant or changing set point. Secondarily controlled variables may be expected to vary more. A change in steam pressure, for example, should be compensated for within the flow control loop without influencing the temperature under control. In most cases, then, the primarily controlled parameter is represented in the process timing diagram as a smooth line, and secondarily controlled parameters are shown as wavy lines.

In an on/off control system, the controller output state may be considered to be secondarily controlled. Vertical stripes are used to represent such a signal (Fig. 9-16).

Forced and Unforced Outputs

At the beginning of a new phase, some control system outputs may be left as they are, whereas others are forced to new values (an option with some systems). If the forced value of an output is the same as the value the output should have had just before the phase transition, it will not be evident from the process timing diagram that it had been forced. To show a variable is forced, use an asterisk to the right of the phase transition (Fig. 9-17).

Checked and Unchecked Inputs

Similarly, the user may wish to check the values of control system inputs (process output) before entering a new phase. Values to be checked are marked with stars: a

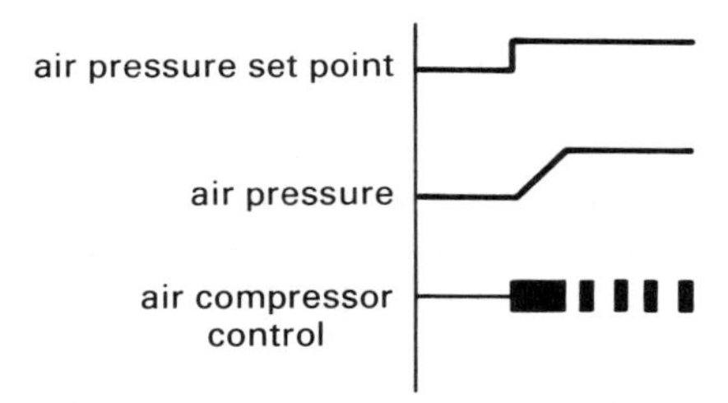

Figure 9-16. Air compressor control as a secondarily controlled on/off signal. *(After Rosenof, 1981; reproduced by permission of Technical Publishing, a company of the Dun & Bradstreet Corp.)*

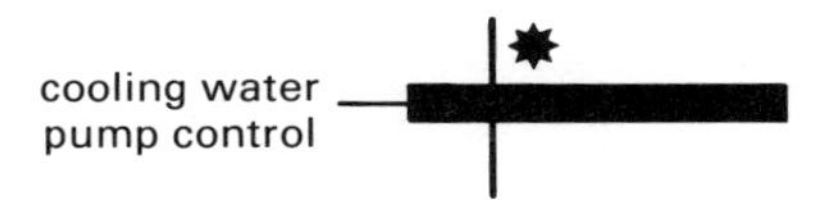

Figure 9-17. Cooling water pump is turned on at the beginning of the phase shown, whether or not it had been on before. *(After Rosenof, 1981; reproduced by permission of Technical Publishing, a company of the Dun & Bradstreet Corp.)*

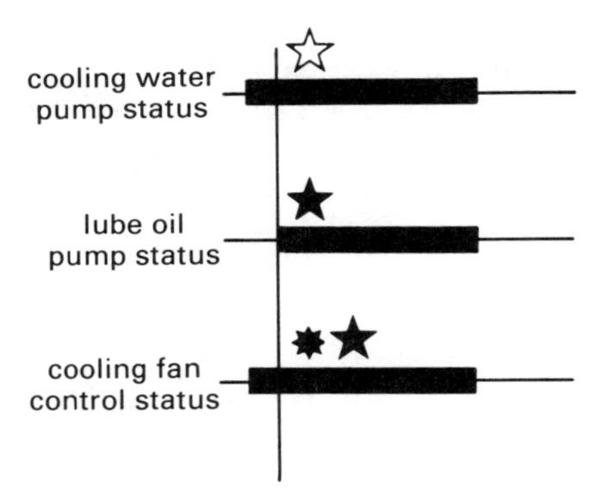

Figure 9-18. All parameters are checked at the phase transition. The lower two would shut down the process if incorrect. *(After Rosenof, 1981; reproduced by permission of Technical Publishing, a company of the Dun & Bradstreet Corp.)*

closed star is used for a parameter that would shut down the process if incorrect; an open star means that the operator is given a message, but no process action is taken. Input checking described here takes place only at phase transitions; checking that takes place continually is discussed under *Alarms*. Parameters that merely cause the process to wait until they are correct are covered under *Combinatorial Logic*. The typical input checked here would be the status signal of a manually operated device (Fig. 9-18).

Combinatorial Logic

Progression of a process to a new step often depends on the conditions of several process and control parameters: Temperature must be above a limit value, a pump must be running, minutes must have elapsed since a certain event. This causes a potential problem for timing diagrams, since, for example, three logic requirements can be met in six different sequences (1-2-3, 2-1-3, etc.). Six drawings would then be needed to show all ways for the process to reach the next step. A solution is to show the step progression once with the sequence the engineer considers the most likely. At the end of the current step all the requirements that must be met are marked with a closed triangle (Fig. 9-19).

The user might also choose to show combinatorial logic explicitly, as explained under *Branching*. Although the example given here allows only the options of waiting or proceeding, branching allows the process to continue at any one of many points.

Alarms

One of the real conveniences of computer automation is the easy control, compared to that of hardwired systems, that it gives over alarms. Limits may be changed, alarms enabled or disabled, and responses to alarms varied to suit changing process conditions. An example is the disregarding of a temperature rate-of-rise alarm when charging a cool reactor with hot material.

Alarms may be documented on the process timing diagram with their own field. The "contact closed" indication shows that an alarm is active. Alarm limits may be

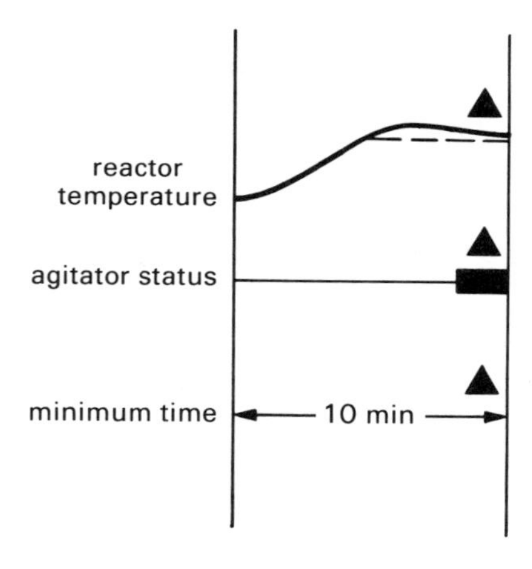

Figure 9-19. All three conditions must be met before the process may proceed. *(After Rosenof, 1981; reproduced by permission of Technical Publishing, a company of the Dun & Bradstreet Corp.)*

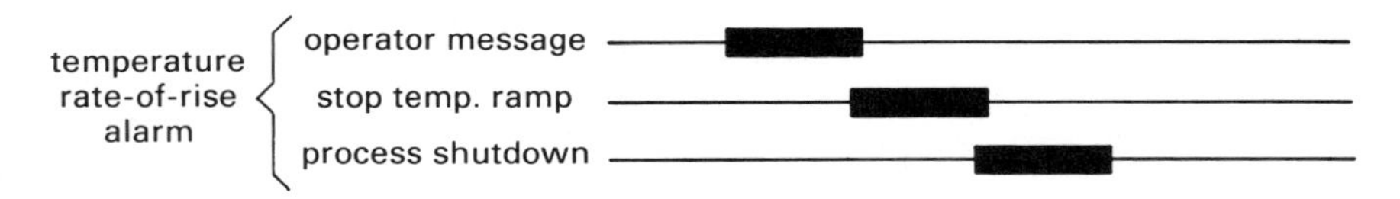

Figure 9-20. Four possible responses to a temperature rate-of-rise alarm. *(After Rosenof, 1981; reproduced by permission of Technical Publishing, a company of the Dun & Bradstreet Corp.)*

shown in the alarm field or elsewhere. If there are several possible responses, they should be grouped together within the field. For example, an alarm condition may cause a shutdown at some places within the process, a branch to service logic at others, and only an operator message at others. An alarm line is used to show time limits imposed when the process waits for an endpoint to be reached. The vertical position of an alarm within the field may be used to indicate its priority (Fig. 9-20). On a separate drawing the user should specify for each alarm the alarm condition, the response desired, and the criteria for return to normal.

Auxiliary Processes

An *auxiliary process* is one that takes place concurrently with the main process, is at times independent of the main process, and at other times interacts with it. Typical auxiliary processes are weighing of ingredients, preliminary mixing, and operator activities performed in synchronization with the process.

Simple auxiliary processes may be shown on the same drawing as the main process. More generally, the auxiliary process is treated from the viewpoint of the main process as an input—for example, "weighing complete," which is the eventual expected response to an output, "start weighing." This is true even if a complex signal protocol is used for interunit communications. In this sense a complex auxiliary process is no different than an on/off valve whose limit switch states should eventually show that it has reached the commanded position. These auxiliary processes, then, need only two contact-type horizontal channels on the main process drawing but may be shown in full detail elsewhere.

Operator interaction is shown on a separate line. A full rectangle denotes a request, made to the operator, that requires a prompt response. If a request is made sometime before the response is needed, the request is shown with the left half of the rectangle, and the response with the right half. Each request/response pair is identified by number and described in detail elsewhere (Fig. 9-21).

Segmentation

Many batch systems are used to make different products at different times, and the procedures needed to make these products can vary in more than temperature set

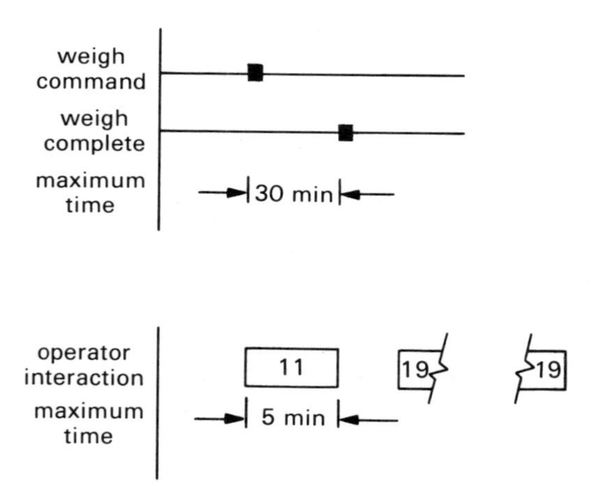

Figure 9-21. Automatic and manual auxiliary processes. *(After Rosenof 1981; reproduced by permission of Technical Publishing, a company of the Dun & Bradstreet Corp.)*

points and material amounts. Flexible batch control systems can hold in their recipes not only set-point data but the sequences in which the various process steps are to be run. With the actual control system as a model, a solution is to show each distinct, schedulable process operation on a separate drawing. Any mandatory combination of segments should be drawn together.

A VS process system may place an additional demand upon the controls—they must be able to drive the system to the desired state without the engineer's prior knowledge of the starting conditions. Reaching a desired temperature, for example, may require an increasing or decreasing ramp. A material to be weighed may not be compatible with the last material to go through the common weighing system, so that the system must first be cleaned. It may be necessary, therefore, to show a segment more than once to illustrate the control system's response to varying starting conditions.

Branching

In some systems it is necessary to be able to modify the process sequence during the process. Changes may be required as a result of operator response to a question from the computer, or failure of a material to meet quality control requirements. To define *branching,* we must specify the starting point, destination points, and logic for selecting the appropriate destination point. Logic may be shown by flowchart, combinatorial logic diagram, or function table. Decision points are identified by number, destination points by letter (Fig. 9-22).

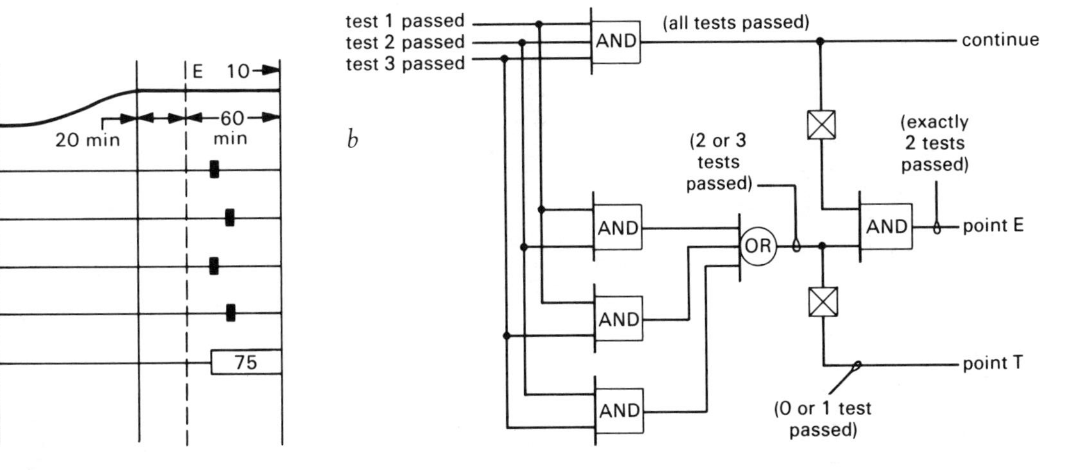

c

if all tests passed, then continue
if two tests passed, then point E
if one test passed, then point T
if no test passed, then point T

d

TEST 1	TEST 2	TEST 3	
P	P	P	CONTINUE
P	P	F	POINT E
P	F	P	POINT E
F	P	P	POINT E
P	F	F	POINT T
F	P	F	POINT T
F	F	P	POINT T
F	F	F	POINT T

Figure 9-22. *a.* A decision to branch may be made at point 10. *b.* Branching logic specified by pseudocode. *c.* Branching logic defined by combinatorial logic. *d.* Branching defined by function table. (*After Rosenof, 1981; reproduced by permission of Technical Publishing, a company of the Dun & Bradstreet Corp.*)

Causes and Effects

The user may emphasize the relationships between process conditions and control system outputs by using vertical dotted lines to connect causes and effects. The lines point toward the effects; thus if a steam valve is closed once the process materials reach a preset temperature, one line may go from the temperature field to the control system's valve output; a second, displaced to the right to show valve travel time, may go from the output signal to the valve position field. If the control system valve output is not shown, the line may connect the temperature and valve position. If one cause has several effects, the vertical line connects more than two points (Fig. 9-23).

Time Scaling

Even if a process would permit the use of a consistent horizontal time scale, there is generally no need for one. The user may ignore the slight time delays inherent in any practical control system, especially if the method described under *Causes and Effects* is used. These delays are easily illustrated, however, by slightly displacing the system's response from the response's cause.

Constant Mass or Constant Volume

A batch system may be able to hold material from several batches at one time. The batch recipes will generally be written to follow each batch's material through

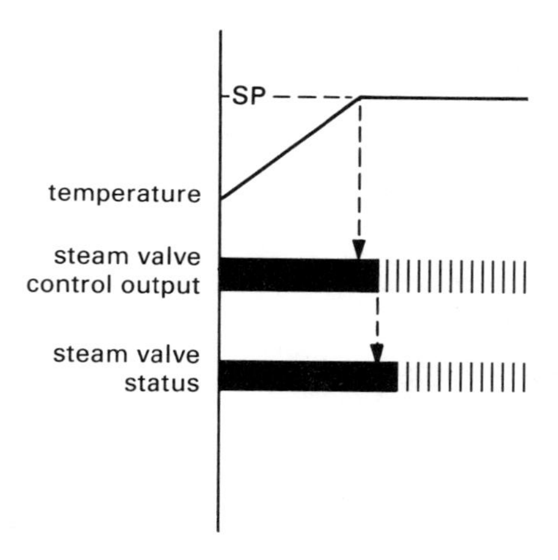

Figure 9-23. Cause and effect. *(After Rosenof, 1981; reproduced by permission of Technical Publishing, a company of the Dun & Bradstreet Corp.)*

various items of process equipment, and it is usual to orient process timing diagrams the same way. It may occasionally also be useful to prepare equipment-oriented "constant volume" diagrams if an item of equipment has batch-independent control for cleanup, material transfer, and so on.

A Complete Process

An example of a complete process timing diagram is given in Figure 9-24.

ENDPOINT PREDICTION AND DETERMINATION

In essence there are two criteria for ending part of a batch sequence: time-out or event. The term "event" is used loosely here: Typically, it means the achievement by a batch of some previously defined condition, such as

```
A GALLONS OF MATERIAL B CHARGED TO VESSEL C
```

or

```
TEMPERATURE IN VESSEL D IS GREATER THAN OR EQUAL TO E DEGF
```

or

```
CONCENTRATION OF MATERIAL F IN VESSEL G IS GREATER THAN
OR EQUAL TO H% BY WEIGHT
```

It may also reference an operator action, such as shutdown command, or confirmation that an activity, external to the batch control system, has been completed. Or it may be an asynchronous event, such as an alarm demanding immediate shutdown of the process.

In batch plants with automation up to about level 5, material transfer processes are virtually all ended by events marking the completion of the operation. However, it is common to allow a drain time after the command to stop the charge has been issued. Processes that modify the nature of the material constituting the batch are ended by event when there are regulatory requirements (sterilization, in the food and pharmaceutical industries) or appropriate industry practices (H-factor for batch digestion in the pulp and paper industry). However, in many other applications, control is defined by time.

Control by time is generally wasteful: Certain changes must take place within the batch, and if the time frame is too short the changes will not have occurred and the batches will be off-spec. Since this is obviously undesirable, batch operations typically are run longer than necessary to insure a margin for error. At best, this

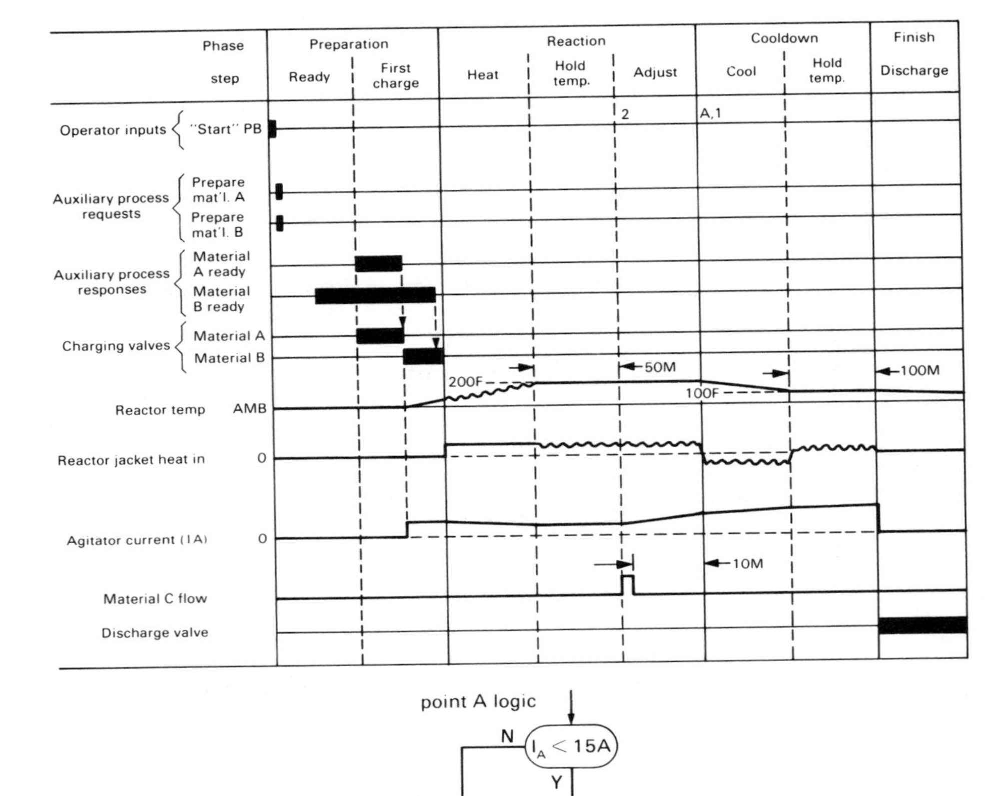

Figure 9-24. A hypothetical chemical batch process. Materials A and B are charged to the reactor and the reaction takes place. The Adjust step is repeated as many times as necessary until the viscosity of the reactor contents is below the preset value, as indicated by agitator current being below 15 A. The reactor contents are cooled and discharged. Note that reactor temperature and jacket heat input exchange their roles as primarily and secondarily controlled. This shows one way to arrange the information on the process timing diagram. Earlier activity is shown at the top, so the chart can be followed along a diagonal from top left to bottom right. *(After Rosenof, 1981; reproduced by permission of Technical Publishing, a company of the Dun & Bradstreet Corp.)*

overprocessing increases batch residence time, thereby reducing the plant's maximum production rate; it also consumes utilities, thereby increasing batch cost and pollution rate. At worst, overprocessing can cause a batch parameter to go out of specification, making it necessary to reject the entire batch.

Consider, for example, the clothes drier. Faced with the potential of an unpleasant damp-clothing condition at the end of the batch, the user is likely to set a time far in excess of that really required to dry the clothes. The added expense due to increases in residence time, utility costs, pollution, and wear and tear are considered to be of secondary consequence or are ignored altogether. (Some driers now use an inferential measurement of clothing moisture content to initiate a process shutdown.)

In an industrial setting batch automation allows the user to discard time-based criteria. These criteria may have been used for operator convenience, to compensate for a lack of suitable sensors, to avoid manual calculation of inferentially measured variables, or because past economic/market conditions simply did not demand tight control. Replacing these time-based measures are endpoint-based criteria that limit batch processing to that required to meet product specifications.

Examples

1. In wood preserving, wood is charged to a retort and treated by exposing it under pressure to a preservative stored in a separate tank. Once the level in the tank stops changing, the batch is said to have reached "refusal": No more preservative is being absorbed by the wood. When the batch is stopped and all unabsorbed preservative has been returned to the tank, a calculation shows the amount of preservative absorbed. At this point the calculation is made primarily for record-keeping purposes; no decision is generally made on the basis of its results.

Under some circumstances industry standards allow the process to be stopped before refusal. In this case the post-batch calculation is required to determine whether the specified retention (weight of preservative per volume of wood) has been attained. A new calculation, run periodically while the batch is active, estimates retention and compares it to the specified value. Tedious for an operator, this calculation is easy to implement in a digitally based control system. This results in less residence time and less use of expensive preservative.

2. In chemical batching processes it is common to follow reaction with distillation. In plants without higher-level automation, distillation is typically controlled by steam flow and time. Although it is convenient to instruct the operator to run the distillation for 90 min, it is less convenient for the operator to run the distillation until the temperature reaches 250°F. However, by doing so, the boiling point of the batch can be used as the source for an inferential measurement of composition. It is comparatively simple to design a computer-based control system this way and thereby decouple batch composition from variations in batch amount, heating rate, and the like. Further, with assurance that distillation will stop when the desired composition is reached, process engineers may be able to specify an increased rate of heating and achieve an additional decrease in residence time.

3. Also in chemical batching the end of an exothermic reaction may be detected as the absence of heat generated from the process vessel; that is, the heat leaving the vessel at constant temperature equals the heat added to the vessel. Periodic calculation of evolved heat may be used to estimate the degree of comple-

tion (conversion of raw materials to product) at any time. This estimate may not only be used for endpoint determination but as part of a closed-loop control system (Lackmeyer and Kempfer, 1975).

4. In sterilization processes adequacy of processing is determined according to time and temperature criteria. The digital control system can regularly calculate the sterilization variable, compare it with the required value, and stop the process when the endpoint is reached. The system can then generate a hard-copy report for the manufacturer and regulatory authorities.

5. In some cases completion of a processing step is determined by laboratory analysis. The computer can help by

Predicting the time of completion, based upon a combination of measurement and process model, and requesting the sample and analysis at an appropriate time
Providing labeling information and test specifications for requested samples
Automating the transfer of data from laboratory instruments to the process controls, or by accepting manual entry of data
Providing limit checks for reasonability of laboratory data
Logging sample times, analysis results, and other relevant data

These measures can all contribute to plant productivity by reducing process time and minimizing operator errors.

MATERIAL TRANSFER SEQUENCES

Tank-to-tank material transfer sequences are pervasive in batch processes used in the chemical, food, and pharmaceutical industries. Although in practice there are many variations, these transfers can be broken down into a few general categories. Once they are understood this way, there may be an opportunity to design and code a very limited number of general-purpose sequences that, with only small variations, are usable through an entire process plant.

Direct transfers simply involve the source tank, shown as storage in Figure 9-25, and the receiving tank, shown as process vessel. The amount of material to be transferred may be measured by level or weight measurements at the storage tank or process vessel, or by flow between them. Direct material transfer systems are simple and economical.

Indirect material transfer systems use intermediate "preweigh" tanks between source and destination. The investment in these tanks is justified by improved weighing accuracy, which comes from using a narrower measuring range than would be available at either the source or the destination tank. (For example, a tank with a 50-lb weighing range and correspondingly low weighing error could be used between a 1000-lb storage tank and a 500-lb reactor.) Batch time may also be reduced this way by allowing the preweigh tank(s) to prepare for the next batch

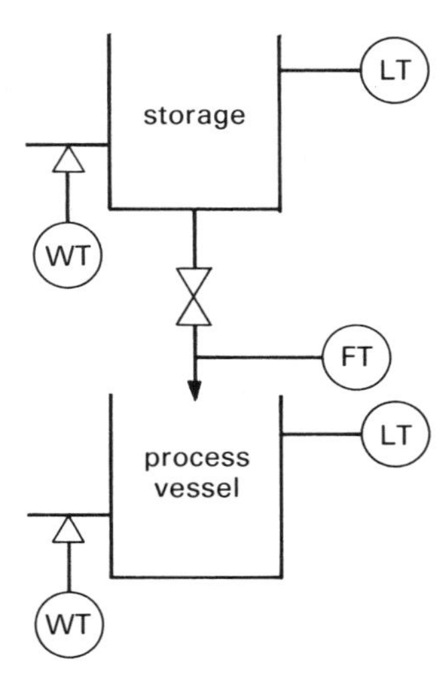

Figure 9-25. The direct tank-to-tank transfer and five options for measuring the amount transferred.

while downstream vessels are working on the current batch. An additional advantage in some applications is that the discharge into the receiving vessel may be arbitrarily fast, since the material has already been measured. The preweigh tank may have been charged at a slow rate to minimize the error arising from scanned operation. (Once the decision is made to stop the transfer, the transfer will continue for a finite time until valves are shut, pumps stopped, etc. Uncertainty is reduced at low transfer rates.)

The usual sequence for a tank-to-tank transfer is as follows:

1. Synchronize units (making certain that both units are at the correct point in their control logic).
2. Start flow.
3. Monitor transferred amount.
4. Stop sequence so that the desired amount is transferred.

High transfer speed with no loss in accuracy is attained by equipping plants for two-speed transfers. When the transferred amount reaches the desired amount minus a predetermined value—the *preact*—the transfer is switched between a large valve and a small one, or from high pump speed to low. In addition, dribble may be used to account for material downstream of the valve but upstream of the sensor and, for the amount of material expected, on the average, to be transferred once the measurement has reached the desired value (Fig. 9-26). It may also be useful

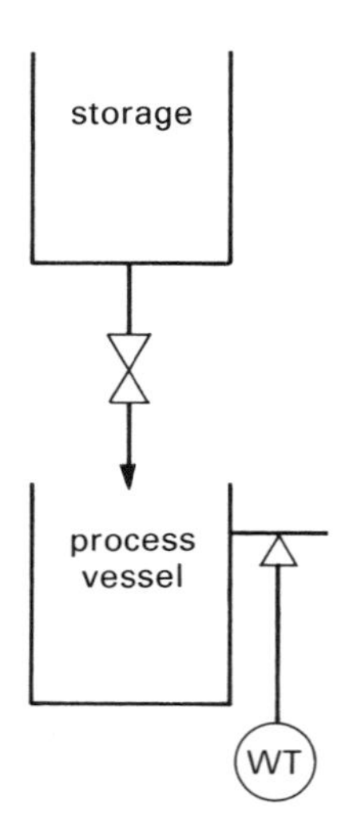

Figure 9-26. Estimating dribble: If 5 gal of material are contained within the volume between the valve and the process vessel, this contributes directly to the total dribble. If the typical flow rate is 10 gpm and the control system measurement-to-output time delay is 1 s, increase the dribble by (10 gpm) (1/60 min) = .17 gal. The total estimated dribble is 5.17 gal.

to record weighing errors and to automatically adjust the preact to minimize average error.

More elaborate transfer systems use control valves or pumps with wide speed variation to permit material addition at controlled rates. Most commonly, several materials under rate control are "blended" into a vessel simultaneously. This procedure may offer several advantages over simpler, sequential "batching":

At any given time the composition in the receiving vessel will be at or near the specified proportions. The controls can be designed, for example, to stop the transfer if one source material becomes depleted. The resulting mix may be immediately usable for a feed to downstream processing, even though the amount available may be below that originally requested.

Processing time may be reduced because less mixing is needed.

The rate of a chemical reaction may be controlled by blending rate.

Where material transfer rates are controlled, blending should lead to shorter cycle times than sequential transfers, simply because several transfers can take place at once.

TEMPERATURE SEQUENCES

Sequences used to change the temperature of a process vessel from one value to another have probably received more attention in the literature than all other sequences combined. The problem is trivial. It is solvable simply by applying full heat or full cooling until a set point is reached, except that in a chemical batch process these sequences are often required to operate for exothermic reactions. Many of these processes, if allowed to proceed out of control (i.e., with generated heat exceeding the capacity of vessel cooling), can create risk to personnel and equipment and certainly cause loss of the batch. Therefore, the problem in temperature sequences is to achieve the appropriate balance between production and safety. Minimization of overshoot can be critical.

The most common sequence is the simple linear ramp. The specification consists of set-point temperature and ramp rate in units of temperature per unit time. In operation the set point to the temperature control loop is ramped up or down until the final temperature set point is reached. Additional specifications may set alarm limits for time (if too long), calculated rate of change (if too high, indicates incipient loss of control), and so on. Heating and cooling are used as required to maintain the ramp. A typical application is to heat a process vessel starting at ambient temperature and then switch to cooling as the reaction becomes exothermic.

As simple as it is, the linear ramp probably offers the best available example of the benefits to be derived from careful use of design-by-level principles. An obvious way to implement a linear ramp is to build a simple loop within the phase logic (Fig. 9-27). The major weakness of this approach is that is prevents the phase logic from doing anything else during the ramp, unless the logic for these additional activities is built into the loop (Fig. 9-28) or off-sequence logic is used.

```
10    LET SETPNT=SETPNT+DELTA

      WAIT 5

      IF SETPNT ≤ GOAL THEN GOTO 10
```

Figure 9-27. A loop that raises the value of SETPNT by an amount DELTA, until SETPNT equals or exceeds the final desired temperature value, GOAL.

```
      LET FLAG=0

10    LET SETPNT=SETPNT+DELTA

      IF SETPNT ≥SPTEMP AND FLAG=0 THEN ACT SAMPLE
      (SAMPLE sets FLAG to 1)
      WAIT 5

      IF SETPNT ≤ GOAL THEN GOTO 10

      LET FLAG=0
```

Figure 9-28. A separate task SAMPLE is required; otherwise, the temperature ramp would be interrupted. SAMPLE sets FLAG to 1 so that it does not receive additional requests once it is initiated.

The first solution tends to make the logic very specific. A different ramp might be needed, depending on whether there were to be a material charge at 150°F, samples at 30 and 60 min into the ramp, and so on. The second solution can be difficult to understand because the individual trying to follow the logic must keep in mind multiple threads of processing. In addition, building the logic into the loop makes it impossible for the operator to access the ramp function without using a comparatively high level of automation.

Design-by-levels suggests that the ramp logic should be built into the regulatory control levels. It is accessed by fairly simple commands, such as INITIALIZE, START RAMP, and STOP RAMP. With the mechanics of ramping implemented elsewhere, the phase logic can be made easy to read and debug (see Fig. 9-29).

The linear ramp, though in common use, does not represent the only alternative in temperature changing. Shinskey and Weinstein (1965) introduced dual-

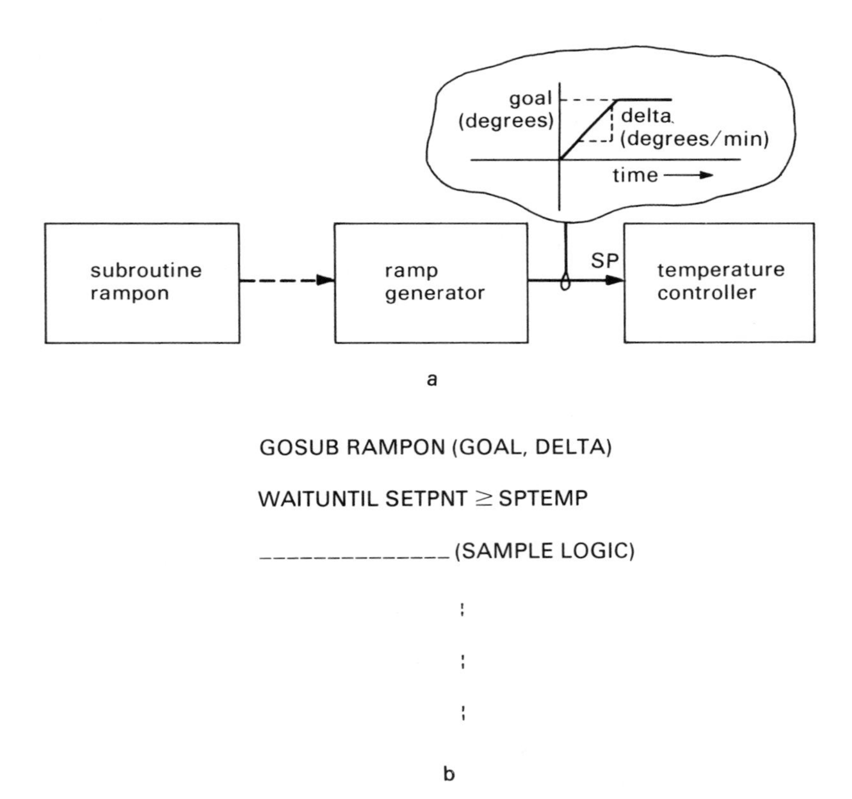

Figure 9-29. *a.* The ramp is initiated by a subroutine call, with arguments GOAL and DELTA defined as for the previous case. The ramp is generated within the regulatory levels, as shown in *b*.

mode control. Dual-mode simply applies full heating at the beginning of the sequence, then switches to full cooling at the end. These are the only control outputs used to elevate the temperature its final set point. Once established, temperature may be maintained by PID control operating in familiar fashion. The authors found that dual-mode control can operate without overshoot or oscillation. No overshoot is particularly important in these applications because reaction rate does increase rapidly (exponentially) with temperature, leading to the possibility that overshoot can cause complete loss of control.

Shinskey and Weinstein describe the implementation of dual-mode control in pneumatics, with higher-level functions performed manually. Their unqualified success with this implementation is another indication of the value of reducing even sophisticated functions to the lowest operable control level.

Ten years after that publication, two European authors compared dual-mode control to suboptimal control and three other methods (Ham and Liemburg, 1976). Although similar to dual-mode, suboptimal control has switching between heating and cooling that depends on a calculated variable that includes the time derivative of batch temperature (Rothstein and Sweeney, 1971). The other methods were variations of regulatory approaches, using no real sequencing at all. Ham and Liemburg found that suboptimal control showed marginally better temperature control performance than dual-mode, but that dual-mode was distinctly superior in the noninteracting relationship between control of unsteady and steady states and in its relative insensitivity to noise in the temperature measurement. (Suboptimal control is sensitive because it uses temperature derivative.) They also present an algorithm for calculating switching temperature for dual-mode control:

$$T(SW) = T(R,SET) - (A^{*}\Delta t[COOL])/2$$

where T(SW) is the switching temperature
T(R,SET) is the desired reactor temperature
A is the heat exchange area
Δt(COOL)is the cooling interval

SYNCHRONIZATION SEQUENCES

A control system's synchronization design, as discussed in Chapter 8, is implemented with synchronization sequences. This section contains process timing diagrams typical of centralized synchronization and distributed, embedded synchronization.

Figure 9-30 shows interactions involving the logic for a tank requiring a material charge, the supervisory unit, and the unit controlling the charge. The tank logic initiates a request, receives a response from the supervisory logic, and opens its valve. The charging unit conducts the charge and signals completion; then the tank valve closes.

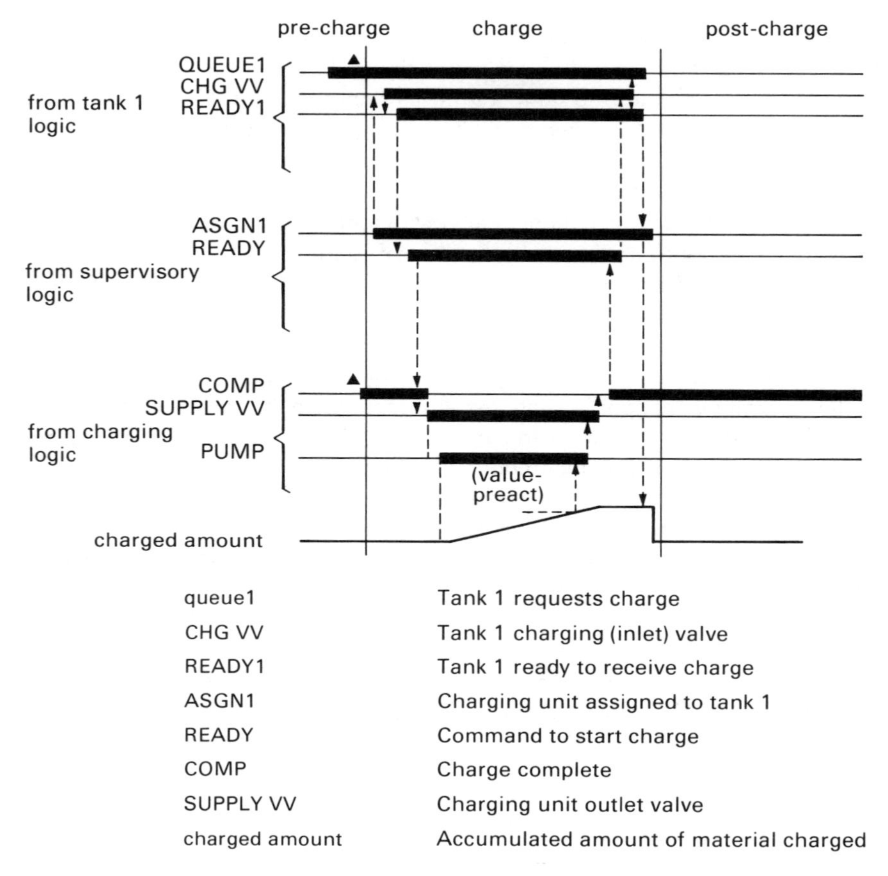

Figure 9-30. Process timing diagram for centralized synchronization, allowing each unit to control its own valve. *(After Rosenof, 1982c; reproduced by permission of the Instrument Society of America.)*

Figure 9-31 shows a synchronization between a charging tank (tank 6) and a shared pipeline, embedded within the synchronization between the tank requiring the charge (tank 4) and the charging unit (Rosenof, 1982c).

SEQUENCE LEVEL OPERATOR INTERFACE

Acting at the sequence level, the operator is able to direct plant equipment to perform such functions as material charging and temperature ramping. The interface devices at a typical material charging system remote from a central computer

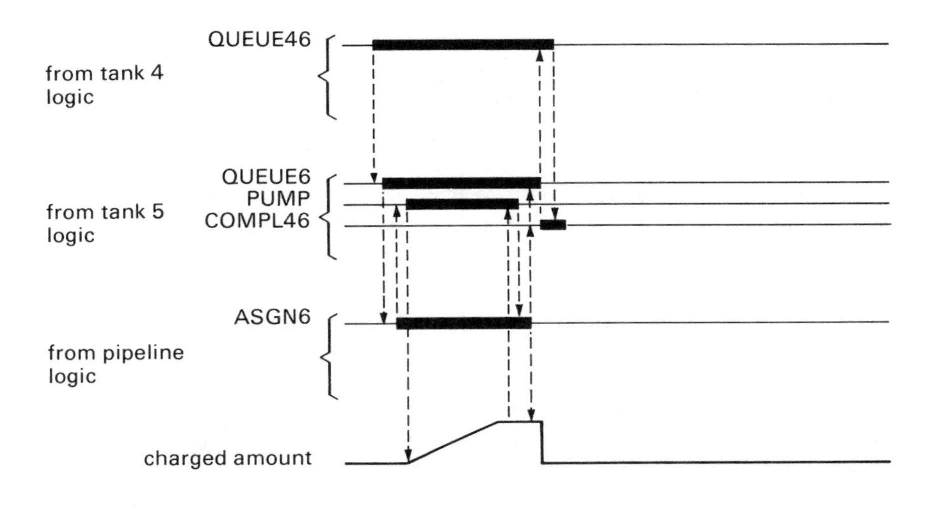

QUEUE46	Charge request from tank 4 to tank 6
QUEUE6	Pipeline request from tank 6
PUMP	Charging pump
COMPL46	Charge from tank 4 to tank 6 completion
ASGN46	Pipeline is assigned to tank 6
CHARGED AMOUNT	Accumulated amount of material charged

Figure 9-31. An embedded synchronization. This diagram is simplified for clarity. In practice, it would resemble that of the previous figure. *(After Rosenof, 1982c; reproduced by permission of the Instrument Society of America.)*

are a good example of the provisions made to support the operator in commanding the system to perform such actions.

The operator can enter a value amount, represented via thumbwheel switch, numerical keyboard, or similar device. For a system with multiple destinations and/or multiple material sources, additional thumbwheel switches, keyboards, or selector switches are provided. The charging action itself is controlled via start and stop pushbuttons. The remote/local switch must also be present to command the system to operate through its panel or its interface to the central computer. There is a simple numerical readout; seven-segment light-emitting diodes (LEDs) are quite common. The seven segments enable each of the figures 0-9 to be displayed as well as several letters, which are often used to display error conditions. Commonly, the numerical display is active even while the system is under direction of the central computer. Various other controls depend on the particular characteristics of the control and process equipment being used.

Remote charging and similar control systems today are often elaborate enough to justify their own keyboard/printers and/or CRTs as devices for local control. These devices become the control equipment's local operator interface.

Finally, local equipment may be network connected to other parts of the system so that local control can be exercised from a location, or locations, remote from the charging controls. These interfaces may be shared with other equipment and may be able to function as interfaces at various levels, but they still fulfill the function of operator interface to the sequential level of control. Operating under supervision from a higher level of control, the control equipment simply accepts similar commands from that higher level.

Where sequence-level control is exercised at the central computer, the computer's CRT facility typically serves as the operator interface, shared with interfaces to other levels and with other functions. A computer with single-step capability offers immediate operator access to the sequence level.

In other cases operator access may need to be designed into the logic to give the operator the option of letting the logic continue to subsequent sequences or stop when the commanded sequence is complete. In any case operator entry of all relevant variables that, under direction from higher levels, would have been derived from elsewhere in the system must be considered.

The computer can report its actions in great detail both on CRT and in hard-copy form. In addition to charged amount, for example, the computer can report the identity of the material charged, its destination, temperature, charge time, and charging error.

ELECTROMECHANICAL SEQUENCE CONTROLLERS

Some simple sequences can be controlled quite effectively with electromechanical sequence controllers. The *cam timer* (Fig. 2-1) is typically used when step-to-step transitions are controlled only by time, not by endpoint. It uses a motor driving a shaft. Cams are on the shaft, and associated with each cam is a two-position switch. By adjusting the cams, the user effects the desired sequence of switch on/off states. By wiring the motor through a cam switch and a switch that closes when an endpoint is achieved, we can make step progression depend on endpoints.

Drum sequencers use intermittently energized motors; each time the connected input changes from off to on, the drum advances one step. Otherwise, they are similar to cam timers. Some versions use cams (large for easy reprogrammability); others use actual drums in which pins are installed to actuate the switches. By using an additional, multiposition switch, we can separately control each step transition (Fig. 2-2).

Analog variables can be sequenced by a variety of *curve followers,* which vary their outputs to match a predefined value-time relationship.

PROGRAMMABLE CONTROLLER SEQUENCING

Programmable *logic* controllers were originally developed as replacements for electromechanical devices used in interlocks. It is always possible to build a sequence controller out of these electromechanical devices, and, PCs have been used this way as well. It is much easier, however, to use some of the specialized sequence functions available in many controllers. Typically, the sequencer has several outputs, each corresponding to a step, and one "change step" input. Each time the change step input changes state from off to on, the sequencer advances one step. The sequencer can be made to advance to the next step only when the present step's endpoint or timing criteria are met, similar to the drum sequencer. *Programmable sequence controllers* are programmable controllers intended specifically for this type of sequence control, with interlock logic serving mainly to determine the step progression.

CHAPTER

10

THE REGULATORY LEVELS

Regulatory controls are responsible for the continuous control of such familiar process variables as temperature, pressure, and flow. The equipment used and its configuration into control loops are much the same as for continuous processes, but some characteristic differences may still be found.

Regulatory controls are subject to reconfiguration according to process conditions. This reconfiguration may be similar to "constraint control" in continuous processes, where control system outputs are governed by the momentarily most restrictive (highest or lowest) of several constraints. For discontinuous control, transition between governing constraints at a more or less predictable point in the process is often an intentional feature of system design. It is common in the control of batch chemical reactors to bring reactor contents to a boiling state by first controlling *temperature* up to the boiling point and then controlling *boiling rate* (measured as heat rejection to the condenser) up to the desired value. In another example, valves used in some material-transfer operations must be treated sometimes as on/off devices and other times as finely positionable control valves.

Reconfiguring regulatory controls is done within the regulatory loops themselves or by directions imposed by the sequence levels. The former is usually best because maximizing the capability of the control loops allows maximum remaining functionality in case of an upper-level failure. Experience suggests as well that internal reconfiguration is the more "robust," less likely to be placed in an unintended state by unanticipated starting conditions or process responses. On the other side is the simple fact that complex loops are more difficult to design and understand than simple ones. The proper separation between regulatory and higher levels may depend on both the process and the preferences of the individuals involved.

Opportunities for reset windup are very common. Controllers with reset (integral) functions can experience reset windup when the measured variable cannot be driven to its set point. The *batch switch* (Shinskey, 1979) may be used with certain controllers to prevent this. In digitally based control systems the upper levels may leave controllers off when not in use and force controller initialization at opportune moments.

Regulatory controls are rarely in steady state. Many continuous processes can withstand a control failure, leaving valves in their last positions, for some time. Process variables merely slowly drift away from their set points until automatic control is restored or operators take control and manually adjust control system outputs. In a batch system a valve may be open to permit a material to flow into a reactor. Failure to terminate the charging action on time may result in wasted material, a ruined batch, or possibly a dangerous chemistry in the reactor.

Detection of off-normal states is complex. For a continuous process under normal operation it is fairly easy to identify one or more alarm values for each measured variable. In a batch process interpretation depends on context: A sudden increase in a reactor's temperature may be normal, when the reactor is being charged with a hot material, or an indication of an impending emergency, when the reactor's temperature is supposed to be increasing at only 1°/min.

The thread that runs through all four of these differences is that the relationships between the regulatory levels and their adjacent ones can significantly affect a control system's success.

ROBUST DESIGNS

We have already introduced the idea of robustness in regulatory controls. We stated that it is generally better to try to "push" functionality into lower levels in order to isolate the functions from outside influences that might adversely affect them.

We return to the initiation of an exothermic reaction. Without adequate cooling, control over the batch can be lost as liberated heat drives up the reaction rate. It is possible to build logic to detect the start of the liberation of heat, then send messages to close steam blocking valves, open water valves, change controller tuning parameters, and so on. It is easier to build a split-range control system that controls the blocking valves exclusively as a function of heat demand (Fig. 10-1). The latter has proven to be easier to start up and is highly reliable. Similarly, comparison operations common to material movement (WAIT UNTIL VOLUME TRANSFERRED ≥ 100 GALLONS) can be eliminated or supplemented in some cases by the use of simple controllers driving the accumulated charged value to a set point.

There is a point, however, at which these loops can become impossibly complex. In practice there is a range of choices that can yield acceptable results. The authors' impressions are that European preference is to make more use of the lower

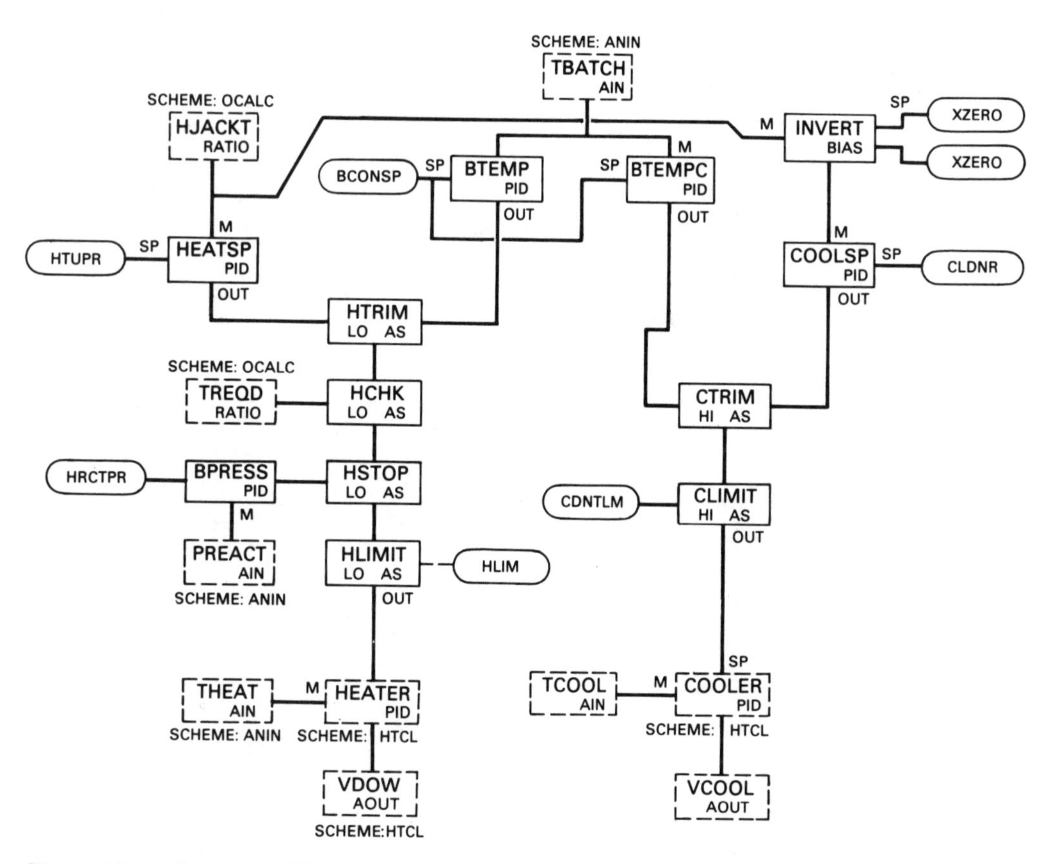

Figure 10-1. Portion of a DDC temperature control scheme used in a reactor for the synthesis of alkyd and acrylic resins. *(After Armstrong and Coe, 1983, p. 61; reproduced by permission of the American Institute of Chemical Engineers.)*

levels than is common in the United States. Supporting this is our observation that there are high-reliability logic cards able to distinguish between the on, off, and fail states readily available in Europe but not here. In the United States the approach has been to stay with a simpler structure while extending the availability of redundant hardware up through the sequence levels, if necessary.

COMPENSATING FOR POOR OR UNAVAILABLE MEASUREMENTS

Unreliable or difficult-to-obtain measurements can be a problem with the control of any process, but batch processes offer their own unique problems: The intermittent nature of the operation of many components of a batch process can produce instrument operation that is satisfactory for one batch but useless for the next. A valve may respond properly during one batch but fail when not being used and prevent satisfactory plant operation for the next batch.

Best (1975*a*) suggests a series of substitutions for measurements that are difficult to make on batch plants or materials that are commonly used in them. Some of these are common practice and have already been mentioned. Use of transferred weight instead of flow (or level) is possible where material characteristics may preclude accurate flow or level measurements by conventional methods. We have also discussed the use of boiling temperature to infer composition. Best also shows how to use a mass-balance calculation to infer moisture in a product.

Batch plants are rife with opportunities for simple instrument misapplication. One of the authors has seen, for example, a level measurement made using differential pressure on a storage tank required to hold products of varying specific gravity (Sp Gr). Specific gravity strongly affects the differential pressure transmitter output, so a correct reading is available for only one Sp Gr value. (An Sp Gr value may be obtained by using a differential pressure transmitter with both high and low pressure connections always below the liquid surface. It can then be used to compensate the output of the level-responding transmitter.)

PROCESS DYNAMICS

The deleterious effect of long time delays in control loops is well known (Shinskey, 1979). Belt feeders are often contributors; spin dryers, being continuously operating units that may be used downstream of one or more batch reactors, can also be. McEvoy (1968) discusses various control schemes used successfully for belt feeders. Best (1975*a*) shows how a feedforward control scheme can be used to compensate for a spin dryer's long lag time. The original and replacement control schemes are shown in Figure 10-2.

The varying materials and amounts handled by a multiproduct batch plant may lead to variations in dynamics that are significant enough to suggest retuning.

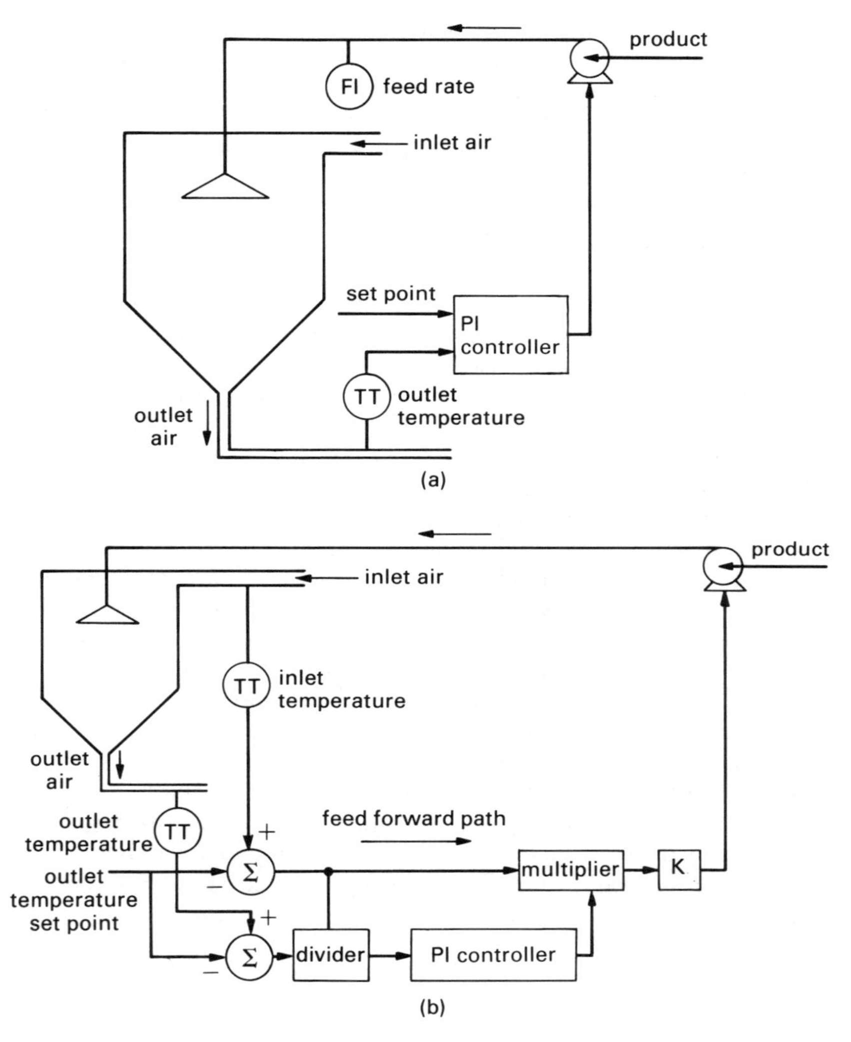

Figure 10-2. *a.* Conventional spray drier is prone to instability and overshoot because of the long deadtime from air inlet and outlet, where exit air temperature is taken as a measure of product dryness. *b.* This feedforward control scheme compensates for inlet temperature disturbances by making pump speed adjustments in addition to those based on outlet temperature feedback. *(After Best, 1975; reproduced by permission of the Instrument Society of America.)*

Today, retuning is rarely if ever practiced; tuning is set to be safe under all anticipated conditions, and then it is left alone. Although we know of no attempt to apply the recently available pattern-recognition self-tuning controllers to a batch process, substantial benefits through better response may be possible. These controllers would, for example, automatically compensate for variations in the thermal time constant, brought about by variations in composition and mass within a vessel, and for variations in the heat transfer surface caused by operating a reactor jacket in sections.

Truëb and Klanica (1978) were able to compensate for vessel inventory variations by controlling the rate of heat removal or addition rather than the flow of heating or cooling medium, as they had done in the past. Their algorithm is shown as a program flowchart, for implementation presumably by a conventional computer language. Figure 10-3 shows the flowchart; it can readily be redrawn for implementation by an analog or distributed control system. Similar algorithms have been implemented with integral action applied to the valve position controllers, and with measurement of flow to support the heat flux calculation.

Lackmeyer and Kempfer (1977) solved a problem caused by heat transfer surface fouling by applying batch temperature control through an overhead condenser rather than through the jacket.

Spruytenburg et al. (1976) describe the operation of a batch fermentation reactor in which process gains, effective for a variety of controlled variables, change with time. Their approach is to allow a controlled variable to vary within a dead band with no control action taken to restore its value to the set point. The time rate of change of the variable, as measured during this time, is used to adjust the control action taken when the variable's value does go outside of its dead band.

Saxon and Glover (1982) describe a control strategy based on an empirical process model. The controller directs the final frequency adjustment of a quartz crystal by which silver is evaporated onto the crystal's electrodes, changing the frequency of an oscillator using the crystal in its circuit. The silver is difficult to remove, so overshoot can be a serious problem. They obtained an overshoot of 5%. This process is fast—from 20 to 30 s per crystal.

SPLIT-RANGE CONTROL

In Figures 10-1 and 10-3, execution of the control logic algorithm proceeds along one path for heating and another for cooling. (In Figure 10-1 there are separate PID controllers for heating and cooling, and in Figure 10-3 there are separate calculations for heating and cooling valve positions.) Another popular strategy is to use one controller and divide its output between heating and cooling. The most obvious way is simply to say that outputs of greater than 50% represent heating, and outputs of less represent cooling.

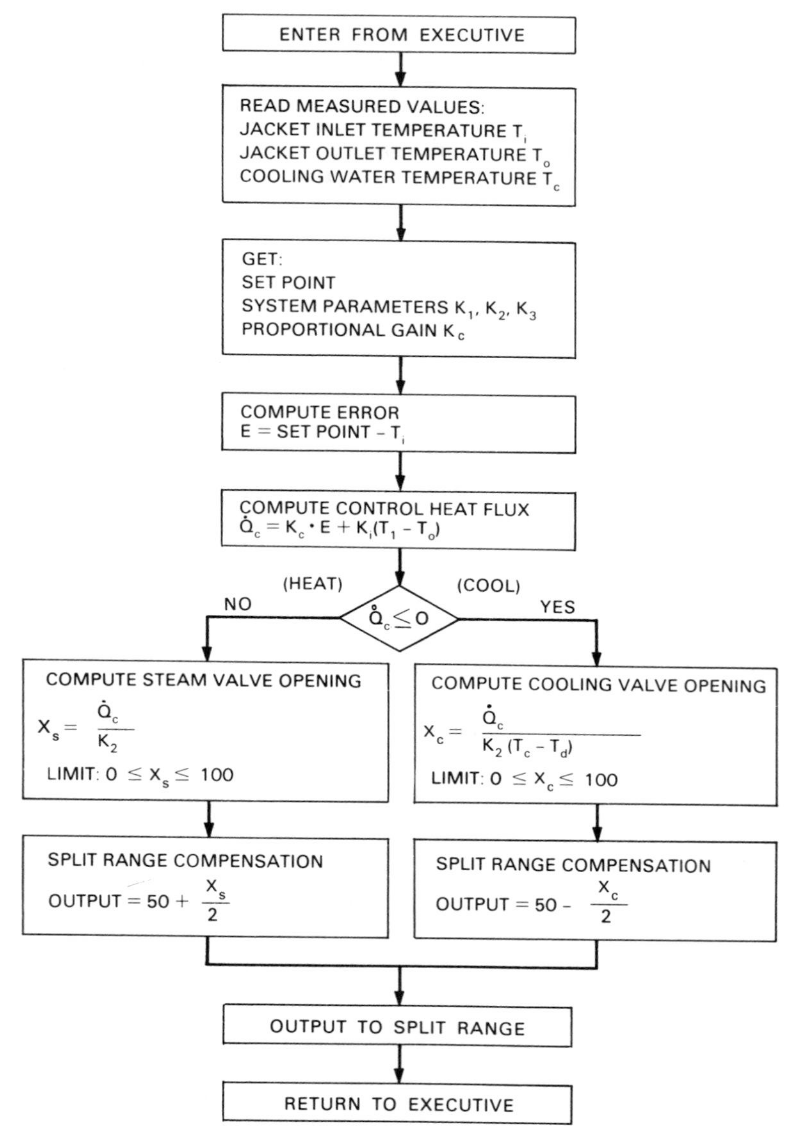

Figure 10-3. Algorithm for control of temperature by heat removal or addition. *(After Trüeb and Klanica, 1974; reproduced by permission of Springer-Verlag.)*

Once some practical problems are resolved, split-range control can operate very well. One problem is due to the reversal in control action that takes place between heating and cooling: When the control valve in the heating medium line is opened, more heat is added. When the control valve in the cooling medium line is opened, effectively, *less* heat is added. Some schemes require a signal inversion (0-100% becomes 100-0%) to operate properly. Another common problem is in the control of on/off inlet and drain valves when more than one medium (steam for heating, water for cooling) share the same piping. These valves may derive their control signals from the heat demand; a heat demand in the cooling range, for example, would cause the on/off cooling valves to open.

The most economical heat transfer takes place when the heating/cooling medium is as close as possible in temperature to that of the process vessel. When a vessel is at 200°F, it is more efficient to cool it with water at ambient temperature than with chilled water. Three-way split-range control would use a heating medium hotter than ambient, water (or another material) maintained at around ambient, and a chilled medium for cooling below ambient or for faster cooling that can be achieved using the ambient-temperature medium.

Three-way split-range control offers an interesting control problem: The system must be designed to treat the intermediate-temperature medium sometimes as a source of heat (when the vessel is below ambient) and sometimes as a source of cooling (when the vessel is above). Whitley (1985) has used a strategy in which the control action of the intermediate-medium valve is effectively reversed to accommodate using that medium to either heat or cool the reactor. With increasing heat demand the control actions are as follows:

1. The cold supply is reduced from 100% to 0.
2. The ambient supply is increased from 0 to 100%.
3. The hot supply is increased from 0 to 100%.

With decreasing heat demand the control actions are as follows:

1. The hot supply is decreased from 100% to 0.
2. The ambient supply is increased from 0 to 100%.
3. The cold supply is increased from 0 to 100%.

The strategy was implemented on a reactor that uses intermediate heating and cooling: A heat transfer material continuously circulates between the reactor jacket and the appropriate heat exchanger. There seems no reason why this strategy would not work on reactors that are directly heated and cooled by pumping the hot, cold, or ambient source directly through the jacket.

There also seems no reason why this technique cannot be extended to another source of heating and cooling, or more than three partial ranges. For example, many chemical batch plants have clean wastewater streams at around 80° or 100°F. In many plants this warm water is simply discarded or pumped through a cooling tower for recirculation. Some have investigated the use of heat pumps to upgrade the heat to a higher temperature. Another approach is to use the heat in these streams to provide initial reactor heating, switching to more expensive fossil-fueled sources only when necessary.

CHAPTER

11

THE INTERLOCK LEVELS

Some simple interlocks are contained within equipment and are so effective that it is easy to forget that they are present. Consider, for example, a reversible, three-phase ac motor, as might typically be used to drive a motor-operated valve (Fig. 11-1). With one connection to the line, the motor runs forward. When two of the phases are switched, the motor runs in reverse. Such a motor can be controlled with two motor starters, as shown in Figure 11-2. Unfortunately, operator error or a short circuit could cause both starter coils to be energized at once, resulting in a direct short circuit across two phases of the line. The solution is to use both electrical interlocks and a *reversing motor starter* with mechanical interlocking to

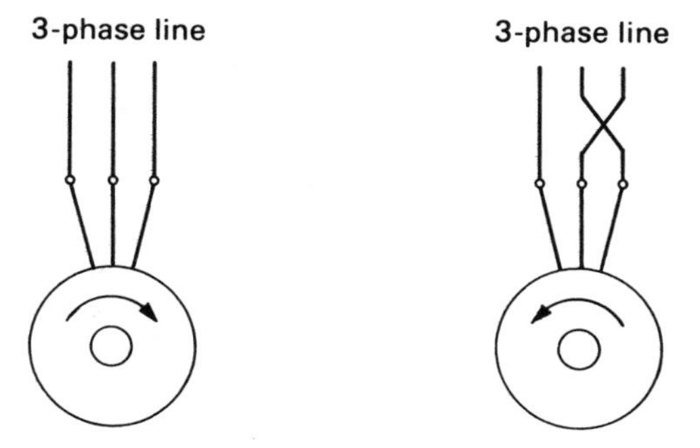

Figure 11-1. Reversing two of the connections changes the direction of motor shaft rotation.

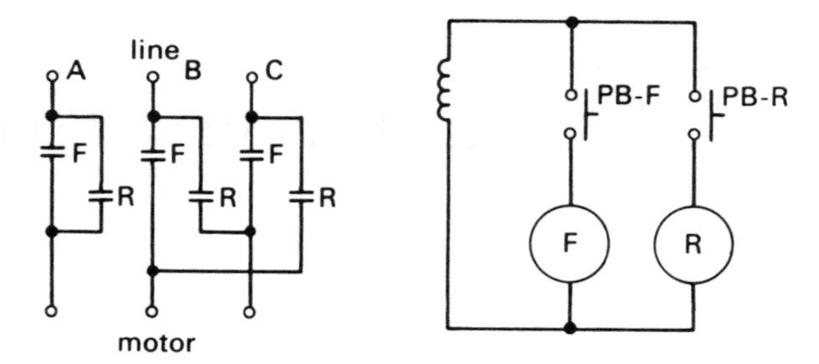

Figure 11-2. Simplified diagram of the use of two motor starters to control a reversing motor. If pushbuttons PB-F and PB-R are actuated at the same time, phases B and C are short-circuited together.

prevent both sets of line contacts from closing at the same time. Here are some other examples of simple interlocks:

A mechanically interlocked pushbutton switch that prevents the short-circuiting of signals with each other (Fig. 11-3).

A three-way solenoid valve that prevents supply-line air from being vented (Fig. 11-4).

Various configurations of key-operated switches. Typically, key-retaining switches are used to prevent multiple valves from being opened at the same time (Fig. 11-5).

Many interlocks serve to detect process malfunctions or dangerous conditions and then act to drive the process to a safe condition. A position switch next to the door of a washing machine, for example, stops water fill or agitator motion when the door is open. The industrial equivalent would be the position switch on an inspection hatch that causes material feed to stop when the hatch is open. Also common are overtemperature alarms and fire alarms, each acting to stop the flow of flammable materials into process vessels and, if necessary (typically for exothermic reactions), initiate process vessel cooling and/or reaction quenching.

More complex interlock implementations (our level 2) can, for example, detect incipient loss of control by determining the time rate of change of temperature in an exothermic reaction, or excessive deviation of measurement from set point, in many processes for many variables. Other examples of these more complex interlocks are as follows:

Detection of a failing sensor/transmitter by comparing multiple inputs for adherence to a predetermined relationship. For example, thermocouples equally spaced at the same elevation around a process vessel should yield outputs that match each other within a few degrees.

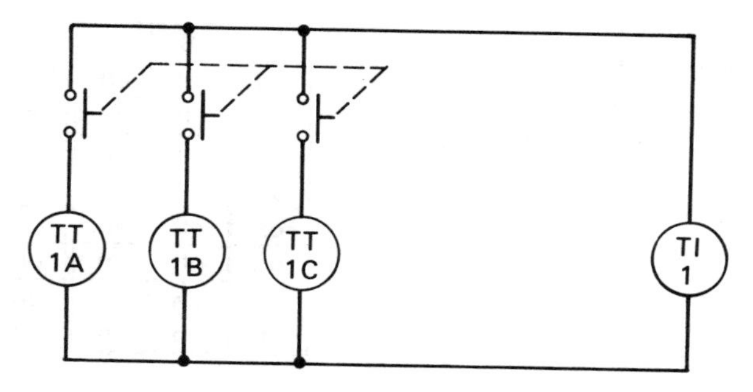

Figure 11-3. A mechanical switch interlock, similar to those used for car radios, prevents temperature signals from being short-circuited together. (When current signals are used instead of voltage signals, short circuits must be maintained across unselected inputs. This requires a different circuit from that shown here.)

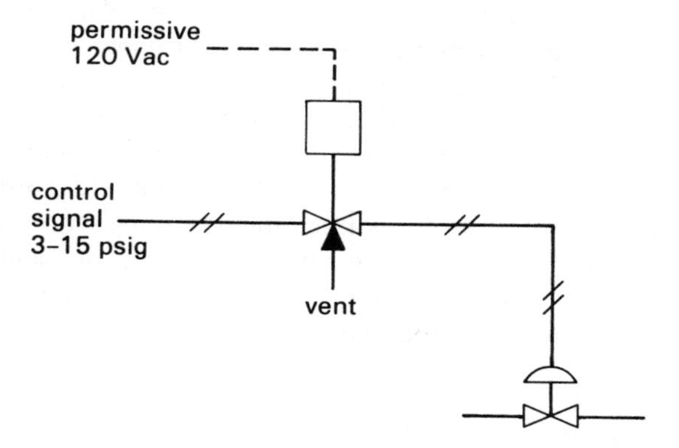

Figure 11-4. When power is removed from the solenoid valve, air vents from the control valve chamber. Otherwise, the valve would hold its last position.

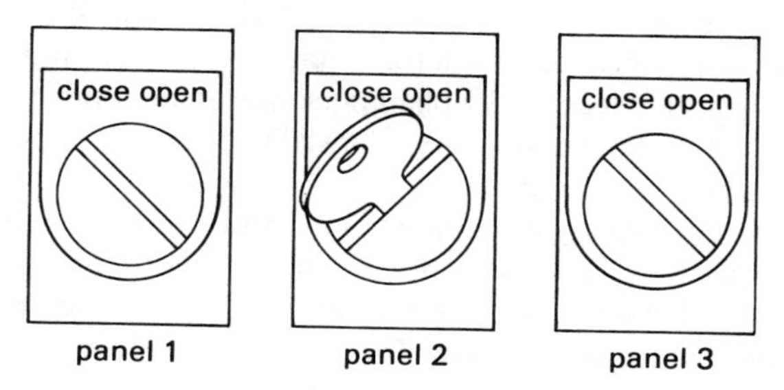

Figure 11-5. Three valves, each controlled by a switch in a different panel, are interlocked by keying. The key is retained by a switch in its open position, so that as long as only one key is available, only one valve may be opened at a time.

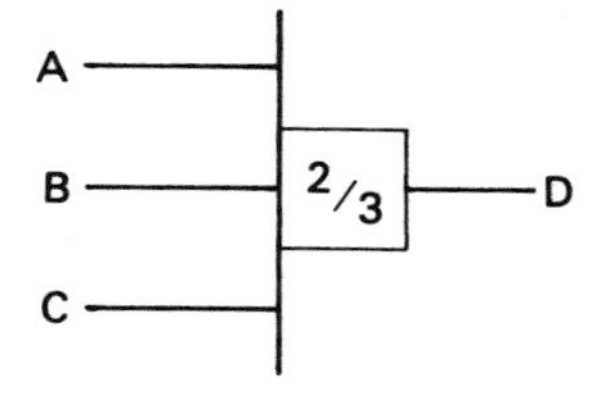

A	B	C	D
off	off	off	off
off	off	on	off
off	on	off	off
off	on	on	on
on	off	off	off
on	off	on	on
on	on	off	on
on	on	on	on

Figure 11-6. Two-out-of-three voting logic. With input A in the on condition for test, the logic simplifies to that shown in the small box: one-out-of-two, or simple OR logic.

The use of voting logic to prevent one monitoring device failure, whether to the alarm or normal state, from causing an unnecessary alarm or shutdown or blocking a valid alarm or shutdown. In Figure 11-6 two of the sensors must show alarm conditions for the interlock output to alarm. If only one sensor shows an alarm over an extended time, a secondary alarm may be initiated. If one sensor fails to alarm under loss of control, the other two will still cause the interlock to alarm. Note also that if one of the sensors, under test, is subjected to a simulated loss-of-control condition, the remaining sensors and the logic form a one-out-of-two circuit and continue protecting the plant.

The use of a timer to enforce a minimum time interval between motor starts, as a means of limiting motor temperature.

INTERLOCK LOGIC

Circuits and systems using combinatorial logic techniques range in complexity from simple interlocks to computers. Here we will discuss combinatorial logic as it applies to typical process (especially batch) interlocks. Interlocks, in general, act to detect actual or incipient conditions detrimental to plant safety and initiate change to those conditions, or at least notify plant personnel of their presence so that they may close the loop and make the corrections. They also act to prevent operator errors from causing such conditions.

For a large piece of rotating machinery, for example, we might require that there be sufficient lubricating oil pressure whenever the machine is operating. If pressure fails, we will immediately deenergize the machine. This is expressed as simple AND logic, as shown in Figure 11-7. The diagram simply says that the control switch has to be in the run position, and the oil pressure must be above the preestablished lower limit for the machine to run.

Consider a multivessel batch plant in which each vessel has a high-temperature

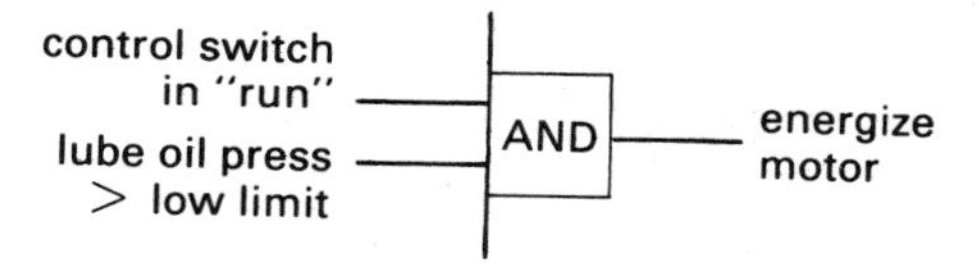

Figure 11-7. AND logic used for motor control.

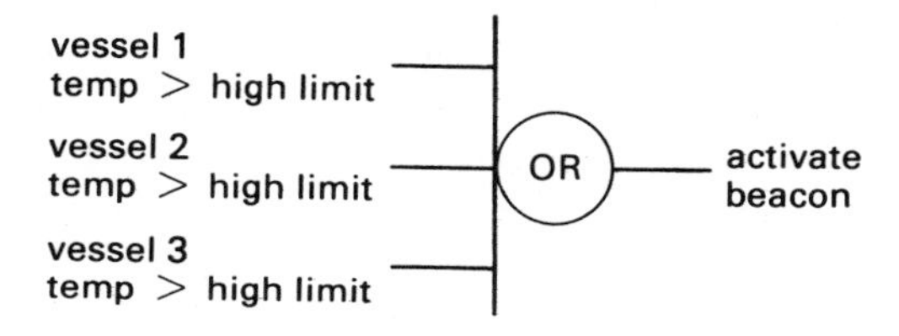

Figure 11-8. OR logic used for a warning system.

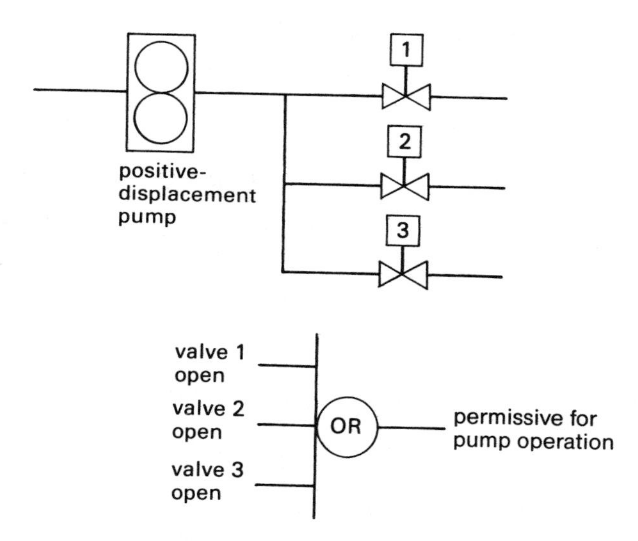

Figure 11-9. Application of OR logic to a positive-displacement pump.

alarm switch. If *any* of these switches goes into the alarm condition, we wish to activate a beacon to warn personnel. This OR logic is shown in Figure 11-8.

Positive-displacement pumps may be damaged if they attempt to pump into a closed valve. Therefore, to protect such a pump installed as shown in Figure 11-9, we might specify that it be allowed to operate only when at least one of the header outlet valves is open.

An additional logic element is the NOT (or negation or inversion), which serves simply to turn a "1" state into a "0" state, and vice versa. Its use is shown in Figure 11-10, where it allows permissives to be logically combined with a NOT alarm condition to allow a pump to be energized.

In many cases the use of the NOT function, as well as the decision whether to use AND or OR logic, is partly arbitrary. Figure 11-11 shows two depictions of the same logic. The authors' general recommendation is to use signal lines to show the user-defined "asserted" conditions—that is, PUMP OK, HIGH PRESSURE ALARM, SWITCH IN RUN POSITION, and so on, and adjust the logic functions accordingly.

More complex logic functions may be built out of the basic ones. For example, NOR logic is OR logic followed by negation. (The NOR gate has been immensely important in electronics for two reasons. It is generally simple to build, even simpler than OR and AND, and NOR functions may be combined to form memory and complex logic circuits particularly valuable in computers.)

The exclusive-OR function (Fig. 11-12) is used to detect mismatches between its inputs. Whereas a two-input OR function asserts its output if either *or both* of its inputs is asserted (it is sometimes called an inclusive-OR function), the exclusive-OR asserts its output if either input is asserted but not if both are. Therefore, the logic

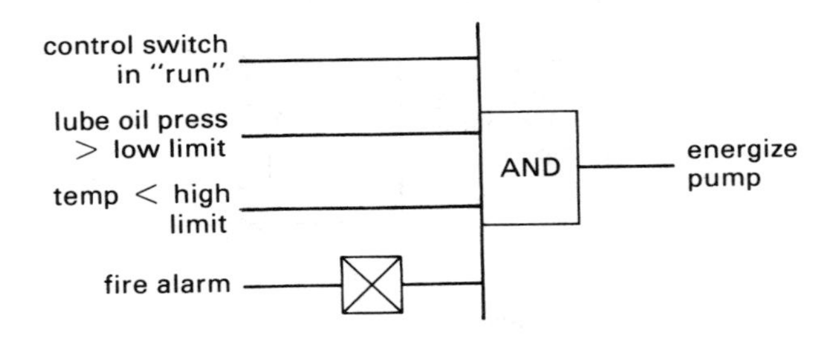

Figure 11-10. Use of the NOT function.

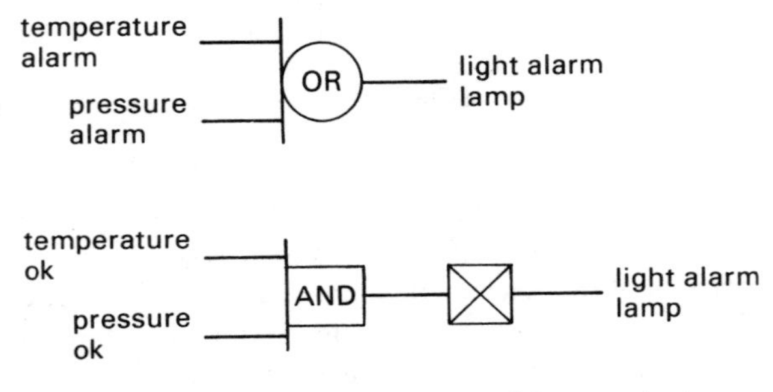

Figure 11-11. Two depictions of the same logic.

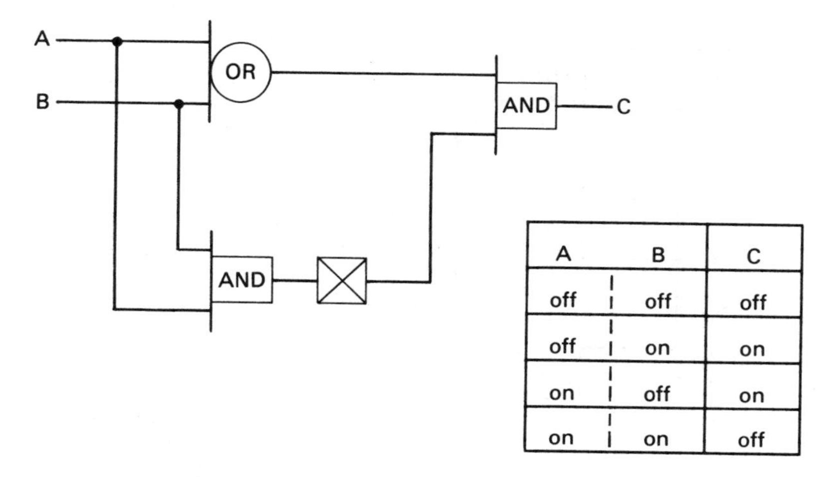

A	B	C
off	off	off
off	on	on
on	off	on
on	on	off

Figure 11-12. The exclusive-OR logic function.

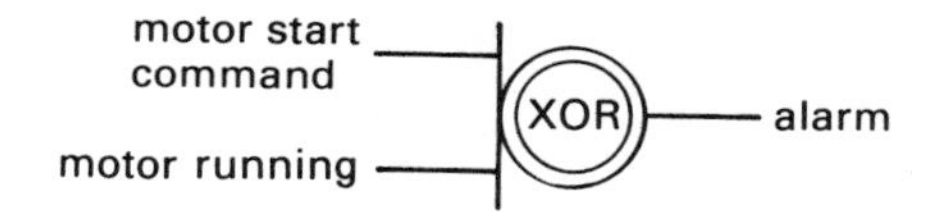

Figure 11-13. A use of the exclusive-OR function.

function shown in Figure 11-13 will assert its output when the motor fails to start after a run command or starts in the absence of a run command. In practice, this output is followed by a timer because momentary mismatches that always occur when the commands are changed should not result in alarms.

Other logic functions are required by certain high-reliability interlock arrangements. It may be desirable, for example, to require multiple sensors to detect an alarm condition before a shutdown is commanded. The simplest of these is the two-out-of-three (Fig. 11-14) arrangement already introduced. The logic combination shown in Figure 11-15 offers further immunity to sensor failure.

Memory elements are used to retain information about the previous operation of the plant or previous operator actions. The simplest memory element is the latch or S-R flip-flop shown in Figure 11-16. The function has two inputs, set (S) and reset (R). A momentarily asserted state at the S input causes the output to remain asserted until the R input is momentarily asserted. When both inputs are present simultaneously, the R input overrides the S input. These functions usually include inverted outputs that take states opposite those of the main outputs.

The S-R flip-flop is the function used to control motors with pushbutton

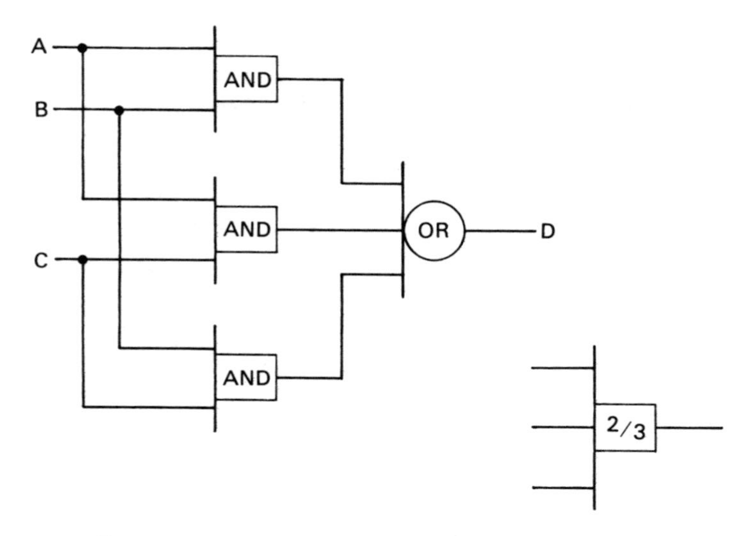

Figure 11-14. Implementation of two-out-of three logic.

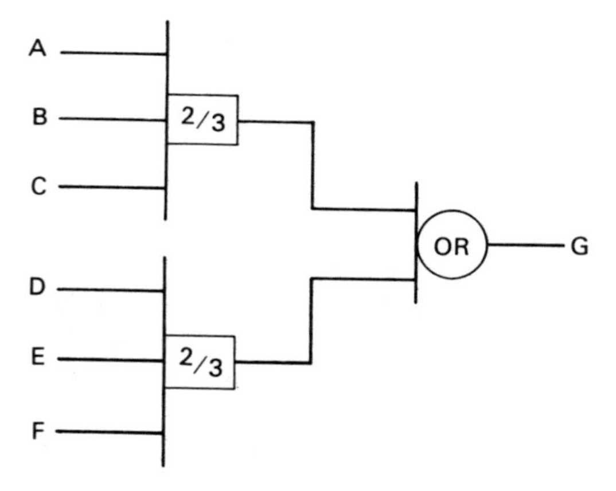

Figure 11-15. This circuit is immune to any three sensors failing to alarm and to any one sensor falsely alarming.

switches. The S input is provided by the start pushbutton. The R input is provided by the stop pushbutton or an overload or a power failure (Fig. 11-17). The function is also used in many annunciator sequences to "capture" transient alarm conditions.

The D type flip-flop retains at its output the state that was present at the D input when a separate strobe input changed state from on to off or off to on, depending on the implementation. The T-type flip-flop changes its output state whenever its

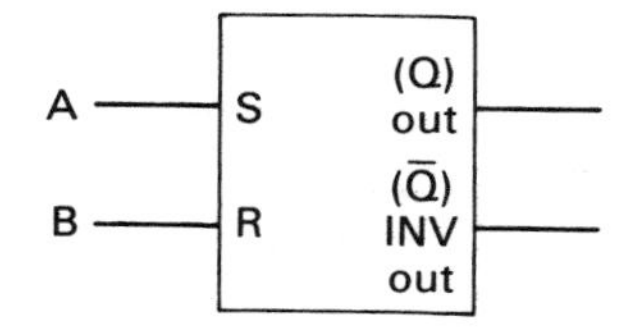

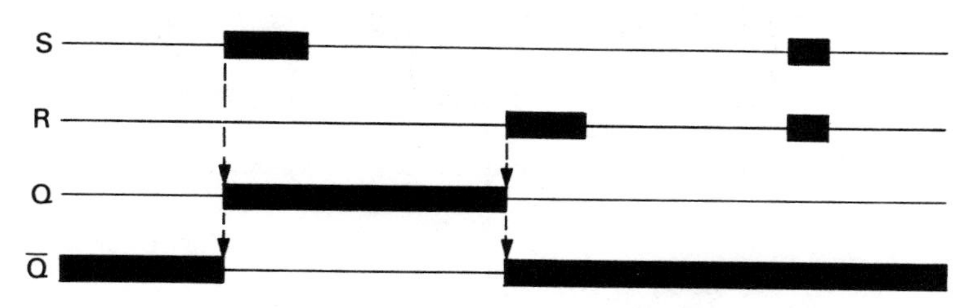

Figure 11-16. Operation of the S-R flip-flop.

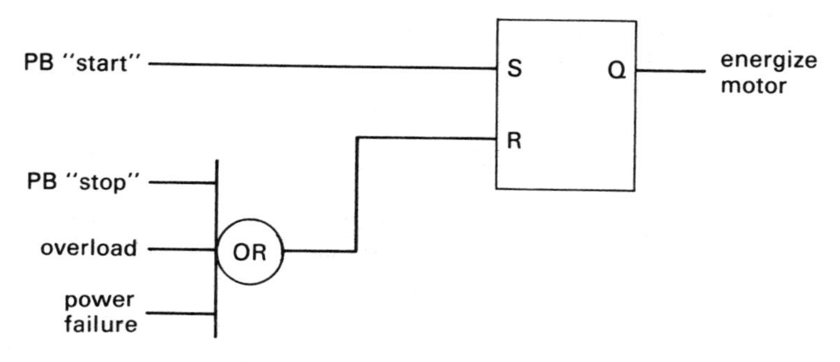

Figure 11-17. Use of an S-R flip-flop. The logic is actually performed by the motor starter and its circuit.

toggle input changes state, again with the direction of state change dependent on the implementation (Fig. 11-18). These and other flip-flops are useful in building digital electronic circuits but are not commonly found in explicit form in industrial interlock applications. The D function is found implicitly in snapshot reporting, and the T function is used in totalization.

Timing elements find a wide variety of interlock uses—some at the off to on transitions of their inputs, some at the on to off transitions, and some at both (Fig. 11-19). The first, for example, may be used in conjunction with exclusive-OR logic to detect equipment failures as we have described. The second is used to maintain

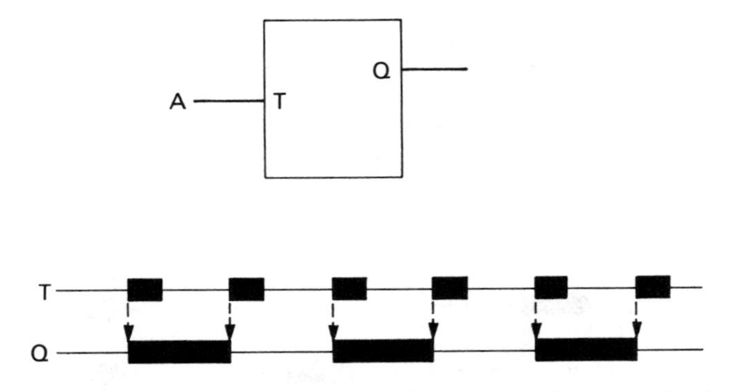

Figure 11-18. The T-type flip-flop functions as a frequency divider.

some service after equipment shutdown, such as lubrication oil pressure after a piece of rotating machinery has been shut down (Fig. 11-20). The third is used in more specialized applications, such as maintaining some predefined synchronization between pieces of equipment. Where endpoint measurement is not available, timing elements are used to determine the durations of steps in a batch process.

A timer is characterized as *retentive* if it accumulates time when its input is on (or off, less commonly) without being reset (losing its accumulated time) whenever the input changes state. A timer is *nonretentive* if it does automatically reset (see Fig. 11-21). Retentive timers must have independent reset inputs—these are optional for nonretentive timers.

Retentive timing elements are used to accumulate equipment run times signaling the need for preventive maintenance. Nonretentive timers are probably more common, being appropriate, for example, in detecting equipment failures. Both varieties are used for timing batch sequences, depending on the process.

Counting elements are widely used in parts manufacturing and occasionally used in process interlock functions. (They are also used to totalize material flows, energy consumption, etc.) A counting element is, by definition, retentive. They are usually equipped with reset inputs. If not reset, some counters will stop at their maximum counts, and others will "recycle" back to their minimum counts.

Counting elements are used to record the number of equipment operations (i.e., motor starts) for preventive maintenance and unsuccessful attempts to perform an operation, such as closing a circuit breaker, preventing further attempts once a limit is reached until the counter is reset.

The most common counters can only add to their present counts. *Up-down* counters can add or subtract, making them able to infer mechanical position from a count, accept "raise-lower" commands from an operator or control system, and compute net totals (i.e., energy consumed − energy rejected for another use = net energy consumed). See Figure 11-22.

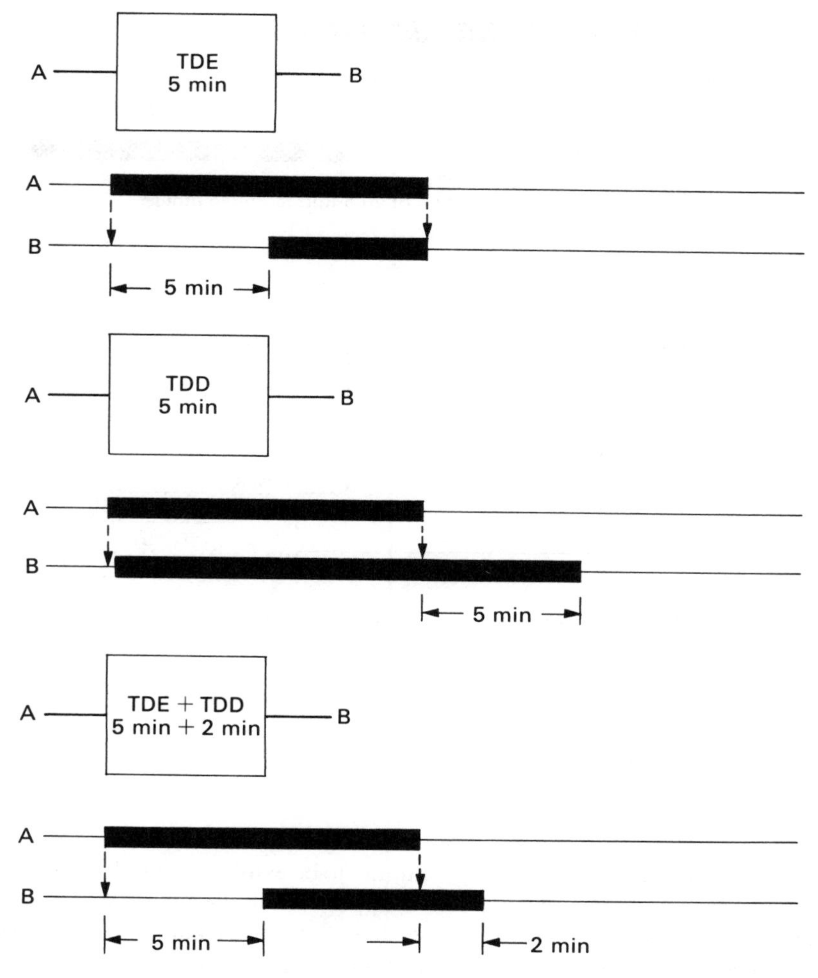

Figure 11-19. Comparison of TDE and TDD functions.

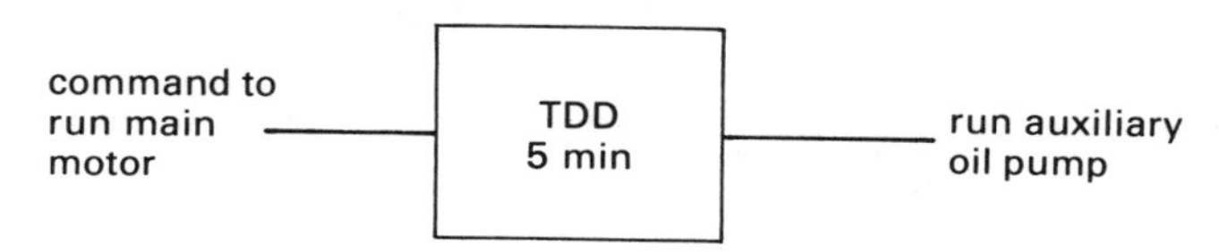

Figure 11-20. The TDD function immediately starts lubrication upon receipt of a command to start the main motor. When the main motor is shut down, lubrication continues for 5 min. The main motor may be interlocked to run only when adequate oil pressure is available.

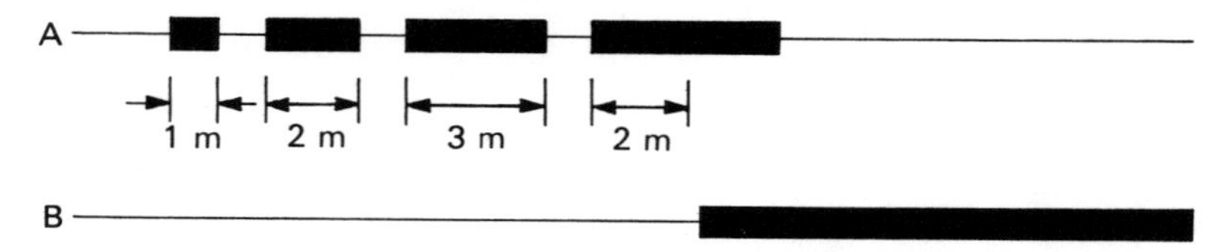

Figure 11-21. Operation of an 8-min retentive timer.

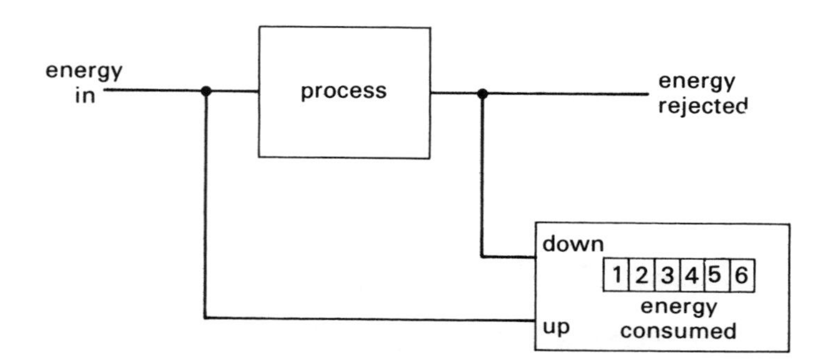

Figure 11-22. Use of an up-down counter to determine energy recovery.

INTERLOCK IMPLEMENTATION

Electromechanical Relays

The familiar electromechanical relay continues to be extensively used for interlock functions, even in this age of computers. Simply, current flowing through a coil of wire produces a magnetic field that causes one mechanical contact (or set) to move with respect to a corresponding fixed one, producing one or more electrical circuit switching actions. Contact "chatter" is avoided in ac coils by modifying the magnetic circuit so that a magnetic field is maintained even when the coil current momentarily is zero.

Relays, in general, provide the control engineer with several important features:

Electrical isolation between input and outputs, and among outputs.
A wide variety of output arrangements.
Wide availability.
Reasonable resistance to typical environmental stresses, including vibration. (Many relay types have met seismic qualification criteria for use in nuclear power plants.)

Easy installation/replacement of the types most commonly used in industrial control.
Ability to tolerate short-term stresses, such as coil overvoltage, without a catastrophic failure.
Availability of compatible devices such as time delay relays and counters.
Ability to directly interface equipment such as motor starters, manual switches and solenoid valves at all voltages in common use with immunity to electrical transients.
Switching speed sufficient for most interlock applications.
Well understood by plant personnel.

Some drawbacks must also be stated:

Limited switching life.
High power consumption compared to some alternatives.
Environmentally sealed devices required in some applications. (Common industrial control relays are open to attack by airborne contaminants.)
Require a large volume of space compared to some alternatives.

Relay Circuit Design: Symbology and Definitions

Many kinds of relay symbols are used in industry (NARM, 1980). The symbols we will use are shown in Figure 11-23. A *form A* contact pair (usually just referred to as a contact) is one that presents an open circuit when the relay coil is deenergized and a closed circuit when the coil is energized. A *form B* contact presents a closed circuit when the relay coil is deenergized and an open circuit when the coil is energized. A *form C* contact has three connections so that one circuit is closed when the relay is deenergized, and the other circuit is closed when the relay is energized. Although many other contact arrangements are available for special applications, forms A, B, and C contacts, usually packaged as several contacts within a relay, fulfill the overwhelming majority of industrial interlock needs.

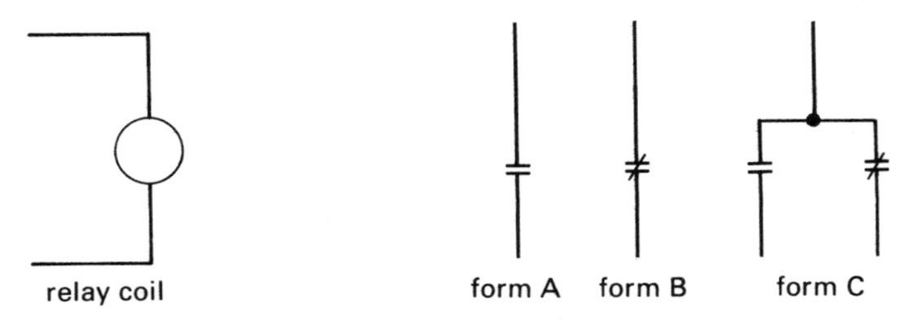

Figure 11-23. Relay symbols.

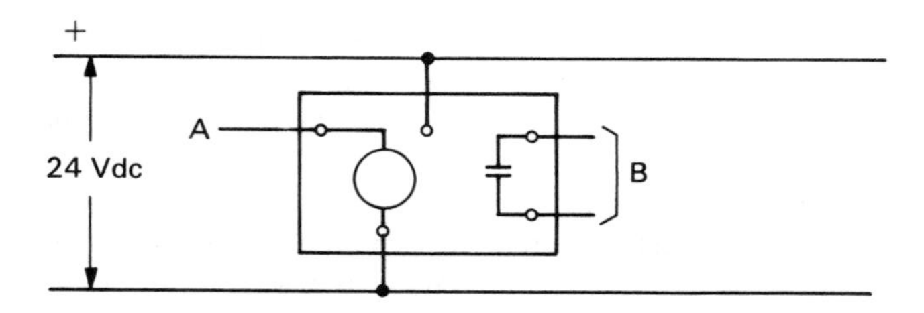

Figure 11-24. This transistorized TDD relay requires a 24-V dc power supply for proper operation. If the power fails, the contact immediately transfers to its deenergized condition regardless of the input.

Most industrial control relays have switching speeds in the range of tens of milliseconds. *Time-delay relays* are those designed to delay their switching actions, either on energization, deenergization, or both. The *time delay on energization (TDE)* relay is probably the most common. Relays designed for *time delay on deenergization* (TDD) are used extensively as well, with the caution that for most types auxiliary power is required to operate the timing circuits (Fig.11-24). Upon removal of the auxiliary power, the contacts transfer to their coil-deenergized positions immediately. Electropneumatic relays store energy as pressure and therefore do not require auxiliary power.

For special applications time-delay relays of both types may be used for a combined TDE/TDD characteristic. Self-contained TDE/TDD relays are also available in some types. We consider time-delay relays to be essentially electromechanical, even though many use electronic, and even digital, timing circuits. Analog electronic, thermal, and pneumatic timing mechanisms are generally limited to fairly short intervals, measurable in minutes. Motor-driven timers can operate over intervals measurable in hours, and digital timers can time for several days.

Some relays do not require electric power to maintain their energized states. These *mechanically latched relays* are usually of the *electrically-reset* type, which means that they have two coils. Energizing the *operate coil,* even momentarily, places the relay in its energized state. Energizing the *reset coil* returns the relay to its deenergized state. Appearing less often is the *manually reset* type. This relay is manually placed in its reset position, where it remains until power is applied to its one operating coil. These relays typically feature very fast operation in comparison with other electromechanical relays of similar size and are commonly used to protect electrical distribution (i.e., switchgear) circuits.

Relay Construction and Installation

Most relays used in interlock functions are designed to plug into vacuum-tube-style or special-purpose sockets, which are in turn installed on sheet metal and wired through screw terminals. These *general-purpose relays* typically contain multiple contacts (i.e., two form C) capable of controlling most pilot-level loads, such as indicator lamps and small motor starter coils, up to 120 or 240 V ac as well as certain dc loads. They are also used to switch low-level loads such as instrument signals. Timing and other accessory-type relays are available in compatible construction. Plug-in relays do not generally offer full environmental sealing, but they do have dust covers.

Reed relays use contacts sealed into glass capsules and so offer a high degree of environmental protection. Sometimes mercury is added to the capsules to improve performance. Most reed relays are too small to be able to switch typical industrial circuit loads. They have solder connectors because they are designed for installation on printed wiring board assemblies. True form C contacts are comparatively rare. Reed relays are available with dc coils only, usually at 24 V or less. They do offer the fastest switching of any electromechanical relay (into the audio frequency range), good switching of instrument-level signals, and small size. They also feature mechanical life into the millions of switching operations. They require very low power and can be driven directly by some solid-state logic families.

Reed relays may be found as contact input and output devices for electronic equipment and as signal switches for analog data acquisition systems. They are also found in some field instruments. Although they are rarely used as interlock logic elements, the designer of an interlock system may well have to accommodate inputs from them and outputs to them.

Larger than the general-purpose relay is the *machine tool relay.* Despite its name, this relay is used in other industries as an interlock device. It is the size of a small motor starter and can accommodate heavy pilot-level ac and dc loads, including switching current into the coils of medium-voltage switchgear. It offers the widest variety of contact combinations, including, in some models, the ability to change these combinations in the field. It is large, and its coil imposes a heavy power load. Since its contacts are designed to switch loads well into the ampere range, most machine-tool relays should be considered unsuitable for switching instrument-level (i.e., 4-20 mA) loads.

A Relay Application

Figure 11-25 is a simplified diagram of a motor starter circuit. The purpose of the motor starter is to control a large load (typically a 480-V, three-phase motor) while allowing pilot circuits operating at lower voltage and current to be used for operator interface and interlocking. For each motor starter we want to light a remote indicator lamp if any of the following happens:

1. The fuse blows
2. Power fails
3. The motor overloads

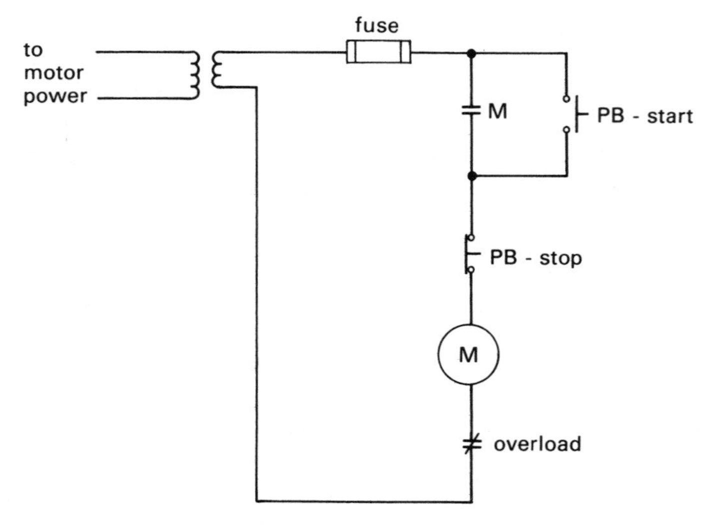

Figure 11-25. The M contact is an auxiliary contact on the motor starter, closed when the starter is energizing the motor. The start pushbutton closes when depressed, and the stop pushbutton opens when depressed. The overload contact opens when the motor is overloaded. Provisions for manually resetting the overload contact are not shown.

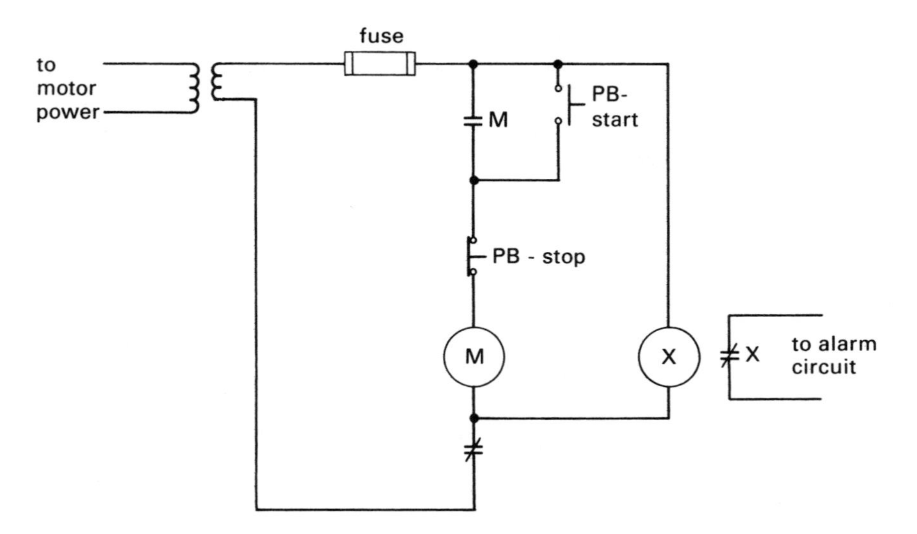

Figure 11-26. Auxiliary relay installed to monitor circuit conditions.

To accomplish this, we install a relay for each circuit and wire it as shown in Figure 11-26. The normally closed contact is held open as long as power is applied to the coil. If any of the problems listed occurs, the relay deenergizes and the contact closes.

Programmable Logic Controllers

We consider here the application of programmable controllers (PCs) only to interlocks. Their use at the higher levels of batch controls is considered elsewhere. Internally, a PC closely resembles a computer; externally it presents itself as substantially different. Instead of being programmed in a conventional, text-oriented language (i.e., with statements like "LET A+B=C"), most PCs use a symbolic language that closely resembles the symbolic language used for specifying relay logic circuits (Fig. 11-27). Virtually any logic design using relays, timers, counters, and memory elements can be realized with a PC—although the user must be aware that the PC "solves" its logic equations sequentially, not simultaneously, so that timing relationships are different.

Programmable controllers have several advantages over relays in interlock applications:

Any input or any output may be used an arbitrary number of times in logic equations. (With relays a separate physical contact is required for each usage.)

They accommodate frequent switching better than relays, since switching does not require parts to move.

In medium-to-large applications they are more compact than equivalent relay systems.

Many models are available with remote I/O, allowing I/O modules to be placed at some distance from the CPU. This can reduce plant wiring costs.

Logic changes can be made by reprogramming instead of by changing hardware.

As with computers, PCs are subject to catastrophic failures caused by failure of a power supply or other subsystem required for overall operation. Redundant controllers are available from some manufacturers.

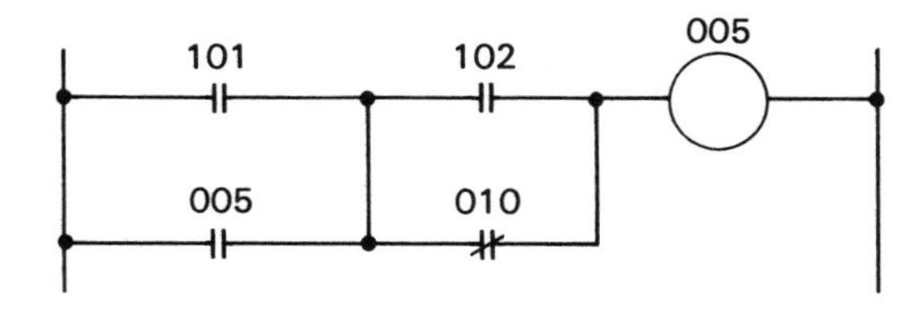

Figure 11-27. A typical programmable controller logic network. In this numbering scheme, outputs have numbers below 100, inputs have numbers greater than 100.

Many PCs are available with serial ports, enabling a computer to determine statuses of input, output, and internal variables and to transmit commands without extensive hardwiring. Serial ports are also used in some models to enable multiple PC CPUs to communicate with each other.

Like computers and modular logic, PCs operate internally at voltages and currents much lower than those commonly used to interface field equipment. Therefore, they require both input and output interface modules.

Most PCs have some mechanism for recording logic programs onto magnetic tape or a similar medium. These recordings can be used to avoid extensive reprogramming as a result of some types of PC failure. They may also be used to transfer programs among controllers performing the same function.

By packaging peripherals that are particularly heat sensitive, such as cassette tape drives, in portable programming units, manufacturers of PCs are able to offer operating temperature limits that exceed those of equivalent computers with such peripherals permanently installed. This enhances the installation flexibility offered by remote I/O. This is not the same, however, as saying that PCs are inherently more reliable than equivalent computers used within specified limits—they are made from the same type of parts and are subject to the same failure mechanisms.

A Programmable Controller Application

An alarm logic system will serve as a challenging, if hypothetical, illustration of a programmable logic controller interlock application. Five temperature switches are installed on a process vessel. They close on high temperature. Three status lamps—green, amber, and red—are provided to give the operators a quick indication of the states of the switches. The lamps are to be lit according to the following rules:

If no switches are closed—green lamp steady
If one switch is closed—green lamp flashing
If two switches are closed—amber lamp steady
If three switches are closed—amber lamp flashing
If four switches are closed—red lamp steady
If five switches are closed—red lamp flashing

A function table corresponding to these rules is given as Table 11-1. The most straightforward logic design is to divide the function into two major modules. The first detects which of the 32 possible logic combinations is present. This module then is the same regardless of the specific lamplighting rules being used. The second implements the rules. An overview design of the logic is shown in Figure 11-28.

Table 11-1. Function Chart for the Example in the Text

Condition No.	*Switch State*					*Lamp State*
	1	*2*	*3*	*4*	*5*	
1	0	0	0	0	0	GS
2	0	0	0	0	1	GF
3	0	0	0	1	0	GF
4	0	0	0	1	1	AS
5	0	0	1	0	0	GF
6	0	0	1	0	1	AS
7	0	0	1	1	0	AS
8	0	0	1	1	1	AF
9	0	1	0	0	0	GF
10	0	1	0	0	1	AS
11	0	1	0	1	0	AS
12	0	1	0	1	1	AF
13	0	1	1	0	0	AS
14	0	1	1	0	1	AF
15	0	1	1	1	0	AF
16	0	1	1	1	1	RS
17	1	0	0	0	0	GF
18	1	0	0	0	1	AS
19	1	0	0	1	0	AS
20	1	0	0	1	1	AF
21	1	0	1	0	0	AS
22	1	0	1	0	1	AF
23	1	0	1	1	0	AF
24	1	0	1	1	1	RS
25	1	1	0	0	0	AS
26	1	1	0	0	1	AF
27	1	1	0	1	0	AF
28	1	1	0	1	1	RS
29	1	1	1	0	0	AF
30	1	1	1	0	1	RS
31	1	1	1	1	0	RS
32	1	1	1	1	1	RF

Switch state = 1 implies alarm, switch state = 0 implies no alarm. GS = green lamp steady, GF = green lamp flashing, etc.

With 32 possible combinations, 32 separate logic equations are required to detect which combination is in effect (Fig. 11-29). The results of these equations are logically combined (Fig. 11-30) to assert one of the six possible outputs. The second module takes these six outputs and uses them to provide control signals to the lamps. A flashing function, as shown in Figure 11-31, uses one timer to determine lamp on time and another to determine off time.

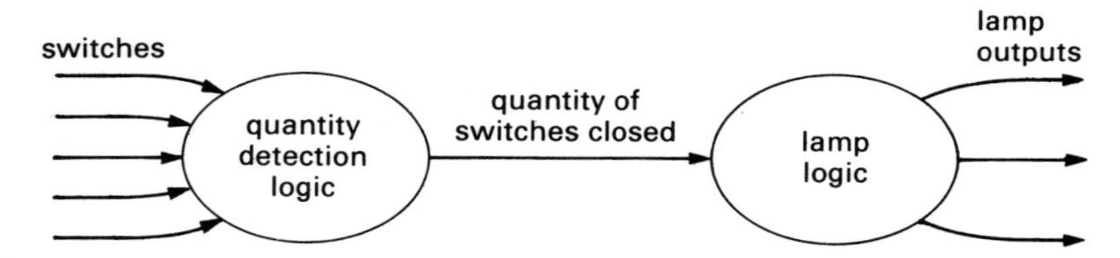

Figure 11-28. Logic overview for lamp control.

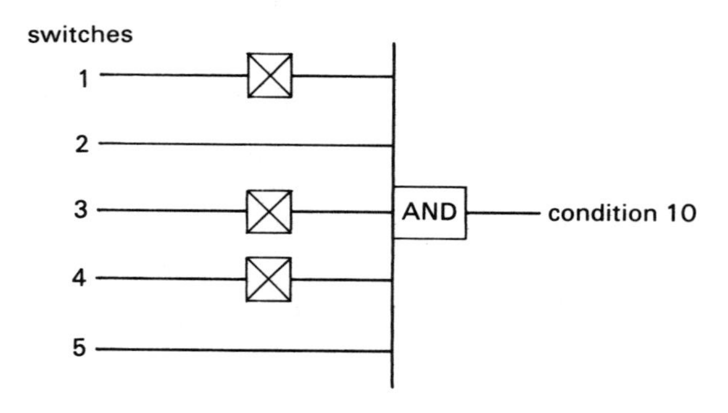

Figure 11-29. Logic to detect condition 10.

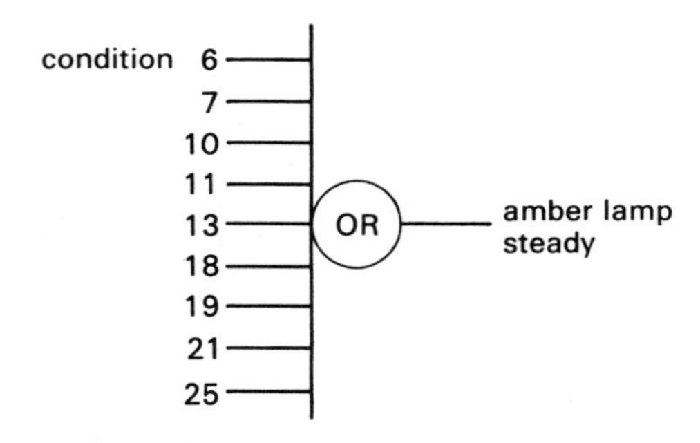

Figure 11-30. Logic for steady illumination of the amber lamp.

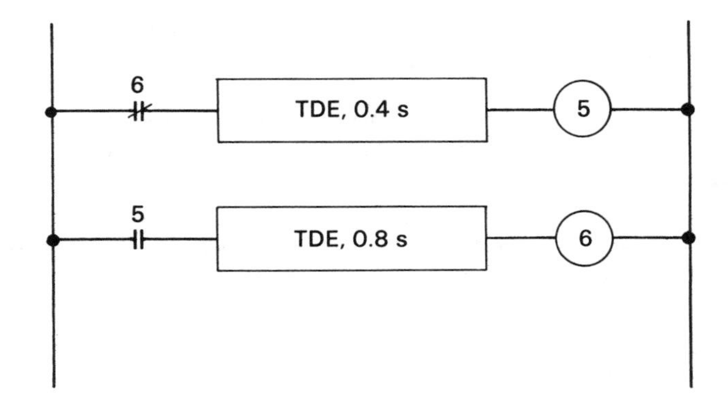

Figure 11-31. At start-up the 0.4-s timer times, so in 0.4 s, coil 5 is activated. Coil 5 remains activated until the second timer completes its timing. Then the first timer resets and the timing begins again. Another contact of coil 6 is on for 0.4 s off for 0.8 s. This is typical for programmable controller ladder diagrams.

An alternative programming approach follows the method that would typically be used in a computer programmed to perform this function. In BASIC, for example, with inputs tagged as TAHØØ1 through TAHØØ5, the code might look like this:

```
LET COUNT=Ø
LET COUNT=COUNT+TAHØØ1 (Assuming TAHØØn=1 if closed;
Ø if open)
LET COUNT=COUNT+TAHØØ2
LET COUNT=COUNT+TAHØØ3
LET COUNT=COUNT+TAHØØ4
LET COUNT=COUNT+TAHØØ5
```

At this point the value of the variable COUNT is equal to the number of switches in alarm. Selection of output conditions, once COUNT is known, would generally be similar to the method used in the PC.

Implementation of the counting technique in the PC depends on the availability of a conditional add function and a compare function. The implementation and depiction of these functions varies among manufacturers, but the logic can be shown schematically as in Figure 11-32. The contact in front of the ADD function indicates that when the contact is closed the function is carried out; when the contact is open, the function is ignored. The relay coil after a COMPARE function indicates that when the COMPARE is satisfied, the coil is turned on.

This method is, in general, preferred over a proliferation of logic gates, but it does require a greater level of user sophistication. The relaylike simplicity that has helped popularize PCs, has been rejected here in favor of a more compact but more

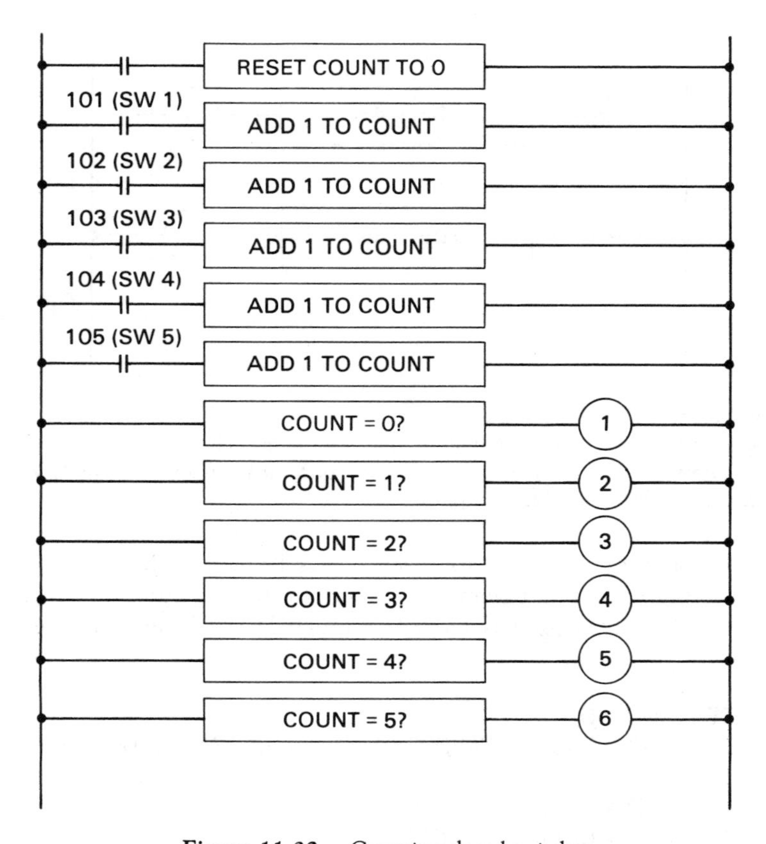

Figure 11-32. Counting closed switches.

sophisticated implementation. The user may find that as these techniques replace ladder logic, the software may become less readily transportable between PCs of different manufacture. Ladder logic is an effective standard even if addressing techniques differ. The more sophisticated functions are likely to show substantial variation among manufacturers.

Modular Logic

Modular solid-state logic is an electronic version of electromechanical relays. The earliest implementations used discrete transistor circuits and were followed by versions using integrated circuits (ICs). Some of these styles used the same ICs that were being designed into computers of the time; others used ICs made specifically for high electrical noise environments. Within the United States these modules are generally restricted in new designs to specialized applications, such as emergency shutdown systems. (Programmable modular logic, described later, is emerging as an exception.) For general-purpose interlocking they have largely been replaced by programmable logic controllers. In Europe, however, interest continues stronger due in part to the availability of high-reliability ac processing modules.

Modular logic offerings in the marketplace can be divided into three groups. *General-purpose* modules provide functions identical to many of those defined as standard logic functions (Fig. 11-33). For example, one such module might consist of 4 four-input OR functions, whereas another might have 6 two-input AND functions. Timers, counters, and memory elements are available in most product lines. *Specific-purpose* modules have their logic organized according to the needs of an end device (e.g., a motor-operated valve). The advantage of the latter type is that, where they are applicable, they require much less wiring between the modules than do the general-purpose types. Furthermore, maintenance is simplified, since it is usually easy to localize a problem to one card. The associated disadvantage is that if requirements change, the modules may not be readily changed to meet them. These modules may have some degree of programmability provided by jumpers. Fully *programmable* types use electronic memory and can provide a wide variety of functions readily changeable. Recent product introductions have made available modules with combined analog and logical functionality, which is excellent, for example, for the alarming of analog variables.

Modular logic elements communicate with each other through dedicated logic level signals and, more recently, serial bit streams similar to those used in communication between computers. For a module sending simple on/off logic signals over dedicated paths, the receiving module cannot distinguish between a sending module failure and one of the valid logic states. (This is similar to the situation with relays.) *Fail-safe* modules use reference ac inputs (at 8 kHz, in one design) to help insure that upon module failure outputs always assume the off state. Thus, when used in properly engineered systems, there is a very high probability that an

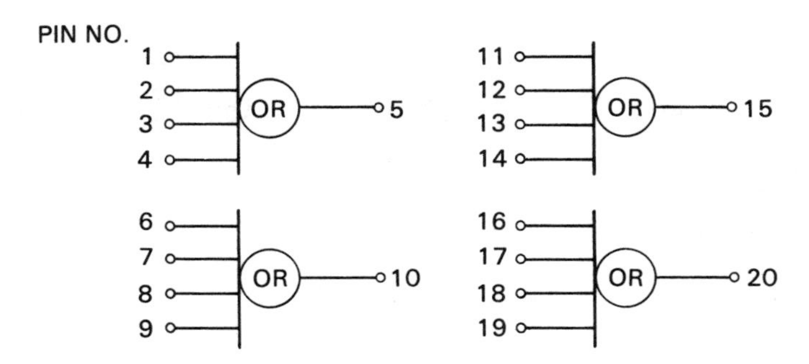

Figure 11-33. Possible organization of a general-purpose logic module. Power, ground, and test connections are not shown.

interlock equipment failure will cause a shutdown or other interlock action rather than failure to act.

Logic modules are usually packaged as small, printed wiring board assemblies. They plug into sockets, with wiring between the sockets determining overall logic functions. They require interfaces to and from plant equipment, since internally they use only low voltages and are comparatively sensitive to electrical noise.

A Modular Logic Application

The five-input, three-output logic function used as a PC application, will also serve as a sample problem for modular logic. A straightforward implementation of the specification shown in Table 11-1 would use commonly available functions but, depending on manufacturer's packaging design, could require about 40 modules.

Once again, the implementation can be simplified by summing the asserted inputs. Unlike PCs and computers, which execute logic sequentially, logic modules are like relays because they operate continuously, with multiple functions performed by multiple pieces of equipment. Sequential operation of logic modules is possible but, for most applications, unnecessarily complicated. Some manufacturers offer "adder" functions, which add binary numbers in parallel. Four of these adders can be combined to yield the sum of asserted inputs (Fig. 11-34). A binary-to-decimal decoder function (Fig. 11-35) yields a distinct output for each possible sum 0-5. Little further processing is required to develop output control signals; logic card families typically offer timers, two of which can be wired together to form a flasher.

With the availability of a larger binary-to-decimal conversion function, even further simplification is possible. A five-bit function can distinguish between all

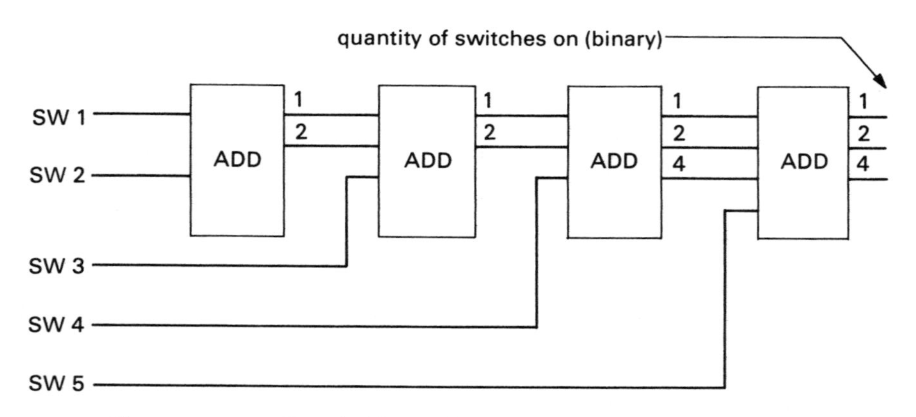

Figure 11-34. Use of adders to determine the number of switches on.

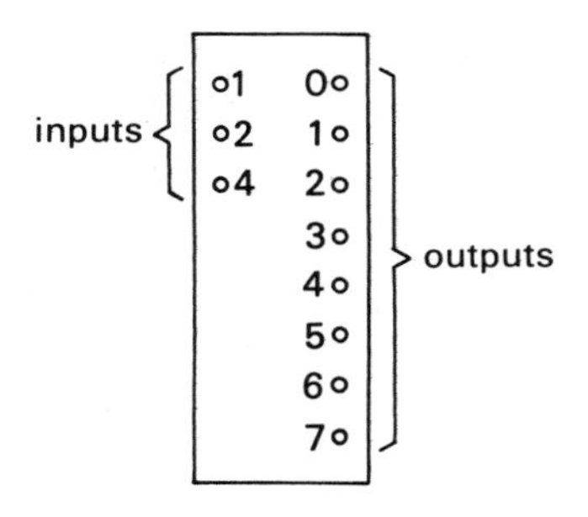

Figure 11-35. A three-bit binary-to-decimal converter (decoder) yields a different output on for each possible combination of inputs on.

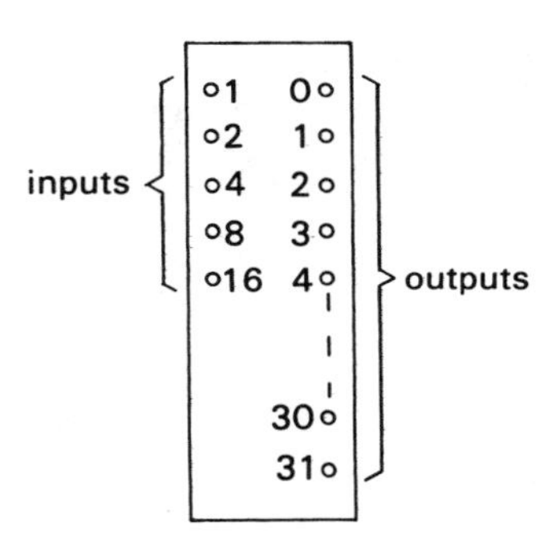

Figure 11-36. A five-bit binary-to-decimal converter immediately determines which input state is present.

possible states; then all states representing each particular sum are ORed to yield the sums (Fig. 11-36). If only a four-bit function is available, some additional processing is required.

Implementation using programmable modules will vary, depending on the particular instructions being used. Generally, though, they should fall into one of the two major categories used here: simple logic or counting.

Continuous-Control Modules

Modular analog control systems are used mostly at the analog control levels; however, their alarming capabilities may be used as an input to the interlock levels. For maximum process security these modules should be separate from those used for control. This would prevent a failure in one module from simultaneously causing loss of control and loss of ability to detect an alarm condition. In many applications, however, the same set of analog modules provides both control and alarm functions.

Alarm functions available range from simple limit comparisons to deviation alarming through alarms based upon complex algorithms. For simple limit comparisons, temperature switches, pressure switches, and similar devices may also be used. Analog equipment has an advantage, though, in that it is easier to test. Sensor outputs may be compared to other (portable or installed) sensors, and test signals may be impressed at the analog inputs to the interlock system to confirm proper alarm operation. To test a temperature switch for operation at the specified value, one must remove it from the process or cause the monitored temperature in the process to reach the alarm value.

A Continuous-Control Application

Figure 11-37 shows the use of analog functions to find the median value of three temperature inputs. (The mean value is not used because an out-of-range signal should have no influence upon the value derived.) The other two signals are compared to the median signal, and an alarm is generated if high deviation is detected.

Programmable Continuous Controllers

The analog world's equivalent to the programmable logic controller is the programmable continuous controller, also called *multiloop controller, distributed*

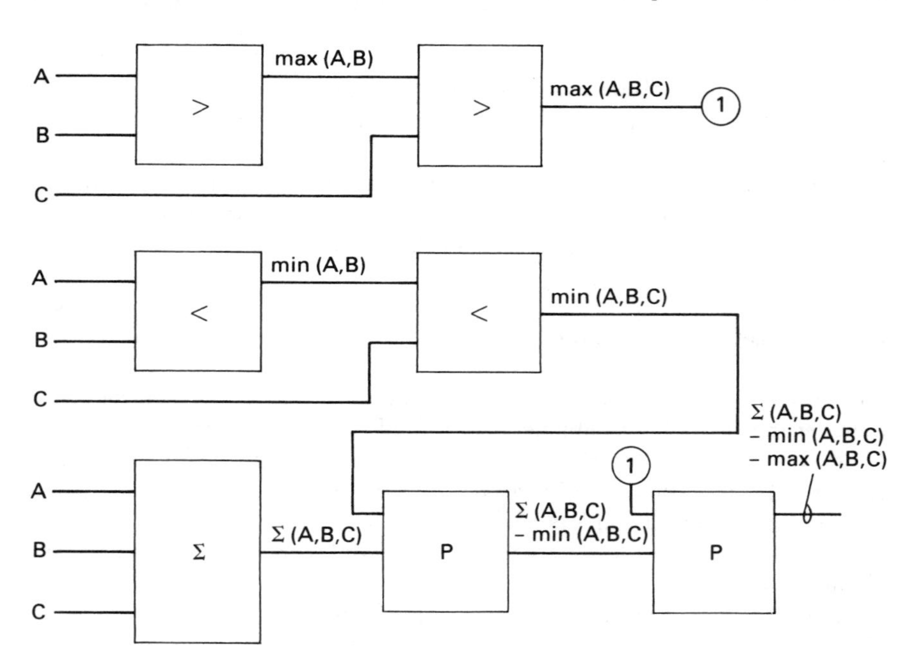

Figure 11-37. Proportional controllers, all with proportional band = 100% (gain = 1), subtract the highest and lowest signals from the sum of all three signals.

controller, or even *microprocessor controller*. Its symbolic language is the language of proportional-integral-derivative (PID) controllers, signal-selection functions, and so forth. These controllers can also be provided with interlock functions similar to those of modular logic. These controllers, therefore, allow analog levels and the higher interlock level (level 2) to be combined in the same hardware and can thereby minimize the number of types of hardware required to implement the batch control system.

Interlock by Computer

A general-purpose control computer may be programmed in various ways, depending on the model, to perform interlock functions. Those described here are typical.

The *continuous-control package* may be used to generate alarms based upon contact inputs and alarm signals derived from analog inputs. These in turn may be used to effect control actions such as those already discussed in this chapter.

The interlock logic facility may be used to condition outputs from the sequential logic. This would commonly take the form of AND logic, allowing a control action, requested by the sequential logic, to reach the process only if all of the permissives are present. Other logic combinations are also possible.

The use of *service and hold*, discussed in Chapter 9, enables the sequential logic to branch out of its normal path of execution to service a process problem.

Finally, the computer's application programming language, such as BASIC or FORTRAN, can be used effectively to establish interlocks. The interlock program is made to run at frequent intervals, typically 1 s or less, and can implement complex logic very simply. The three-lamp example given in the section *Programmable Logic Controllers* requires six lines of simple BASIC code to determine the number of input switches on, even fewer if the tags can be represented by subscripted variables. Output logic for the three lamp-steady conditions is similarly simple:

```
LET GREEN = OFF
IF COUNT = Ø THEN LET GREEN = ON
LET AMBER = OFF
IF COUNT = 2 THEN LET AMBER = ON
LET RED = OFF
IF COUNT = 4 THEN LET RED = ON
```

where GREEN, AMBER, and RED are tags associated with output points, and OFF and ON are variables with values 0 and 1, respectively.

To cause the lamps to flash for COUNT = 1, 3, or 5, define the variable MARKER with an initial value of 1. Then the statement

```
LET MARKER = - MARKER
```

will cause the value of MARKER to alternate between 1 and -1 when executed with the remainder of the logic. Then statements like

```
IF COUNT = 1.AND.MARKER = 1 THEN LET GREEN = ON
```

may be interspersed with the code shown previously. Since on alternate executions the logic equation will be satisfied and not satisfied, the lamp will flash at half the execution rate of the software.

Maintaining Reliability

The user thinking about implementing interlocks in the same computer to be used for control should carefully consider the reliability and safety implications. There is, however remote, a chance that the same computer failure that causes erroneous control will also make the computer unable to detect or respond to the process conditions caused by the erroneous control. Interlock logic has two functions: protecting the process against process failures, and protecting the process against control failures from higher levels and operator errors. Interlock logic installed within the control computer will provide the same level of protection against process failures, but somewhat less protection against control failures and errors than logic installed externally. Finally, the designer should assume that all computer-based interlock functions can fail at once and should provide sufficient external interlocks to protect the process in that case.

OPERATOR INTERFACE

Switches

Selector-type switches have served industry for years by providing the interface through which an operator may start and stop a motor, open and close a solenoid valve, and perform other tasks. The two-position switch (Fig. 11-38) allows on-off

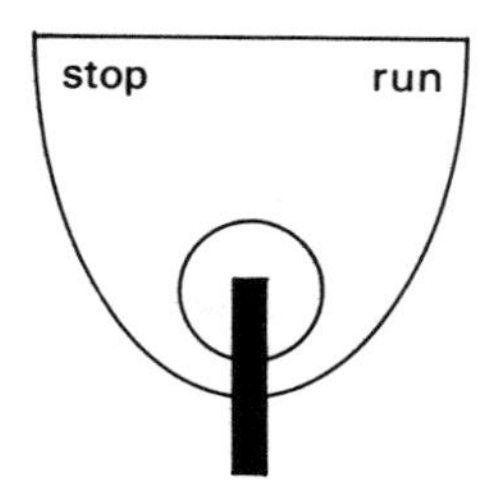

Figure 11-38. A two-position switch for on/off control. Escutcheon plates in these illustrations are shown marked for motors; solenoid valves and other types of equipment may be similarly controlled when appropriate markings are used.

control by the simple circuit shown in Figure 11-39. The power industry and, frequently, the process industries as well use three-position switches (Fig. 11-40) with the circuit depicted in simplified form in Figure 11-41. This implements R-S flip-flop logic. The third position enables the controlled device to receive only a momentary "start" signal, rather than a continuous "run" signal. This is sometimes referred to as *three-wire control.* If the motor starter (or equivalent device) is deenergized for any reason (e.g., line power loss), it will not be energized again without specific operator action. Additionally, local switches at the equipment are more easily accommodated with three-position central switches.

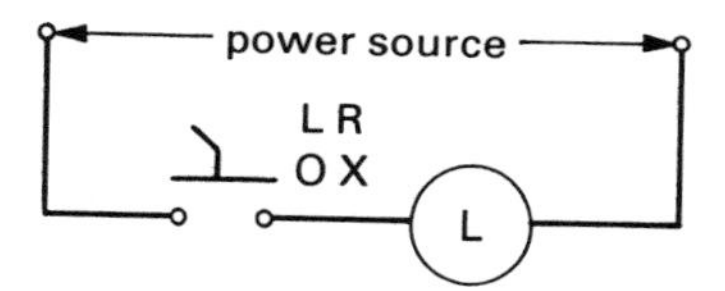

Figure 11-39. Circuit for the switch shown in Figure 11.38. Load L may be an intermediate device, such as a motor starter, or the controlled equipment itself, such as a solenoid valve.

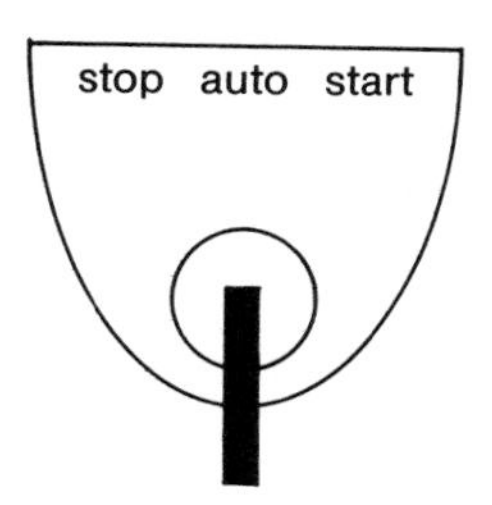

Figure 11-40. A three-position switch for mandatory operator reset.

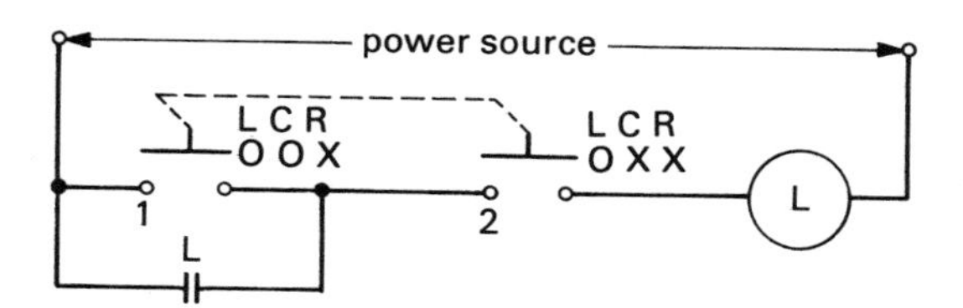

Figure 11-41. Circuit for use with the switch shown in Figure 11-40. Additional momentary-action start contacts may be wired in parallel with contact (1) and stop contacts in series with contact (2). Contact (L) is an auxiliary contact that follows the condition of load L.

Application

These arrangements are satisfactory when "automatic control" consists only of thermal overload devices that require local reset and similar means of equipment protection. When manual control is to be combined with more complex automatic control in a batch plant, there are problems with these conventional methods. The power industry commonly labels the center position as *automatic* and allows the automatic system to control the equipment when the switch is in that position. It is possible to prevent the equipment from being automatically energized by using a "lockout" position to the left of the "stop" position or by making the stop action continuous (maintained position) rather than momentary (spring return from position to center). Allowing the equipment to remain energized despite higher-level commands to stop (while accepting interlock commands to stop) is more difficult. Changing the "start" position to "run" would make implementation of mandatory operator restart action impossible, or at least much more complex, and require additional hardware.

When mandatory operator restart is not required, the process industries use the "hand-off-auto" configuration shown in Figure 11-42. In the "hand" position the equipment is continuously energized. Because the "off" position is in the center, it is not practical to go from "auto" to "hand" without turning the equipment off.

When mandatory operator restart is required, a switch with four defined states is needed:

1. Start (momentary)
2. Stop (momentary or continuous)
3. Manual (continuous; start and stop in response to switch states 1 and 2, stop in response to interlock signals)
4. Automatic (continuous; start and stop in response to higher-level commands, stop in response to interlock signals)

(Some equipment may require starting as well as stopping from interlock signals.)

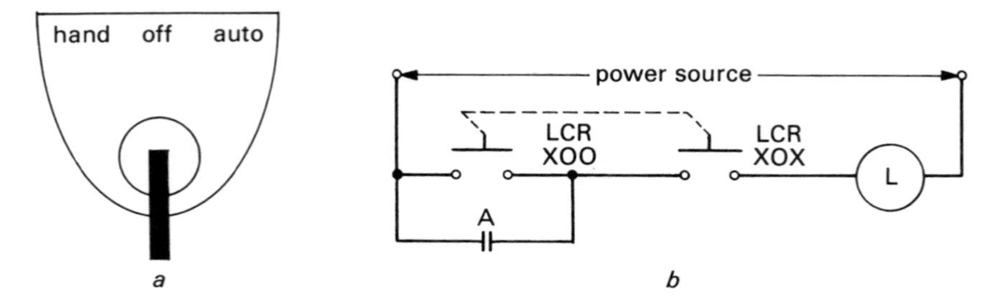

Figure 11-42. *a.* Configuration of a hand-off-auto switch. *b.* Circuit for use with switch shown above. Contact (A) is a run command from the automatic control system.

Transitions

The following direct transitions between states are required:

1. Manual to start (and return automatically)
2. Manual to stop (and return automatically)
3. Manual to automatic
4. Automatic to start to manual
5. Automatic to stop to manual

With transitions 4 and 5, when the operator takes action to start or stop the equipment, it is evident that he or she no longer wishes to have the equipment under automatic control. Therefore, the equipment must be removed from automatic control after an operator start or stop command. Note that if the stop state is continuous, at least one additional transition—from stop to manual—is required. Placing the device in automatic from stop would require two transitions or an additional two-step transition. It should be possible to accomplish transition 3 without a change in equipment state.

Switch Types

What control switches are available to perform these functions? Three-position selector switches are available with center pushbuttons. When configured as illustrated in Figure 11-43, with the circuit shown in Figure 11-44, they come close to meeting the stated requirements.

To energize the equipment, the operator rotates the switch to the "push start" position and presses the button. With the button pressed the switch is in start state; with the button free it is essentially in the manual state. To reach the stop state, the operator rotates the switch through automatic to stop, which is maintained (otherwise, there would be no way for the switch to prevent the higher levels from starting a stopped motor). This violates the requirements for state transitions.

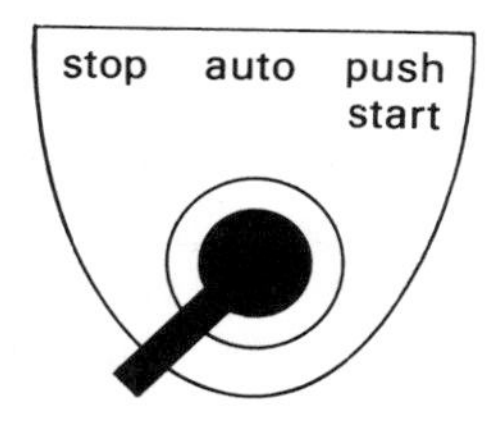

Figure 11-43. This switch is rotated to the push/start position, and the center button pressed, to energize the load.

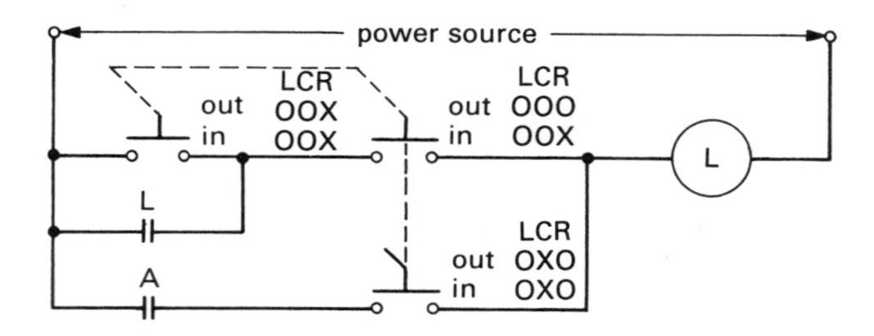

Figure 11-44. Circuit for use with the switch of Fig. 11-43. The automatic control system output (A) is a continuous run signal; a similar circuit with two automatic control system outputs, one for start and one for stop, may be used to achieve "fail-inplace" action on failure of the automatic system.

There is no problem in the manual-to-stop direction unless the equipment has been deenergized by another (usually local) momentary-action switch or momentary-action protective device. Then the higher levels could energize the equipment as the switch is rotated through the auto position. To enable another external switch to energize the equipment, the operator must rotate the switch back through auto to push start and again, in auto the equipment might be automatically energized when this is not desirable. When all protective devices require deliberate reset and no other switches are used, this switch should be satisfactory, provided that operators do not rotate it to the push start position from stop until they are ready to energize the equipment. It is assumed that the motor starter (or equivalent) will remain in the energized condition through the short power interruption that could result as the switch enters the push start position. Therefore, the equipment does not receive a momentary energization.

By adding a fourth choice, we can improve the switch action. The positions are now identified as auto, push start, and push stop. With the center button not pressed, push start and push stop are manual. This switch and a possible circuit to use it are depicted in Figure 11-45. Such a switch, however, is not known to be commercially available.

Simple pushbutton switch circuits are of little help—a pushbutton switch is a binary device and at least four states are required. With some assistance from the interlock levels, the switches may be logically connected so that they operate as required in this example. A switch arrangement is shown in Figure 11-46, and the interlocking logic in Figure 11-47. Note that this is a departure from a condition held to previously. There is no way for the front-panel operator to disconnect the equipment from the automatic control system; the disconnection depends on the interlock logic itself. Furthermore, failure of the interlock can render the equipment uncontrollable from the front panel. In systems where this lack of true manual backup does not constitute a hazard, it should prove economical and

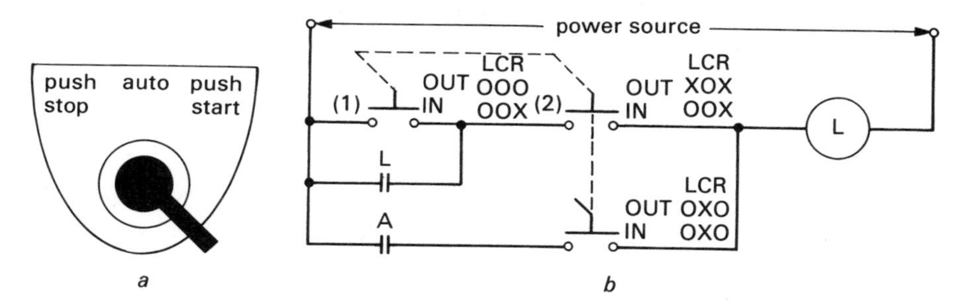

Figure 11-45. *a.* Push-to-stop operation is added to the switch of Figure 11-43. *b.* Circuit for use with this switch. Now additional momentary-action "start" contacts may be wired in parallel with contact (1), and "stop" contacts in series with contact (2).

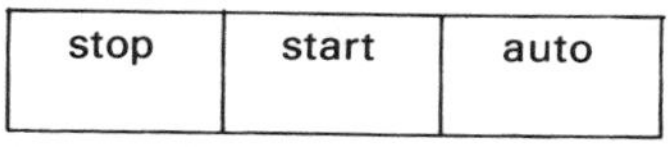

stop	start	auto

Figure 11-46. Pushbutton arrangement for manual control through the automatic system. A fourth switch, independent of these three, may be added to lock out other switches in the circuit.

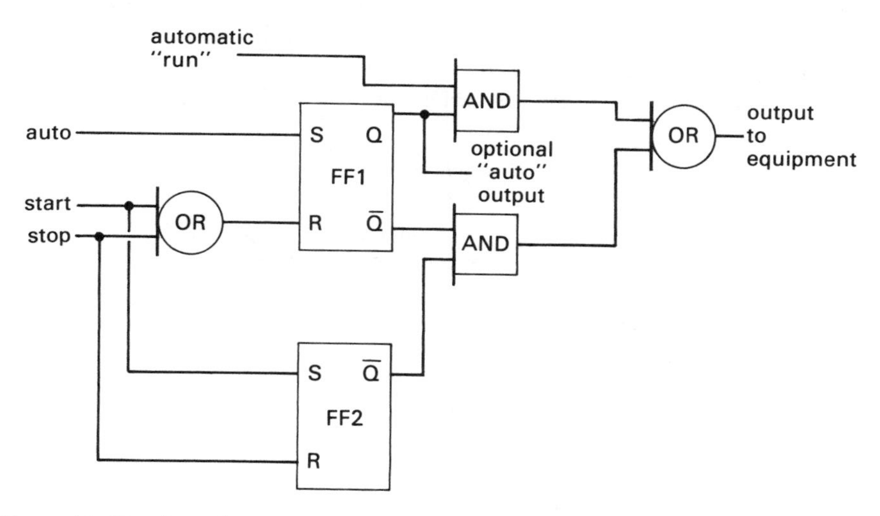

Figure 11-47. Logic for pushbutton control. For mandatory operator restart when in the manual mode, replace flip-flop FF2 with an AND gate combining the start signal and a signal indicating that the equipment is energized. Similar logic may be used for the automatic mode.

simple to operate. A lockout function may help alleviate the problem by forcing the equipment off; this is done with a fourth switch.

Operation

To energize the equipment, the operator presses the start button. The start signal is conveyed to the equipment by the interlock logic, which then does not process subsequent change-of-state requests until the auto button is pushed. The advantage of this arrangement is that exactly one operator action—pressing a specific button—is required for each transition listed.

An advantage of processing commands through the interlocks is that all or a selected part of the controlled system may be placed in manual by the system itself. This might, for example, be an appropriate response to sensor indications that the process does not respond to automatic control. Or the equipment might be placed under manual control after the automatic system processes a request for automatic shutdown.

For applications requiring completely separate backup control independent of any other controls, the switching circuit in Figure 11-48 utilizes a solenoid-held manual/automatic pushbutton switch to provide the same operator interface while allowing direct control of equipment. By controlling the solenoid power supply power line from an output of the interlock system, the system may still place all or some of the equipment on manual control. Once this is done, the equipment is disconnected from the automatic system, and deliberate operator action is required to reestablish the connection.

An intermediate approach is a switch with the same front-panel appearance

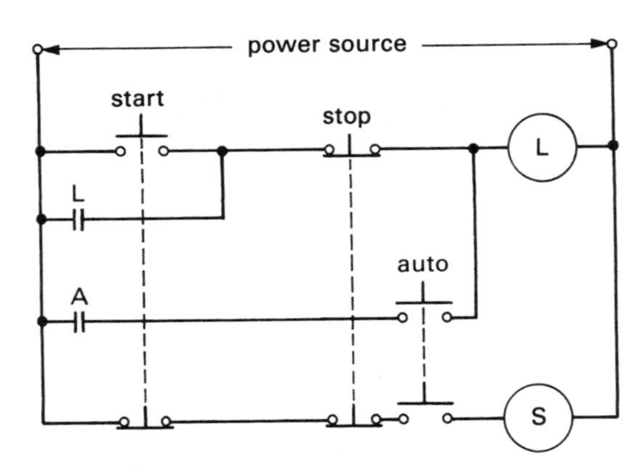

Figure 11-48. Use of a solenoid-held pushbutton switch to hold the equipment in an automatic mode. Activating either the start or the stop pushbuttons interrupts current to the solenoid S and causes release of the auto switch.

that achieves its interlock mechanically rather than electromechanically. With such a switch the auto pushbutton would remain in its depressed automatic position until either the start or stop button is pressed. Then the auto button would be released by the start or stop action, disconnecting the automatic control system from the equipment until a deliberate operator action restores automatic control. Such a switch does not seem to be listed in any catalog we have seen, either explicitly or by making use of available interlock features. One of us has, however, received a working sample, specially constructed, from the manufacturer.

Taken together, the three pushbutton switch assemblies provide a broad level of security: dependence on the automatic system; independent of it; and independence with remote switching to manual. The assemblies have the same front-of-panel appearance, so they may be combined within a system, as required by economic and performance considerations, without adding a source of confusion for operators. Widespread acceptance of the two high-security choices may allow manufacturers to achieve production economics and panel fabricators to simplify inventories; both lead to lower panel costs.

Annunciators

Annunciators are illuminated indicators driven by electromechanical or solid-state logic so that they do more than simply represent the on or off condition of a field device. *Annunciator systems* incorporate audible alarms and one or more pushbutton switches. Although numerous sequences are in use (Ref. ISA Standard S18.1), the following is a description of typical annunciator operation.

A contact input, previously defined as having its open state represent an alarm, changes state to open. The corresponding annunciator flashes at full brightness, and an audible alarm is sounded. The operator *acknowledges* the alarm by operating a pushbutton, audible alarm ceases, and the annunciator itself (the *annunciator lamp* or the *annunciator window,* because many use translucent plastic panes illuminated from behind) stops flashing and remains at full brilliance. When the alarm "clears," as indicated by the contact closing, the window flashes at half brilliance until the operator *resets* the alarm with a different pushbutton.

Usually one set of buttons is used for each group of annunciator windows. It is tempting to minimize the number of windows by combining inputs (i.e., wiring three open-to-alarm contacts in series to an open-to-alarm annunciator). With this arrangement, however, one alarm can *mask* either of the other two, preventing the operator from knowing about a new alarm condition. The *reflash* module causes the window to flash every time a new alarm condition occurs and so prevents masking. Obviously each contact sharing a particular window must be wired through the window's reflash module.

Auxiliary contacts follow the input contact states and sometimes annunciator internal logic states. They allow alarm information to be processed by other equipment, such as a computer. The *ringback* feature calls the operator's attention to the return of a contact to its nonalarm state by an audible signal. The *first-out* feature uses a different flashing pattern for the first annunciator point to go into alarm. *Analog input* annunciator points accept analog signals directly as inputs and compare them against preset limits to detect alarm conditions.

Sequence-of-Events Recorders

Sequence-of-events recorders provide hard-copy records of contact changes of state. Their extremely high time resolution (to 1 ms) enables the engineer to distinguish cause and effect after an emergency shutdown or similar occurrence.

A Sequence-of-Events Recorder Application

A reactor receives a simultaneous feed from three preweigh tanks. If one of the control valves loses its air supply and shuts down, the other two will shortly follow to maintain correct material proportions in the reactor. A sequence-of-events recorder will enable plant personnel to determine which valve closed first and, therefore, imply the first items to check.

USING SOLID-STATE INPUT AND OUTPUT DEVICES

Input and output interfaces of solid-state hardware are, themselves, usually solid-state devices. A few guidelines will help the inexperienced user achieve success.

1. Like relay contacts, solid-state outputs are two-state devices. The two states of a contact are on and off. The two states of a solid-state output are almost on and almost off. The almost on state is indicative of a small, nonohmic voltage drop across the power terminals of a solid-state device. "Nonohmic" means that the voltage drop is not proportional to current; voltage drop is, usually, virtually constant over a wide range of currents. In the overwhelming majority of cases this voltage drop may be ignored by all except the equipment manufacturer, who must account for the heat produced. In certain cases, for example the use of multiple solid-state outputs in series, especially near the low-voltage limit of the devices' normal operating ratings, there may be a problem. Similarly, solid-state devices exhibit a small leakage current when they are supposed to be turned off. Obviously, it can be very dangerous to work on plant wiring under the assumption that a solid-state output device has deenergized it. Even if normal leakage current is far below the level that poses a danger to personnel, a device short circuit could not be detected by any method except measurement. (We do not advocate the use of potentially unsafe practices with relay circuits.) In addition, connecting these devices in parallel, especially with a high-impedance load that does not need much current to operate, can lead to an inability to effectively deenergize the load.

2. Industrial power wiring exhibits frequent electrical transients, during which, for a short time, line potential may exceed several thousand volts. This can cause damage to or false operation of electronic devices. Industrial equipment is usually specified to meet the requirements of American National Standard C37.90a. This should be adequate in most cases. Specific additional precautions to control these transients may, for example, be required when switching direct currents to inductive loads.

3. Incandescent light bulbs (those with filaments) show a much lower resistance when cold than when hot. Therefore, a solid-state output device rated at twice the light bulb's design current may seem appropriate, but this rating can be temporarily exceeded every time the bulb is energized. Solid-state outputs are less tolerant of overloads like this than are their electromechanical counterparts.

4. Be knowledgeable about wiring requirements. Some solid-state inputs and outputs are similar to their relay equivalents in that they offer electrical isolation between their own inputs and outputs. Others do not, and they impose wiring requirements upon the circuits they are connected to.

CHAPTER

12

THE ECONOMICS OF BATCH AUTOMATION

Throughout this part of the book we have divided the functions of batch automation into levels. Each level represents a unique investment in design, including software, and a unique or shared investment in equipment. Each level should separately justify its associated investment. In the first part of this chapter we discuss the economic contribution to be anticipated from each group of levels. In the second part we consider the corresponding investments required.

INTERLOCK LEVELS

Interlock levels exist to protect work in progress, equipment and buildings, plant personnel, and the public. Measurements of the economic effects of investment at the interlock levels are not as direct as, for example, measurements of reductions in energy consumption. The interlock levels deal with probabilities—what is the probability of a particular failure or error? What is the investment required to prevent it or mitigate its consequences? This indirectness does not reduce the real economic effects that interlock deficiencies can have. We will not speculate on what additional interlocks, if any, would have prevented the release of a highly toxic chemical into the environment, reduced the amount released, or at least allowed local residents earlier warning, in a case like the process plant accident in Bhopal, India. It will not be necessary to speculate on the direct economic effects of the accident, including direct payments to survivors, victims, and their families, and the extended shutdown of similar plants in other places. Interlock deficiencies can surely allow such accidents to happen, whether or not they had any influence on the Bhopal plant. At the Three Mile Island plant accident in Pennsylvania, which fortunately has not been demonstrated to have injured members of the public (although charges to this effect have recently been made) but did cause the extended shutdown of a nuclear power plant, improper placement of indicators was blamed for encouraging an erroneous decision by operators (Rubenstein, 1979).

Specific economic benefits can be associated with various interlock functions. Work in progress, as well as raw materials and finished products, must be protected against contamination, excursions of temperature, or other conditions that might make additional processing necessary or cause irreversible changes, thus making disposal the only available alternative. The costs of additional processing are discussed in more detail later. Disposal cost is directly measurable and is likely to be high if the material is considered hazardous. There may also be a tankage cost while the material awaits disposal.

Protection of materials also involves protection against the materials. Here the costs of failure to provide appropriate interlocking can include public liability, lost production, equipment damage due to release of acids, corrosives or similarly active materials, and Workmen's Compensation costs. Adverse publicity may also accompany a release of dangerous material.

An important class of interlocks protects the plant staff from exposure to process materials during the normal course of their jobs. Included are detection and warning of toxic gases or of gases displacing air, agitator shutdown when charging hatches are opened, and blocking of flow to vessels under maintenance.

Protecting plant equipment helps avoid replacing expensive apparatus, as well as other plant problems that might result from equipment failures. Interlock equipment can detect failures of services such as coolant and lubricating oil before they damage equipment, mechanical problems that cause vibration of rotating equipment, motor overload, and electrical supply problems (undervoltage/overvoltage, underfrequency/overfrequency, single-phasing).

Vibration monitoring ranges from fairly simple displacement measurements to sophisticated frequency spectrum analyzers capable of suggesting incipient problems not yet resulting in large displacements.

Motor overload, which if unchecked could result in motor destruction, is detectable as excessive current or as an excessively high motor winding temperature. A low-load condition, which might indicate shaft breakage, is detectable as a low-current condition.

Electrical supply quality can be monitored at the plant service entrance or at the feeder to particularly large, expensive, or critical motors.

For maximum value the engineering associated with plant interlocks needs to be on par with that applied to plant process operations. One of the authors has had experience with an automotive cooling system failure: by the time the high-temperature switch closed and the warning lamp came on, high cooling-system pressure had evidently caused the rupture of a hose, making the resultant repairs substantially more extensive than they otherwise would have been. So even for an application as simple as a high-temperature switch, an incorrect operating point can render any subsequent warnings and interlocks useless. These switches must also be properly installed, in the right place, and make the right measurement of the right material or circuit.

One place where engineering judgment is required, and one which can directly affect a plant's overall economic performance, is in the decision between giving the operator a warning of a plant equipment problem and bypassing the operator's judgment to automatically shut the equipment down. The latter might seem to be the more prudent; after all, what can the operator do in this case except delay the shutdown and allow the equipment to be damaged? The answer is found in the job that the affected equipment is, or can be, doing at the time. Is it prudent to automatically shut down the single engine of a light airplane for low oil pressure or to allow the pilot the maximum opportunity to safely land, albeit with a damaged engine? The operators at Three Mile Island were criticized for shutting down some important pumps to prevent possible damage to them, based on the mistaken belief that these pumps were no longer necessary to maintain the safety of the plant. In a batch plant containing the strongly exothermic reaction we have discussed so much, this logic might be applied to the cooling medium pump. The economic consequences of pump or motor damage have to be considered against the consequences of a runaway reaction caused by insufficient cooling.

The interlocks considered so far have been of the simpler level 1 variety, with the possible exception of sophisticated air monitoring and vibration analysis equipment. Interlocks at level 2 can help protect the plant against problems more subtle than the unconditional excursion of a variable outside of preset limits. The effect they have, generally, is to allow relaxed limits when appropriate but severe limits when necessary. Reactor temperature time rate of change, for example, can only be used as a measure of incipient loss of control if the high values that might occur normally during the process can be ignored.

The two techniques can be applied to the same variable: Agitator current can be compared to a nonvarying absolute limit and to a varying limit that is high when the agitator starts and lowered as the agitator is presumed to reach normal operating conditions.

Interlocks can be used to anticipate problems rather than merely react to existing ones. An attempted transfer of material into a vessel with insufficient capacity can be prevented from the start, instead of being stopped as the receiving vessel approaches its capacity. The economic benefits here are in reducing disruption of plant operations.

REGULATORY LEVELS

The major economic contribution of regulatory level automation is in the consistency of products from batch to batch. More common than a simple mistake is the typical batch-to-batch variation observed in the production of what should be the same product. Some of this may be explained as variation between operators—in response times, the ways they read analog indicators, their willingness to run a process near its stability limits, for example. Some may result from not taking into account variations in "secondary" variables, such as cooling water temperature.

The economic effects depend on the resolution of those batch-to-batch variations. The simplest case is where the market accepts these variations. This acceptance can be deceptive, though: We have been told of one case in which a manually controlled plant regularly sold the same product to a particular customer. The company started up a new, computer-controlled plant to make the same product and received congratulations from the customer for the improvement in consistency. The acceptability of variations, then, needs to be continually reevaluated in a competitive marketplace. Where the market accepts variation in one direction but not the other, the processor's response is normally to adjust the manufacturing process to ensure that all, or the overwhelming majority, of the batches are acceptable. A processor weighing a food item into containers, for example, may choose to give away product in most containers to insure that the least net weight that is reasonably likely in a container still is not less than the container label's printed net weight. Although some such giveaway may always be present, better control equipment can reduce measurement uncertainty and hence allow the average to be brought closer to the minimum. The cost of giveaway is simply the selling price of the product given away.

Another strategy is to simply downgrade the product specifications when selling into a market that accepts more than one grade of a product. There is a cost here too, because the selling price of the downgraded product will generally be less. Some processors, selling into a one-grade market, hold material that is slightly off-spec for blending with a batch or batches that exceed the spec. This may at first seem like an ideal answer—for a small tankage and inventory cost all of the off-spec product is brought back into the product stream and sold. However, more careful analysis of a particular process may show the hidden costs, because of the nonlinear processing costs that may be incurred to make better-than-spec product for blending. Shinskey (1977) shows energy costs as a function of purity for continuous distillation (Fig. 12-1). Clearly, if a batch process operation follows a cost structure like this, the least expensive way to make 10,000 gallons of 98% pure product is to do just that, not make 5,000 gallons of 97% pure product and 5,000 gallons of 99% pure product.

SEQUENCE LEVELS

Automation, even without higher-level functions like production scheduling, has been shown in many cases to substantially reduce the time necessary for a plant to prepare a particular batch. There seem to be two main reasons for this, both closely

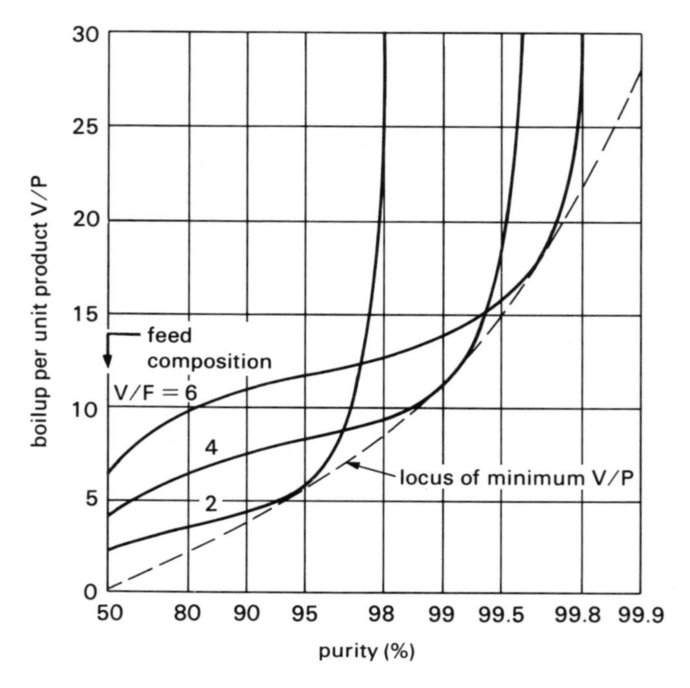

Figure 12-1. The dimensionless ratio V/P, vapor flow per unit of product flow, is an indicator of the energy required for continuous distillation. For various values of ratio of vapor flow to feed flow (V/F), the energy required is shown to increase rapidly as higher purities are demanded. *(After Shinskey, 1977; reproduced by permission of McGraw-Hill.)*

related to the improvement in consistency between batches. First, regulatory level automation allows the process engineer to specify operating points, and the engineer may choose points closer to heat removal and other limits than some or all of the plant's operators might normally choose. Second, batch times reported are usually arithmetic averages. If 90% of the batches of a particular product are within nominal times, but 10% are longer than they should have been, the number reported for all batches will be higher. The reason for some of the batches being longer than others may be the inconsistent level of attention that batches receive from operators. If a batch has reached an endpoint, or a point at which the achievement of an endpoint should be tested for by off-line analysis, and the operator's attention has been diverted to another part of the plant, then the batch may "sit" for a while. Steam and similar costs incurred during this time are counted as overprocessing costs. The excessive residence time of the batch in a process vessel—during which the investment in the vessel pays no return to its owners—must also be accounted for. In a plant with 10 parallel process trains, a reduction of residence time by 10%—not at all unheard of—represents the equivalent of the 11th process train. Generally, the same bottlenecks apply to this artificial 11th unit as would apply to a real one.

In a plant under manual control some proportion of operations will simply be in error—the wrong material charged to a batch, for example. The economic consequences can include the unrecoverable loss of raw materials, energy, and plant time, and disposal cost of the resultant unsalable product. Properly designed automation systems are very unlikely to make such "mistakes"; their failures tend to be manifest as I/O failures or large-scale "crashes," both of which typically cause the process to halt in its tracks. (This applies after sufficient testing. It is certainly possible for an undetected programming, data base, or wiring error to cause a process error.) Many computer-based automation systems have run for years without making an incorrect charge.

Sophisticated controls can also lead to increased production by helping to eliminate bottlenecks. They can help avoid errors in operator-performed procedures by putting requests from contending units in order of importance, by calling the operator's attention to procedures that need to be performed, and by presenting requests unambiguously without extraneous information; that is,

```
CHARGE 204.5 LB OF MATERIAL EX12 TO VESSEL 17
```

Probably their most important contribution, though, is simply allowing plant operators to direct their attention to unusual process conditions, problems, and so on, by automating routine sequencing duties.

EXECUTIVE/SUPERVISORY LEVELS

Compared to the effects of lower-level automation, the benefits of supervisory level automation are just beginning to be quantified. At this writing the industry that has probably done the most to automate supervisory level batch functions is the pulp and paper industry in many of its batch digester plants. The economic motivation is the reduction in savings in steam demand, an achievement that requires unit-to-unit coordination beyond the sequencing required for a particular batch. Large paper mills often generate their own electricity, but when process steam demand peaks electricity may have to be purchased. Eliminating the peaks by scheduling batches so that their individual peaks do not coincide minimizes the need to buy electricity. Concurrently, valleys in demand are also minimized so that boiler operation tends to be at higher average efficiency.

Additional economic gains are suggested by the power of computers available to be applied to the problem and by practices in other industries. The most economical way for a utility or region to generate electricity, for example, is by "equal incremental cost." If increasing the output of one generator increases the utility's cost by X¢/kWh and decreasing the output of another decreases the utility's cost by 1.1X¢/kWh, then it makes economic sense to do just that. Similarly, a multireactor or multiplant operation should be able to schedule production for minimum overall cost.

Record Keeping

From an investment point of view, the operator interfaces associated with the various control levels simply represent additional costs, with associated benefits intermingled with those of the levels themselves. Automated reporting, however, should be treated separately, because it represents a substantial, but optional, expenditure. For many food and pharmaceutical processes where extensive record keeping is mandated by regulatory authorities, automated record keeping can allow the processor to satisfy the regulations at lower cost and with greater accuracy.

Where the need for record keeping is determined from within an organization, there can still be a substantial economic benefit from prompt and credible reporting. It enables a full-scale production plant to function almost as a pilot plant, fine-tuning the process for optimum performance. The folklore of batch processing offers the story of a processor that, having long made the same material from the same plant, made a "perfect" batch. Samples of this batch sit on the desks of the plant's process engineers and managers. Unfortunately, they have never been able to duplicate it. They simply do not know what happened to the batch to cause it to achieve this perfection.

In more mundane circumstances, good record keeping can help a processor's quality control activities by allowing product problems to be traced back to processing differences—perhaps variations in timing of a procedure that has to be performed by an operator. Some users, in addition to printed records, use computer storage media to enable review of temperature histories and similar time-dependent data from past batches.

Some processors also perform statistical analyses on data representing numerous batches. This enables them to detect long-term trends, such as material variations or equipment degradation, possibly long before they would become evident as malfunctions or off-spec batches.

By tracking utilities consumed, a computer-based control system can help plant management accurately determine the cost to manufacture each product. This information can then be reflected in the price charged for each product and can contribute to justification studies on methods to reduce consumption.

SAVING LABOR

Reduction in labor per unit of production is one benefit commonly claimed for automation of all kinds. Our experience in batch automation, however, suggests that labor cost reduction is a secondary component in many justifications. Of course, it is possible to imagine a plant with no automatic control at all, one that requires an operator at every valve. We doubt that any plant has been built that way since James Watt invented the governor. The limited pneumatic automation designed into plants long before the common use of electronics probably accounts for most of the labor savings ever to be achieved by conventional control, although there does remain, especially in many plants handling solids, the opportunity for effective use of automated material handling systems.

In many processes labor is simply not a major contributor to the cost of the final product. Increasing the number of operators will not improve the accuracy with which they weigh out ingredients—an operation at which automatic control excels. And it would be grossly impractical to have a large enough operating staff to easily handle peak demands, perhaps occurring a few times a day—for most plants it is far more practical to accept the inefficiencies that result from peak demands, but once again they can be readily minimized by automatic control. At one plant, for example, an electromechanical control system in good condition and capable of running the entire batch process was replaced with a minicomputer system. The major economic justification was the minicomputer system's capability of performing an endpoint calculation repetitively. Under electromechanical control it was only possible to determine adequacy of processing *after* the batch, so the plant's tendency was to waste money by running the batch too long. The alternatives of asking the existing operators to perform the calculations at, say, 60-s intervals or of hiring additional staff for this purpose were, if considered at all, quickly dismissed. There was simply no practical labor/investment tradeoff; some form of automation was the only realistic solution. The plant management chose to replace the electromechanical controls with a system offering the significant, if not critical, advantages of consolidated display, automatic reporting, and capacity for additional software and I/O to extend its domain to other process areas.

SAVING CONSTRUCTION COSTS

Common practice during the 1970s was to wire field devices to a central marshalling area from which signals were distributed to (and from) control room instruments. With modern data communications technology incorporated in or appended to the control system, it is possible to substantially reduce the overall length of multiconductor cables in favor of much shorter lengths of twisted pair, coaxial, or fiber-optic cable.

CASE HISTORIES

The performance of process automation systems has a significant effect on a company's position in a competitive environment. It is not surprising that specific financial results are only occasionally released. Manufacturers of automation systems also sometimes release details on performance improvements, attributed perhaps to an industry but not to a company.

Mehta (1983) recounts three attributed and nonattributed case histories. Probably the most interesting is the experience of Chemische Werk Huls, as recorded by Amrehn (1967). This processor used one of the earliest computer-based batch control systems to operate a plant containing 80 polyvinyl chloride (PVC) reactors. Dr. Amrehn reported "an 11.5 percent increase in production, uniform product quality, no lost batch, improved plant safety, and lower manpower requirements." Results from undisclosed plants include eliminating a 5% out-of-spec rate for completed batches, eliminating release of monomer vapors to the atmosphere, and reducing batch time from 8 to 6 h, all at an alkyd/acrylic resin plant, and a 16% increase in production, reducing the rate of bad batches from 10% to near zero (all bad batches were caused by operator error), and reducing the adjustment frequency in downstream processing equipment from 12-16 adjustments per month to 1-2 per month, all at an epoxy resin plant.

Yamada et al. (1981) report on the benefits of computer-controlled start-up of a distillation column: A procedure that took between 8 and 24 h, and the attendance of skilled operators now requires only 6 h and no skilled operators.

Several authors have published results of batch digester automation: Lemay (1979) published Figure 12-2 showing a marked decrease in variation in Hypo number, an indicator of the degree of cooking. Norris (1980) stated that a 10-20%

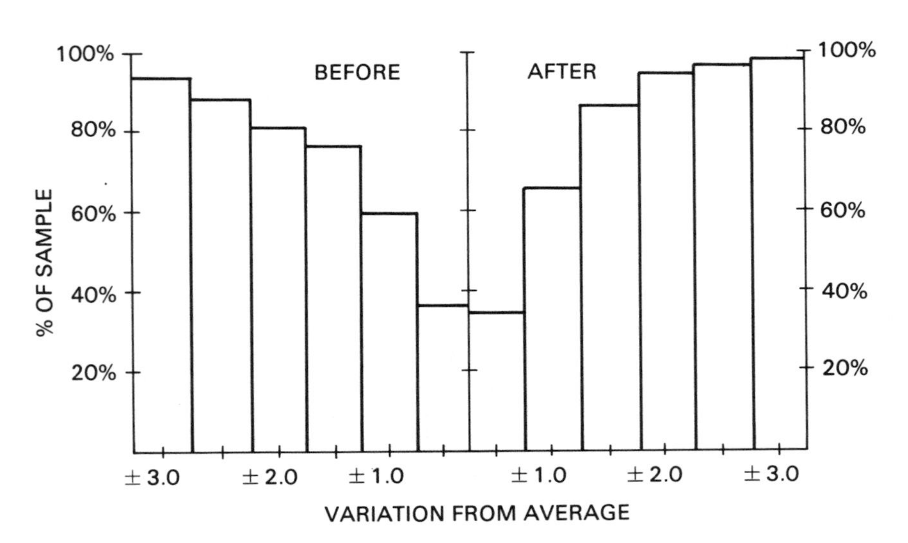

Figure 12-2. Hypo number distributions, before and after the installation of computer control. *(After Lemay, 1979; reproduced by permission of the Canadian Pulp & Paper Association)*

improvement is possible by combining digital control and on-line analysis. Canup (1981) found "batch digester production was increased by 4% and steam usage/ton of pulp was reduced by 6%. Hardwood and pine K-number standard deviations were decreased by 30% and 2%, respectively." The K-number is an indicator of further processing required; the processing (chlorination) must be adjusted for the highest K-number so that reducing the standard deviation reduces the use of chlorine.

Hochkamp (1980) reported on the savings achieved by computerizing the operation of a series of blenders, ranging from 20 hp to 200 hp, in a batch plastics production process: By only operating the blenders when they were needed, the plant reduced their energy consumption by more than half, corresponding to an overall plant reduction in energy consumption of 8%. Production capacity was increased by an unspecified amount, and overblending was eliminated.

Löhteemöki, Jutila, and Paasila (1979) have stated as "typical benefits" of automation a 10-36% increase in pulp production, a yield improvement of 0.2-1.5%, reduced steam energy use of 8-32%, and a 2-7% decrease in chemical usage, the latter two per unit of pulp.

Ellingsen (1985) predicted a savings of up to $278,000 per year for one reactor by optimizing reactor charge amounts, batch time, and other variables. Twenty percent represents energy savings; the remainder represents savings in materials. The study found several other scheduling strategies that would accommodate various specific plant conditions, like maintenance shutdowns, while maintaining annual savings of over $200,000 per year compared to the base case. These numbers do not include the potential benefits of reducing waste treatment and disposal costs by increasing yield.

THE INVESTMENT

Control systems for batch processes are asked to do more than equivalent systems for continuous processes—continuous controls generally have no equivalent to the extensive sequencing and off-sequence logic required for automatic batch control. This suggests first that batch controls are likely to be more expensive than continuous controls, and, second, that I/O count and loop count, the usual units of measure of system size, will be of only limited usefulness as predictors of batch control system cost. A single-product plant and a multiproduct, multisequence plant may each have similar process equipment lineups and therefore similar I/O and loop counts, but the latter will require a more expensive control system for automation at the sequence levels and above. Barona and Bacher (1983) estimate instruments and controls to account for 2-7% of total plant cost, with 5% the recommendation for preliminary estimates, including installation. Plants at the top of the range will tend to have multiproduct processes and extensive sequencing. A portion of the plant's engineering and start-up costs is also attributable to controls.

For a more detailed estimate of the size and cost of the control system, the design-by-levels approach once again is useful. Each level of control and its associated operator interface use some combination of dedicated and shared hardware and dedicated software. Hardware costs can indeed be estimated on the basis of I/O count and some "one-liner" equipment specifications—redundancy requirements, backup panels, shared versus dedicated displays, centralized versus distributed control, quantities of CRT displays, printers, and other peripherals, and so on. Field equipment is estimated the same way as is usual for continuous plants—by counting instruments and assigning a value for each type (direct-wired RTD, turbine flow transmitter, etc.). Installation costs are estimated by using one of the construction industry's guides, requiring estimates of lengths of conduit to be installed, and so on.

Engineering and programming are estimated by first applying a data base cost representing the work needed to program and/or design the system to acquire and tag inputs and distribute outputs. Then for each level engineering and programming charges are applied by using appropriate estimators. For interlocks a common basis is the quantity of logical functions—decisions, counters, timers, and so on—in turn based on counts of the items being protected. For the regulatory levels the familiar method of loop counts may be used. Sequence levels are characterized by the number of process steps, with an upward adjustment for extensive logic not directly driving the process functions. Recipe handling, if not provided by a vendor-supplied software package, and recipe-directed choices between processing methods are examples. Operator interfaces are estimated separately, with numbers of CRT display pages and report pages being common indexes of work.

Specific practices vary widely within these general guidelines. Some may increase the estimates for individual tasks to account for such activities as project team meetings, and others may prefer to leave these separate. Some may combine specification and design for estimating purposes, and others, especially if they plan to separately contract for the latter, prefer to separate them.

Engineering services are required for specification, design, implementation, and start-up. These services may be obtained, in various combinations, from plant staff, corporate staff, control equipment manufacturers, and third parties.

We believe that too little attention is given to the specification portion of the engineering services, because of the common belief that those details remaining unspecified will become clear as design and implementation progress. Our experience is that design and implementation proceed much more smoothly if based on a clear and comprehensive specification. We recommend that the engineering hours be divided approximately into thirds: specification, design, and implementation. Additional hours for programming, during design and implementation especially, may be required for special operator interface, reporting, and similar functions.

Purchasing Software

The prospective purchaser is often faced with a "make-or-buy" decision regarding the software required to establish an environment for batch application software—contact scanning, for instance. Manufacturers offering standard batch software generally try to build it to meet the needs of a variety of processes and industries, so for many users the investment in this software is an excellent one. Where the process deviates significantly from those that the manufacturer assumed, or simply requires few of the additional features the software offers, it may be better to forego the batch software and build application software as needed. A middle ground, available with some batch packages, is to use only those parts deemed useful and make whatever small changes are required.

Phasing of Implementation

Capital budgets may permit only the gradual implementation of a modern batch control system. The design-by-levels technique favors one sequence of implementation—bottom-up. Although it is not impossible to provide interlock and regulatory controls along with automated production scheduling, but without unit-level sequencing, the most obvious solution is to purchase hardware and purchase and/or write software only for those levels to be implemented at a particular time. With the discipline imposed by design by levels, it is reasonable to try to anticipate the eventual system and purchase all the hardware required at once. There are advantages that will help justify the additional expense: The purchaser will be assured of having all of the elements mutually compatible (provided they were all purchased from one source, or that the manufacturers have made a clear commitment to the appropriate compatibilities), the entire system will be available for subsequent application development, and decisions to proceed with subsequent development will not be delayed pending equipment installation.

Enclosures

Batch plants designed in the past for simple, local control were often built without control rooms. The operators were required by the limitations of the control equipment to remain in close proximity to the process vessels. When control is exercised by a system with centralized operator interface (whether control itself is distributed or centralized), it is common to provide a reasonably comfortable, officelike area for the operators to work in. In any case minicomputers and mainframe computers are frequently specified only for use in such environments.

For applications over a widespread area, in which local control equipment is not specified to meet anticipated environments, or the environments can be severe enough to interfere with maintenance and emergency operation, satellite instrument houses offer the needed protection at a reasonable cost. They house marshalling cabinets, electronic controllers, and other equipment, protected by heating, air conditioning, smoke detection, and other means, as required. They are generally manufactured by a company specializing in them, then shipped to the control system vendor, where equipment is installed and tested. Once in the field, the satellite instrument house is treated as a large panel, with field connections being made only at designated terminal areas.

CHAPTER

13

THE HARDWARE AND SOFTWARE ENVIRONMENT

This chapter first discusses the various hardware configurations, followed by the possible architectures for batch control. It then considers the aspects of hardware reliability and the configurations that may enhance or degrade it. The chapter concludes with a discussion on software reliability.

HARDWARE SYSTEM ARCHITECTURE

With the availability of various types and sizes of controllers for batch systems and the means of communication between them, the hardware configuration options for controlling a batch system are virtually unlimited. This section describes the more common types of architecture used today and discusses their essential advantages and drawbacks. This discussion is largely on the hardware aspects and is applicable not only to batch processes but also to process control systems in general. The next section deals with the functional aspects of architecture for batch control systems.

Historical Note

The first manually controlled batch processes had essentially distributed control. Indicators for level and temperature and the control of inlet and outlet valves, pumps, and motors were largely local. Operators for these processes were either stationed locally near each process vessel or had to move from one process vessel to another to effect control. With the advent of pneumatic and electronic signal transmission systems, centralized control stations with indicators, recorders, and control devices mounted on large panels came into vogue. Each indicator or controller in these panels usually indicates or controls one or a very limited number of process variables, thus the need for a large number of indicators and controllers. In this type of environment the batch control is essentially manual or automated in a very limited way. With the introduction of digital computers, programmable controllers (PCs), and other programmable devices, where one piece of hardware controls multiple inputs and outputs, the batch process started to get centrally automated. The human-machine interface for these devices is multifunction panels and/or CRT-type display units with keyboards (Fig. 13-1). These interfaces are far more compact than the large control panels and thus require much smaller control rooms and are more economical. At the beginning, however, process engineers and plant management had little confidence in the long-term reliability of digital electronics, so the large instrument panels were usually retained as backups even when the control functions were normally entrusted to programmable devices (Fig. 13-2).

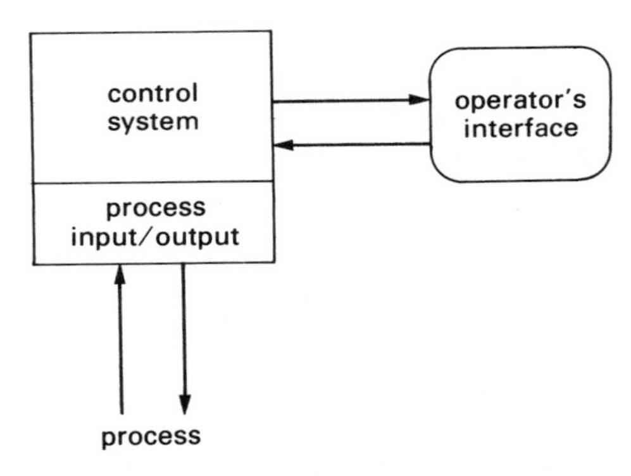

Figure 13-1. Centralized control system configuration.

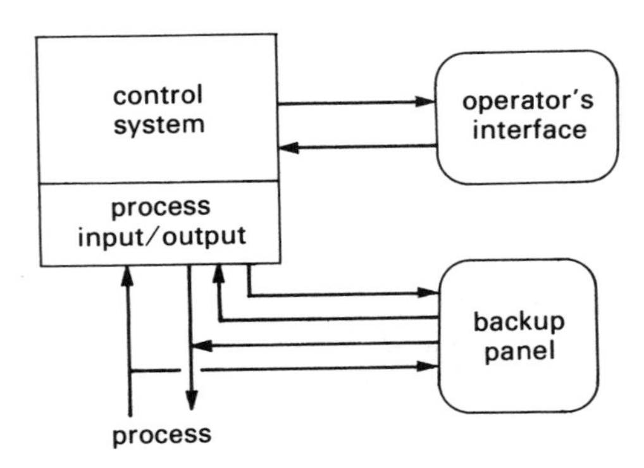

Figure 13-2. Centralized control system with backup panel.

In addition to being expensive to build and maintain, backup panels pose other serious problems. A CRT-based human-machine interface is very dissimilar in operation to a conventional control panel. An operator using a CRT-based interface most of the time loses the proficiency of using a conventional control panel. So on those infrequent occasions when a programmable control system fails, an operator may have considerable difficulty in controlling the process via a conventional control panel. Also, if the backup panels are not regularly tested, there is no guarantee that they are in working order. These considerations have made conventional backup panels less popular in recent years.

Dual Control System

A common approach of recent years is the use of two essentially similar control systems, one backing up the other (Fig. 13-3). In this type of configuration the two systems start with the same control logic, and the inputs/outputs to the plant are duplicated. The two devices communicate either via the process I/O ports or through independent high-speed channels. Typically, the primary controller scans the process inputs and controls the outputs while the backup controller waits for the failure of the primary controller. While waiting, the backup controller is updated at regular intervals for all changes in the primary. This type of arrangement, while providing increasing reliability, is not entirely foolproof. Changes

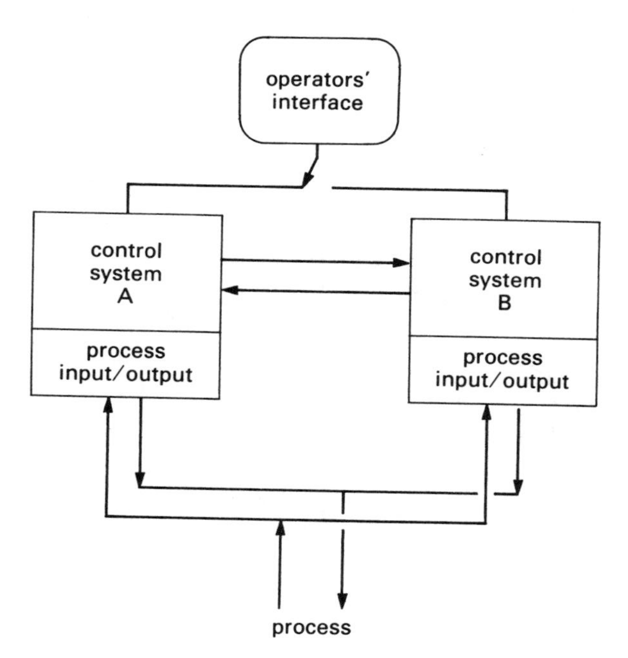

Figure 13-3. Dual control system for backup.

made by the primary controller between the last update and its failure are lost, and the result can be serious in certain circumstances. For example, if a pump has just been started and the desired state is changed, with loss of this information, the backup computer will find the state of the pump in discrepancy and may generate an alarm. If the input has been designated as critical, the discrepancy may even shut down the unit. There are further discussions on reliability of this type of system in the section *Maximizing Reliability.*

Remote Multiplexing

So far we have considered control devices like computers, PCs, and so forth, where each input and output is directly connected by a pair of wires to the I/O bay of the controllers. The increased expense of wiring and cabling can become a major part of the total system cost. In recent years remote multiplexing has become increasingly common because it can reduce considerably the cost of cabling.

In a typical application, remote multiplexers are distributed throughout a plant, and the plant's inputs and outputs are connected by pairs of wires to the nearest multiplexer. Usually, the multiplexers are connected to the controller(s) by dedicated copper cables, but they may also be connected through telephone lines, wireless or microwave links, or optical cables. These multiplexers are usually passive devices; that is, they do not initiate the communication of messages but only respond to the request from active devices, like computers, that receive or supply process data (Fig. 13-4). The transmitted or received data must be error

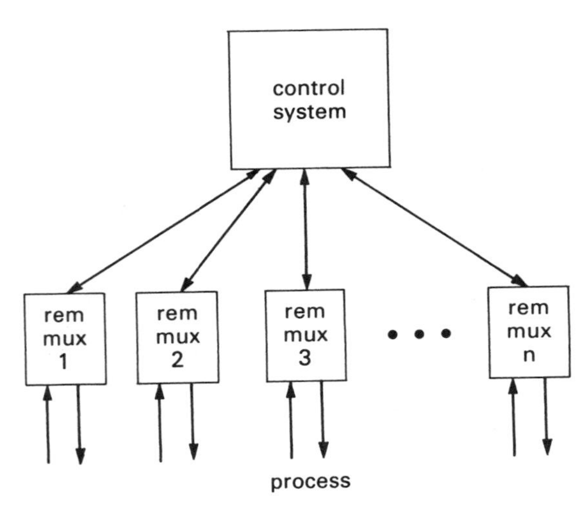

Figure 13-4. Control system with remote multiplexer units.

free. Various types of error-checking methods, from simple parity checking to more complex redundancy checking are available on standard vendor equipment. When a transmission error is detected, the message is usually retransmitted. When an error-free message is not received after a fixed number of retries, the link is normally disabled with an alarm message to the operators.

Networks

With the increase in the number of functions performed by a single batch controller, the size of a controller in a traditional configuration (Fig. 13-1) had to grow. The designing of a single large controller with enough reliability to economically control a large batch process is an enormous task; particularly difficult are the procedures for handling the process on control system failure. These problems can largely be simplified if the control is distributed among multiple controllers, where the failure of one may not be a great problem in terms of the total process. Then the strategy of manual or automated backups for a subset of the total system may be much simpler. These considerations, combined with the increased sophistication in digital transmission technology, have given rise to the idea of networks, where a number of controllers and process I/O interfaces (multiplexers) may be linked together to form a distributed yet unified control system.

Some control system vendors now supply controllers and remote I/O interfaces along with data communication channels, which makes network configuration easy. Interfaces to large data processing computers and control devices made by other manufacturers are also usually available. Thus, unique networks appropriate for each particular application can readily be designed.

The basic network types are (Fig. 13-5) star, multiconnected, ring (or loop), and multidrop. In these illustrations a box represents a discrete piece of hardware that could be a process controller, a process I/O device, a human-machine interface, a large data processing computer, or a communication port to which other devices may be connected. These boxes are also called *nodes*. A line between these nodes is called a *link*, which is a communications channel, such as coaxial cable, telephone line, microwave link, and so on.

The star network (Fig. 13-5*a*) consists of a central node to which is connected all other nodes by individual links. The central node may be a computer and the other nodes may be satellite controllers, I/O devices and human-machine interfaces. In this configuration the central computer may communicate with other devices slowly while maintaining a high overall data transmission rate. There is also good channel security, because no third processor directly affects the communication between the central computer and another device. However, the central computer provides the communications between other nodes, if any, and the failure of the central computer will cause the suspension of all communications between them. Because of individual links, the wiring cost for large distances may be high. Star

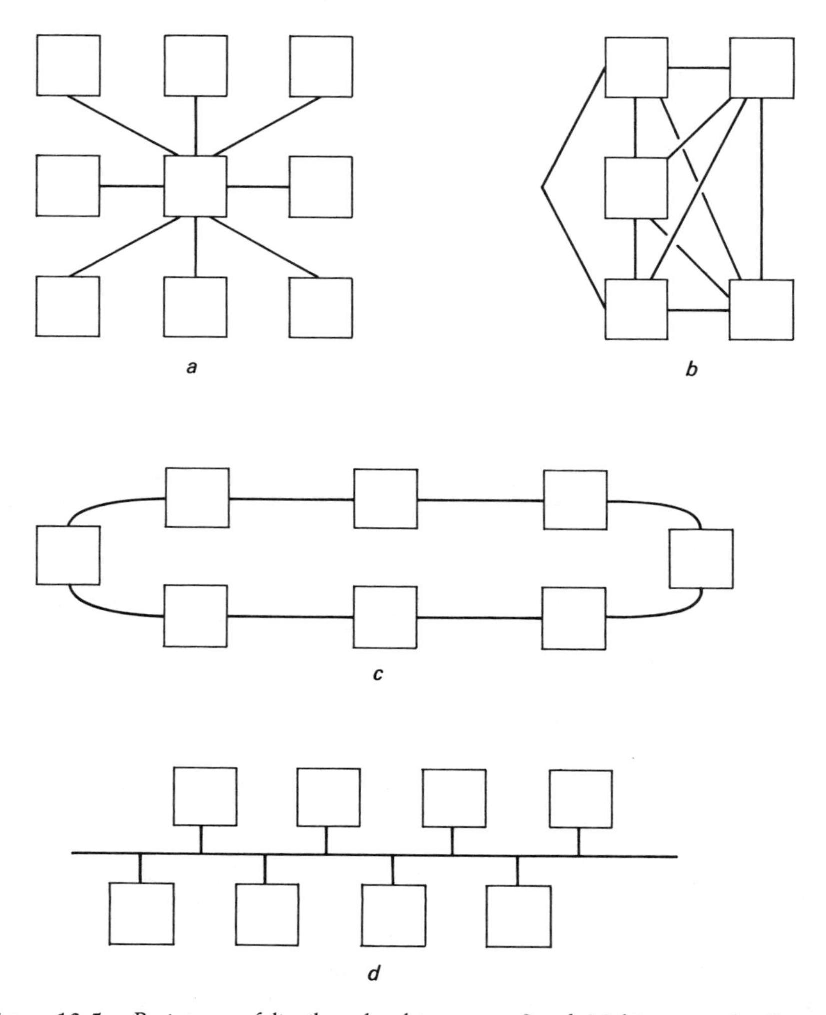

Figure 13-5. Basic types of distributed architecture: *a*. Star. *b*. Multiconnected. *c*. Ring. *d*. Multidrop.

configuration may be appropriate for a large batch control system where different areas in the process are largely independent and little communication is required between these areas. The central computer may be for collecting historical data and generating management information, so its failure may have little effect on the process control in the short run. Star configurations are commonly implemented with nonproprietary communication standards, so it is comparatively simple to use equipment from many vendors.

In a multiconnected network (Fig. 13-5*b*) each node that has to communicate with another is connected by an independent link. This type of arrangement allows the flexibility of using links of different types and speed within one network. This also has high channel security, as in the star configuration, without burdening a central computer. If one link fails, other paths might be found for communication by a third node. However, this architecture is expensive and generally not used in process control environments; instead it is commonly used in city-to-city telecommunication applications.

In a ring configuration (Fig. 13-5*c*) each node is connected to two other (usually adjacent) nodes. A message received by a node is relayed to the next node, and so on, until received by the source node. Any node can normally initiate a message, and since all messages go through all the nodes, communication between any two nodes is possible. Because each node acts as a repeater, long distances can be covered, but because each node has to receive and transmit a message, the effective transmission rate is slow. The major disadvantage with this configuration is in the failure of communication on failure of a node. Failure of two nodes will isolate the system into two parts. This type of architecture is suited for long-distance communications because the signals are amplified and retransmitted at each node.

In a multidrop configuration (Fig. 13-5*d*) a single link is connected to all the nodes. This is also called a *data bus* or *data highway* and is widely used in process control environments. The multidrop configuration allows communication from one node to another, but generally only two nodes can communicate at one time. However, with the increased communication speed of dedicated links (1 Mbps or more), that is not usually a serious problem. This architecture is also usually the most economical.

Another type of configuration is a "cluster," where devices are connected together by a communication port. It is a variation of the star configuration (Fig. 13-5*a*), where the central node, instead of being a computer, is a communication port. The communication port, unlike a computer, does not process any information but acts as a coordinator for the communication between the satellites. Each cluster usually contains a limited number of nodes, and multiple clusters may be connected in a multidrop configuration (Fig. 13-6). The maximum number of

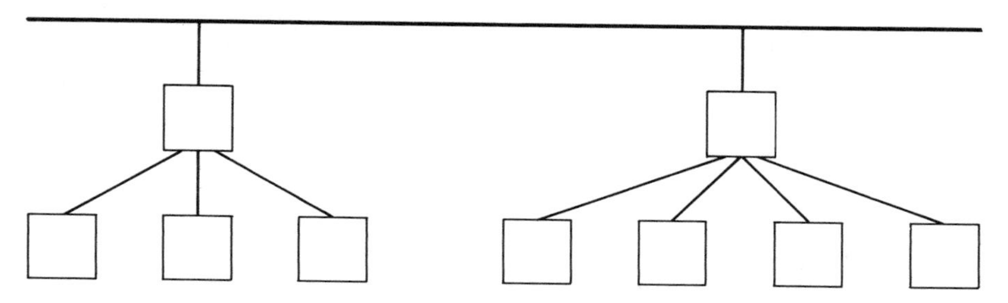

Figure 13-6. Multidrop configuration with clusters.

devices and their distance from a communication port, within a cluster, is usually more limited than that for a multidrop link. Thus, devices physically close to one another are usually put in the same cluster, and clusters separated from each other are connected by a multidrop link.

If a link fails in a multidrop configuration, then all communications between the nodes may stop. This can be avoided by having a backup link (Fig. 13-7). A backup link is generally identical to the primary link and is connected to the alternate ports of the nodes. Normally a backup link does not handle any data; it waits for the failure of the primary, at which time it automatically takes over. Sometimes, however, a system is configured in such a way that both links handle data under normal conditions. This reduces the load on a single link. Thus, when a link fails, the other takes over the entire data communication, possibly in a degraded mode. Redundant links may be used for a multidrop configuration with clusters (Fig. 13-8) and other types of configurations with similar advantages. The flexibility of network design allows other network types and variations to be configured.

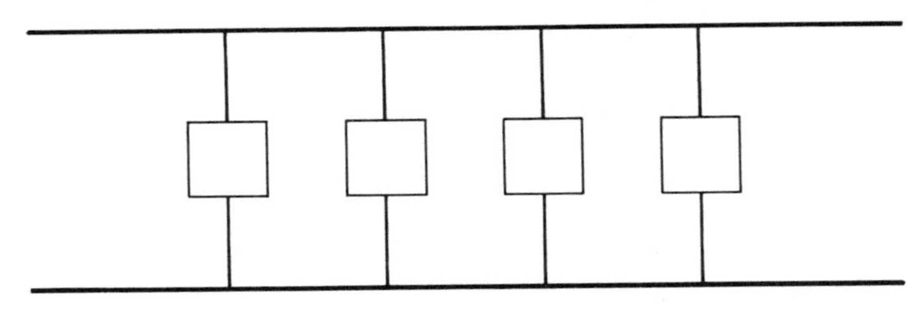

Figure 13-7. Multidrop configuration with redundant links.

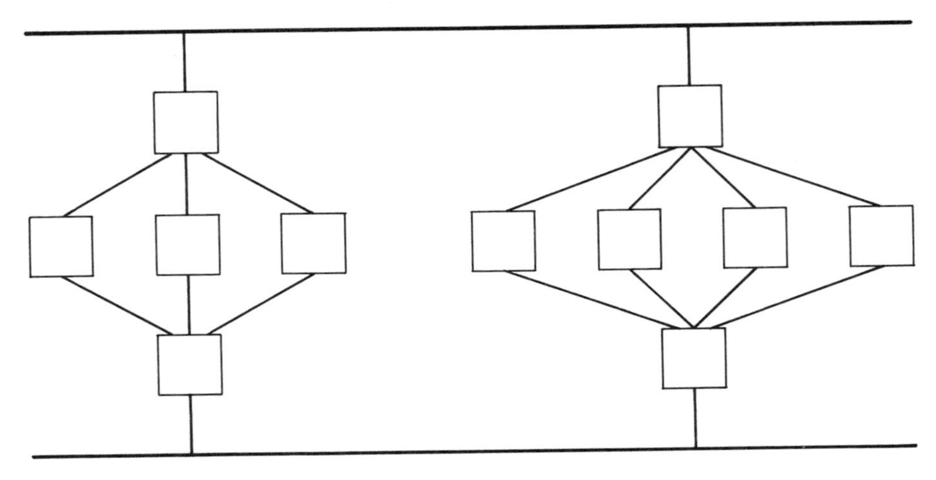

Figure 13-8. Multidrop cluster configuration with redundant links.

As stated earlier, a node in a process control network can be one of several devices—for example: host computers, process control computers, programmable controllers, local analog and sequential control devices, centralized and distributed display devices, remote multiplexer units for interfacing with field inputs and outputs, and communication ports. All of these devices may broadly be classified into two types: active and passive. *Active* devices are those that can initiate a data communication with another device, either active or passive. Host computers, process control computers, controllers, and interactive display devices are usually active devices. *Passive* devices are those that do not initiate any data communication with other devices but only respond to requests from the active devices. However, one thing is common to all the devices in a network—they all need interfaces to communicate within the network. For a network with a common link, such as a multidrop type, there has to be a common protocol. The interfaces for different types of devices must all be able to input and output messages in the same language. However, the protocol does vary between networks of different vendors of control systems. Thus, though it is usually easy to configure a network of devices supplied by a single vendor, custom interfaces generally need to be designed when devices supplied by different vendors are configured in a single network.

So far we have considered the redundant links in a network. Redundant devices (nodes) may be linked to increase the reliability of a control system. The reliability considerations for redundant devices are discussed in detail in the section on maximizing reliability.

Hierarchy in a Network

In a simple multidrop network (Fig. 13-5*d*) all the nodes are connected as peers. However, functionally they need not be. One of these nodes may be a host computer that may only be required to communicate with a few process control computers and PCs. The process control computers may, in turn, be communicating with remote multiplexers for process inputs and outputs. Thus, several levels of functional hierarchy in a network system can be set up easily.

A network lets dissimilar pieces of hardware be configured to work in a unified manner for controlling a process. Redundancy and backup provisions can be readily incorporated in a network system. The network allows these devices to function in multiple levels of hierarchy, which may be altered, when required, without changes to the hardware configuration. With intelligent I/O devices connected to a network, the operator-process communication functions may be centralized to any required extent without altering the control architecture. Thus the control system architecture has gone full circle from the original decentralized manual control system to a fully centralized control system to distributed network control.

ARCHITECTURES FOR BATCH CONTROL

Chapter 7 made the case for dividing the batch control system horizontally by function. There is also a vertical division to be made by process unit or area. The controls for a small batch process area, depicted in Figure 13-9, can be represented as shown in Figure 13-10. At the lower levels, control for each process unit is distinct. At level 5 the possibility of coordinating control loops across process unit boundaries is shown by the dotted line. The basic sequential control level, level 6, is characterized by the need for each reactor to send service requests to, and receive acknowledgments from, each shared unit. The ability of the finishing tank to receive material from the charging unit is also shown.

Batch control, level 7, exists across all units but communicates only with the reactors since they have the responsibility to request shared unit service when they

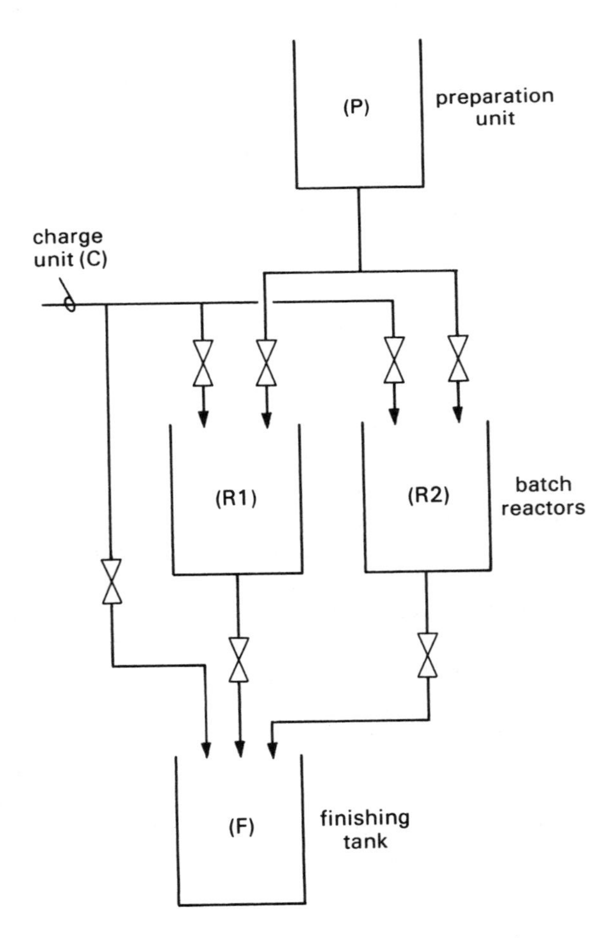

Figure 13-9. A small batch process plant, referred to in subsequent examples.

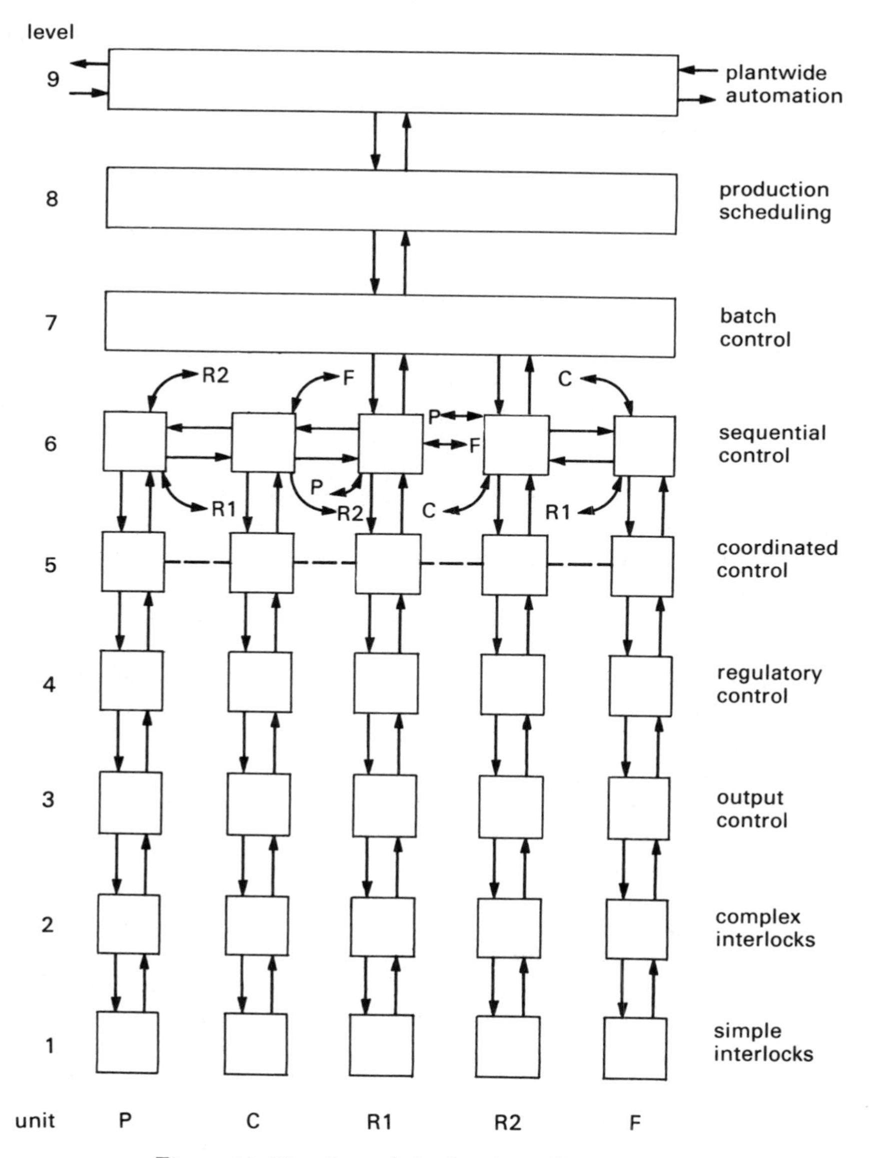

Figure 13-10. Controls for the plant of Figure 13-9.

are running a batch. Alternatively, the first unit to receive a request to start a batch could be the preparation unit. In this case, though, there is a possibility that the preparation unit could prepare material for R1 and find R1 unable to receive it. If the batch cannot be run in R2 instead, all production ceases. The method in use here requires reactor logic to be operational before the preparation unit receives a request, so it minimizes this possibility. Production scheduling, level 8, is associated with all of the units as a group. Although not used here, production scheduling might have interfaces to transportation scheduling and raw material inventory functions, for example. The plantwide automation function is assumed to have interfaces to quality control and marketing information, for example, each of which might affect product recipes.

A wide variety of hardware and software products are available for batch control. Some of them can do all of the jobs dictated by Figure 13-10; many do not. There are safety considerations, as discussed in Chapter 7, that suggest it is not advisable to try to put all of these functions into one "box."

As a starting point for analyzing the various possible methods for distributing responsibility, Figure 13-10 suggests that some of these methods may require cutting numerous communication paths, whereas others require comparatively few. Those requiring few paths are more "natural" because they more closely conform to the organization of the process and its controls. They can generally require less engineering work and are more reliable than their more complex alternatives.

The penalty for using an architecture other than a natural one depends on the equipment selected—especially its support for horizontal communication—and the process, which determines how much of these horizontal communications are required.

Basing our analysis upon communications alone, we would conclude that the largest possible subsystems would generally be the easiest to apply. However, due to the need for independent safety functions to minimize common failures, we are required to keep level 1 controls separate from each other. At levels 2, 3, and 4, combining controls for multiple units is permissible, but it retains some penalty, process dependent, because of the possibility of failures affecting more than one unit. At level 5, coordinated control, processes vary widely in their need for communication across process unit boundaries. With increasing interunit linkages, there is incentive to centralize coordinated control. Similarly, the sequential control of level 6 may be better implemented in a centralized or distributed setting, depending, as always, on the needs of the process and the equipment's ability to make horizontal communication easier. At levels 7 and above, the controls are associated with whole plants or plant sections, rather than individual units. There may be specific exceptions, but it is generally not practical to functionally distribute these activities among multiple computers or other items of equipment.

The foregoing discussion established that there is a penalty for any batch control architecture, and this penalty is high at the extremes of full distribution and full centralization. If we use a measure of distribution versus centralization as the horizontal axis and the penalty function as the vertical axis, our evaluation of practical alternatives is likely to yield a relationship like that of Figure 13-11. Procedures used to measure both the degree of centralization and determine the penalty value are somewhat subjective, depending on, for example, one's perception of the difficulty of writing software. Since these procedures will rely strongly on context, we will not suggest any but leave their formulation to the user organization. In the next section we do discuss some of the most popular equipment configurations.

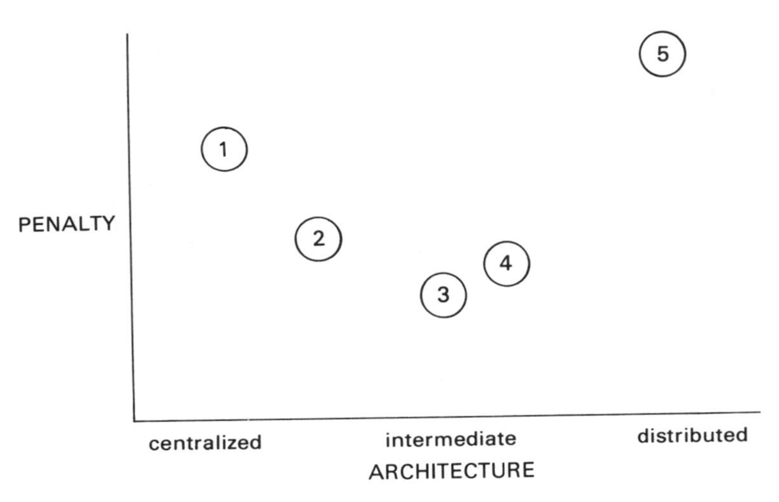

Figure 13-11. Six possible control system architectures, ranging from fully centralized (1) to fully distributed (6), are plotted with their corresponding penalties. Architecture 3 is shown to impose the minimum penalty.

Selected Equipment Configurations

Configuration 1 (Fig. 13-12) uses individual interlocks only at the lowest level; the rest of the automation functions are within one piece of equipment. Programmable controllers (PCs) with regulatory control capability are available and can be considered as candidates for levels 2 and up, but the reliance of most of them upon ladder logic and its derivatives precludes them from all but the simplest activities at levels 6 and above. In addition, PCs tend to offer less extensive display and reporting capabilities than minicomputers applied to process control. Also personal computers, bolstered by real-time software packages and analog and contact I/O capabilities, have emerged as contenders in the automation market. At least at this writing, personal computers are not yet able to control a batch plant of the size shown here, while driving multiple CRT displays independently, printing out reports, and controlling interfaces to laboratory instruments. Some are available in ruggedized packaging; most are intended only for office environments.

The higher-level controller performs several tasks and needs a variety of languages: a hardware-emulation language for regulatory control, a ladder or Boolean capability for interlocks, an I/O-oriented batch language for sequences, and a general-purpose programming capability for the higher-level functions. A batch-oriented reporting package and access to low-level commands for I/O are also useful.

In most cases, then, the equipment of choice for levels 2 and above in this configuration will likely be a process control minicomputer. (Process control mainframe computers are available for larger plants.) Communication with level 1 and the process units themselves will be by discrete I/O.

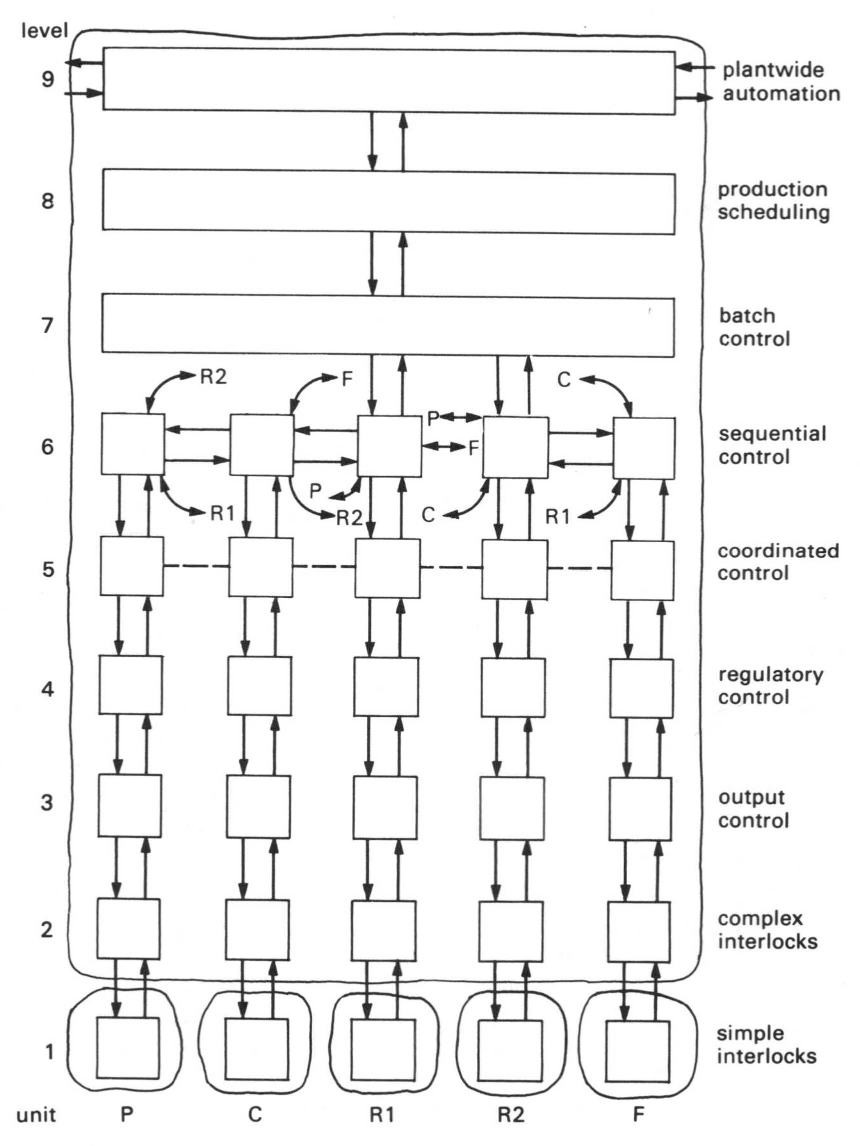

Figure 13-12. Configuration 1.

This configuration is, where applicable, probably the easiest to apply. Communication problems between the levels are minimized, since any signal or other value can be readily made available to any function that needs it. This capability can make system start-up easier by allowing the engineer to access these values from a CRT or keyboard/printer regardless of their level or function. Simple test programs can provide automatic recording or data display not required during normal operation but useful during start-up. Small changes necessary during testing and start-up can be made within the confines of one box, except where they involve I/O.

The major limitation of this configuration is its absolute dependence on the condition of the computer's hardware. Proper design can help isolate software failures, but a serious hardware failure can render the plant incapable of making product, and even leave it without level 2 interlocks. This configuration, then, is not generally used to control processes such as exothermic reactions, where an inopportune failure can deprive a reactor of adequate coolant flow for removing the heat evolved and possibly cause thermal runaway.

The economic penalty for exposing all of the plant's production control to a single point of failure is limited in this plant by the interdependencies among the process units themselves. No matter how distributed the risk, failure to control P, C, or F will soon stop all production. It is true that with distribution of reactor control, the failure of one control would not necessarily affect another, so limited production might still be maintained. This configuration is usually used to control processes that are inherently reasonably safe and in which process units are tightly coordinated.

Configuration 2 (Fig. 13-13) substitutes a PC for the process control computer's interlock logic. This off-loads the computer from some high-speed interlocking and helps make it practical to use computer equipment of less capacity. Placing level 2 interlocks in the PC should, all else being equal, result in a safer system, since all interlocks are now isolated from computer failure. The disadvantages, compared to configuration 1, arise from the increasing variety of items of equipment and the need to communicate between them. Training, maintenance, and application costs are all likely to be higher because of the need to deal with two types of digitally based equipment. Although the chances for a dangerous failure may have been reduced, the chances of *any* failure have probably been increased, simply because there is more equipment to fail—two sets of digital memory boards instead of one, and so on. In practice, this configuration is more likely to be found on systems larger than that of our example.

Configuration 3 (Fig. 13-14) is one type of distributed control system. Complex interlocks and regulatory controls are grouped within one piece of equipment per unit. With an operator interface independent of the computer—mandatory if the full value of the configuration is to be realized—a plant controlled this way can usually be operated satisfactorily without the computer. With proper planning (which is recommended in any case!) controls need be installed initially only for level 5 and down, so buying the computer can be postponed.

It is not necessary to use the same equipment for each unit's control; this configuration accommodates specialized controllers for different units. For the reactor units, depending on the number of loops, control might be implemented with single-loop controllers with either individual (panelboard-mounted) or shared (CRT-based) operator stations, or with distributed multiloop controllers, usually with shared stations. The preparation unit P might be controlled with a specific-purpose batching controller designed to charge it with a specific sequence of

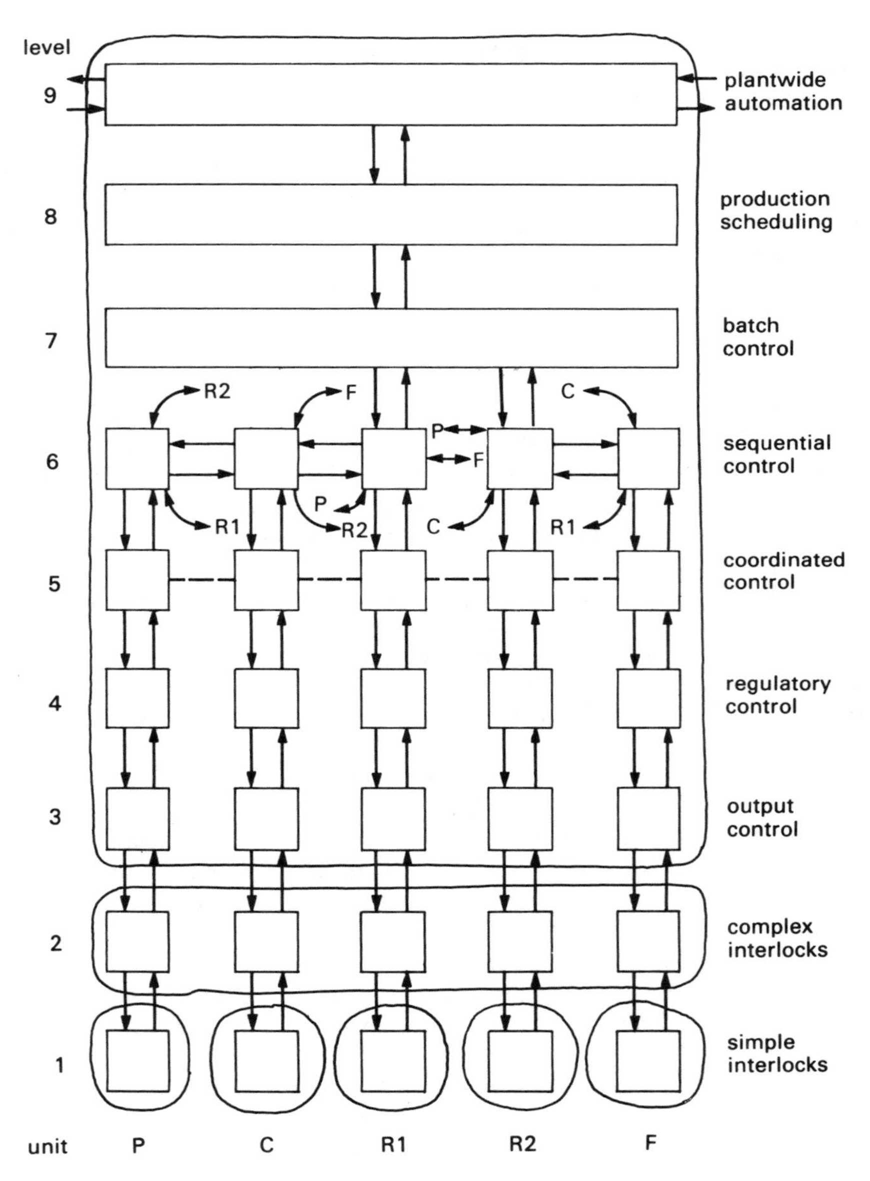

Figure 13-13. Configuration 2.

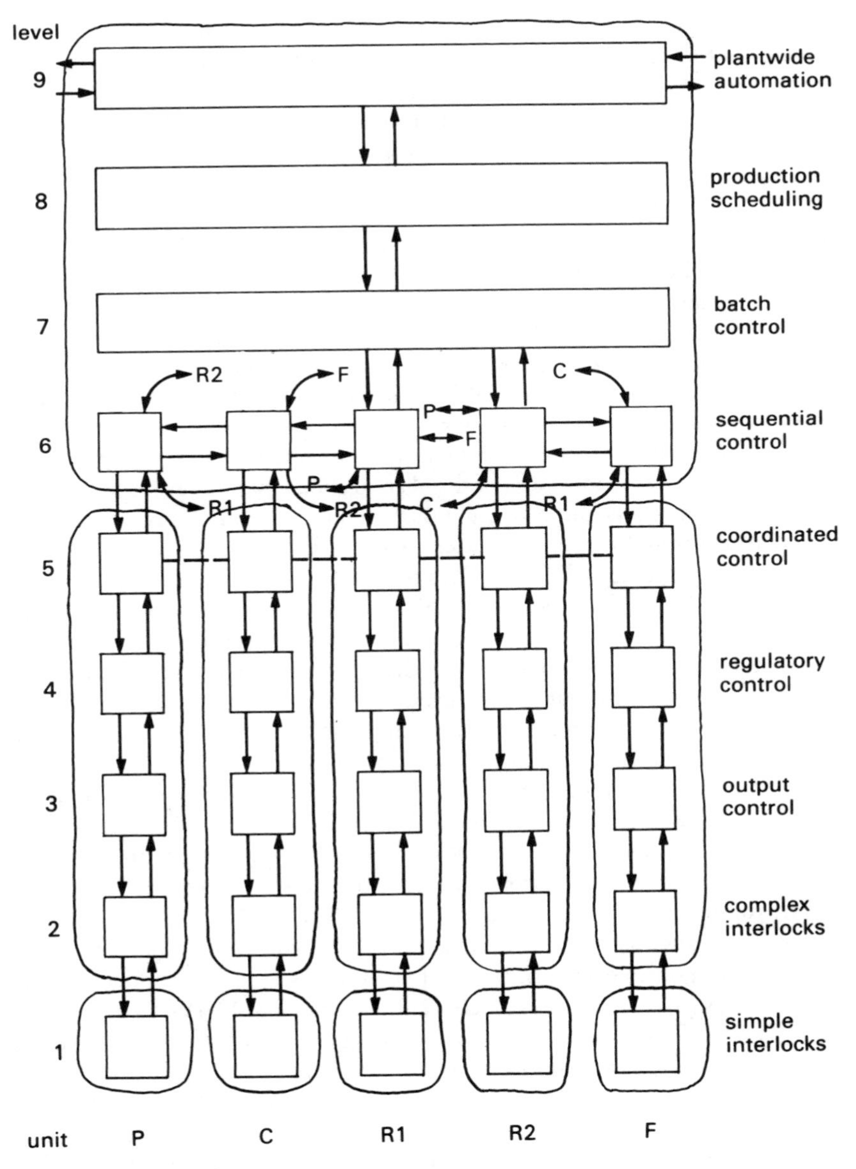

Figure 13-14. Configuration 3.

specified amounts of raw materials from multiple raw material sources. The liquid charging unit C requires a totalizer for the amount charged, which might be a specific-purpose device, or be programmed into a distributed controller or PC or, more recently possible, into a configurable, microprocessor-based single-loop controller. The finishing tank F might also be controlled by these newer single-loop controllers or by a PC. Obviously, units to be controlled by the same type of controller may be combined into fewer pieces of equipment, depending on controller capacity limitations, although this compromises the unit-to-unit independence that can be one of the major attractions of a distributed architecture.

The economic value of this unit-to-unit independence, though, relies strongly on the process to be controlled. In our example, as stated, failure to control P, C, or F will soon stop all plant production regardless of the control system used.

The electronic equipment available for a configuration 3 system is, relative to the first two systems, less likely to be available from one vendor. The user may have the responsibility to make this equipment work together, although many vendors will agree to take over some or all of this responsibility by purchasing appropriate equipment from others and providing the system engineering services necessary to build an integrated system. The extent of the system engineering should not be minimized. It is not enough for vendors to note in their specification sheets that their equipment has ports for RS-232C (an Electronic Industries Association standard insuring electrical and mechanical compatibility) and uses ASCII (American Standard Code for Information Interchange) coding. These standards do not, for example, specify the handshake protocol used to establish communication between two items of computer-based equipment.

Industry is proceeding to resolve this problem by promulgating more comprehensive communication standards, such as MAP (manufacturing automation protocol), advocated by General Motors among others. In the meantime, potential users of multivendor systems should be aware that there is a history of problems in interfacing digitally based control equipment from different sources.

Configuration 3 is often seen in a modified form, in which for some units, some, but not all, sequential control resides within the computer. The logic in the computer is responsible for synchronization with other units. The effect that this can have is best shown as a series of concentric circles (Fig. 13-15). We will assume that the preparation unit controller is a free standing batching controller with serial computer interface and that charge unit control is from a PC. Both of these are microcomputers (microprocessors with memory and other devices)—the innermost envelope. What makes them a batching controller and PC, respectively, is the software packaged with each—the personality envelope. With different personalities, there is little system-level compatibility between the controllers. The interface envelopes take instructions from the remainder of the control system and format them for their respective controllers. The synchronization envelope controls synchronization with the remainder of the system.

Proceeding outward from the microcomputer and personality regions, the two systems present interfaces that become progressively less idiosyncratic and more standardized. The benefit is that this allows the remainder of the system to be

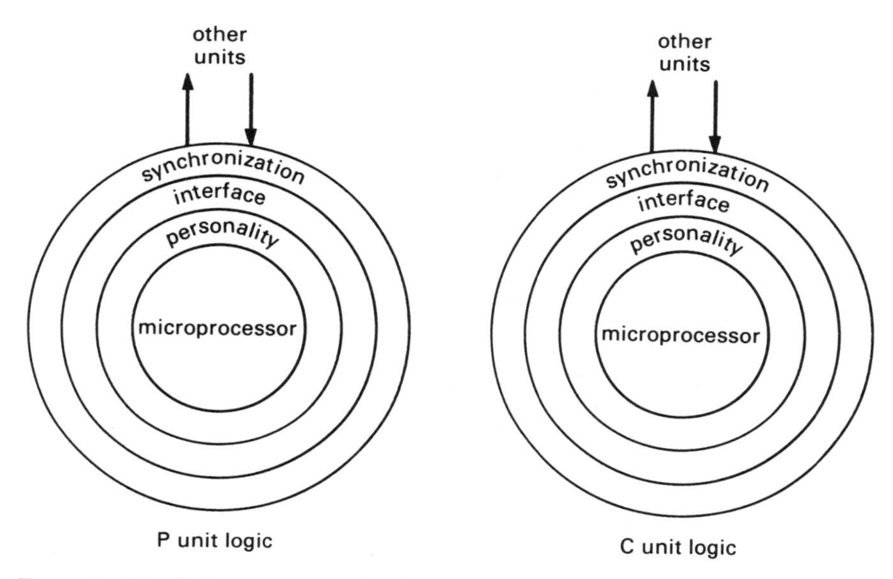

Figure 13-15. Software structures for different devices may be seen as concentric "envelopes."

designed with very little consideration for the communication requirements imposed by, or the sequences and processes used by, each particular item of equipment.

There is a general relationship between this envelope model and the method of sequence structuring in Chapter 9. The entire charging operation typically constitutes a phase. The synchronization and interface envelopes correspond to steps, and statements are executed by the microcomputer according to its personality.

Configuration 4 (Fig. 13-16) shows all sequential control distributed among controllers dedicated to each process unit. Synchronization between units is implemented by some combination of hardwiring between controllers' I/O modules, peer-to-peer communication over a network, and peer-to-master-to-peer communication involving the computer. The latter adds communication delays and a point of failure and is rarely used. The other two are more common, with hardware peer-to-peer practical for simple situations (Rosenof, 1980) and network peer-to-peer used for more complicated situations such as recipe transfer.

The potential purchaser of a networked peer-to-peer system is advised to check carefully the communications capability offered in order to determine how much engineering and application programming, or even custom programming (modifications to software supplied as a standard part of the product), will be needed to meet a process's synchronization requirements.

Configurations presented so far have been symmetrical: Each process unit's control was structured the same way. In Configuration 5 (Fig. 13-17) simple interlocks of level 1 operate independently of the other levels. A batching controller is used for the preparation unit, for levels 2-5 and part of level 6 sequential control, as described above. A PC provides complex interlocks (level 2) for the remaining

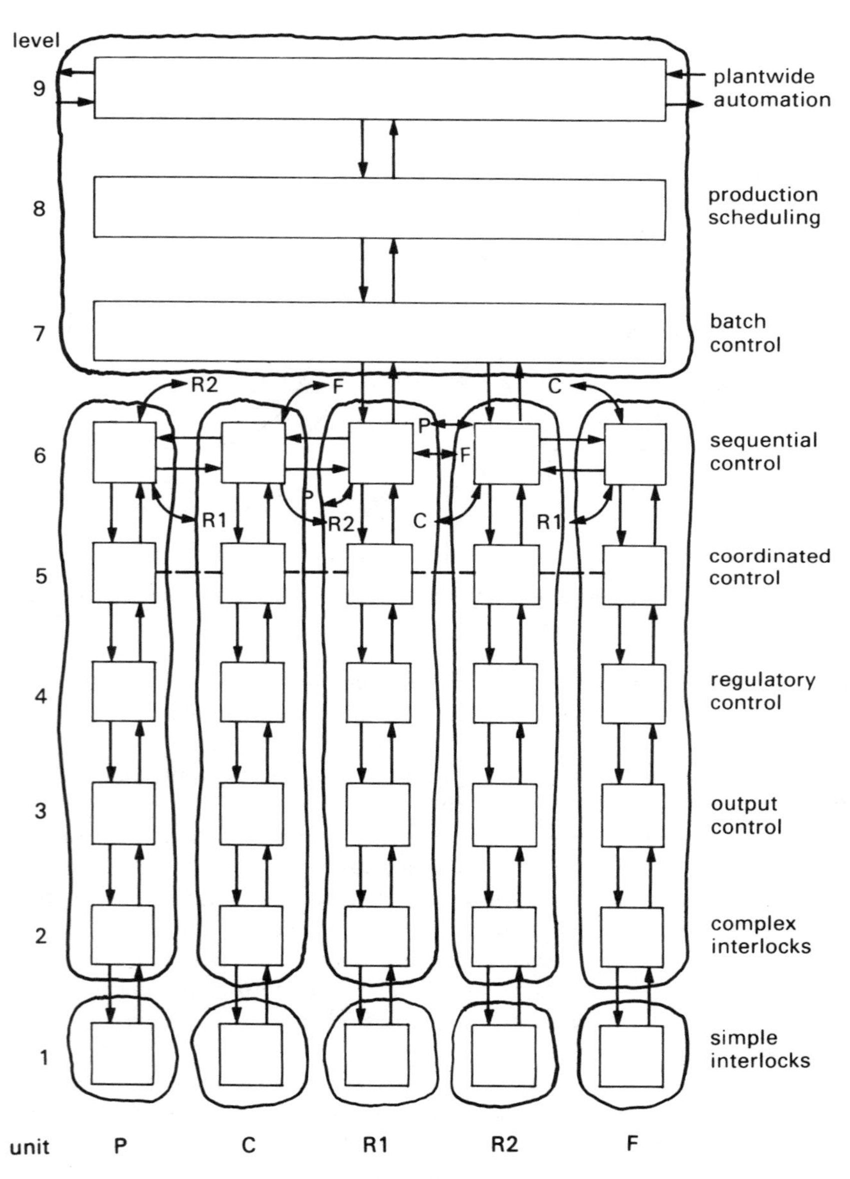

Figure 13-16. Configuration 4.

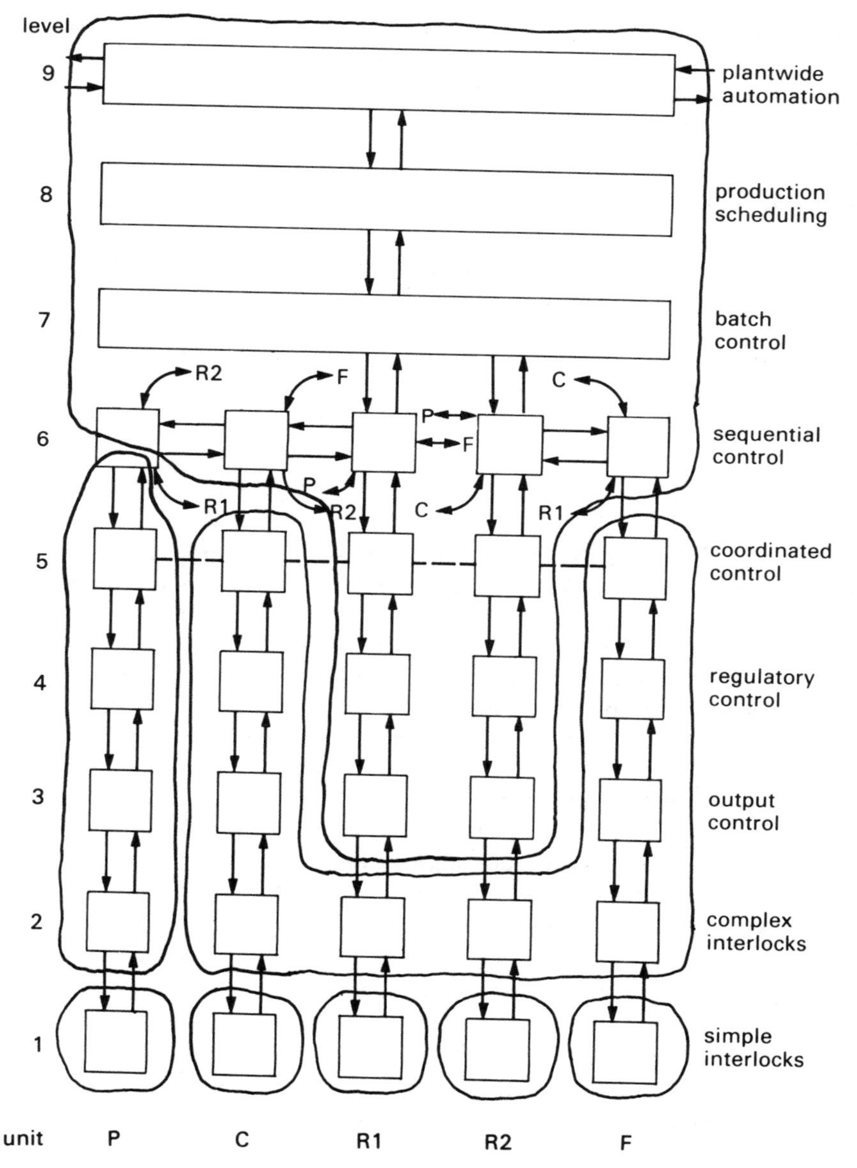

Figure 13-17. Configuration 5.

units and the bulk of control for the charging unit and finishing tank. A process control computer is responsible for control at levels 7 and above as well as for direct digital control of the reactor regulatory loops.

MAXIMIZING RELIABILITY

Reliability is a key factor in specifying and designing a control system. When a control system is not working or is generating wrong outputs, the result could be loss of production, decrease in product quality, loss of equipment, and even danger to life. Ensuring the reliability of batch control systems is generally more difficult than for continuous control systems.

Hardware Reliability

On failure of the primary control system for a continuous process, the backup system, whether automatic or manual, generally needs only to continue controlling at appropriate set points with little or no loss to product quality. The backup system in this case generally does not require any knowledge of the immediate past history of the process. For a batch system, however, the fallback system must be able to either continue the control or bring the process to a safe state. Therefore the system must know the process state at the time of its takeover. In a chemical batch process, for example, if the reaction has not been started, then preventing further addition of reactants or energy may suffice to keep the batch in a safe state. However, if the failure occurs in the middle of a reaction, then the backup system should be able to continue executing appropriate sequence steps to complete the reaction, or it must cool the reactants quickly or add materials to inhibit further reaction so that the process can be brought to a safe and stable condition.

Definitions

A measure of system reliability may be obtained by calculating its availability. *Availability* is defined as the probability that the system will operate to an agreed level of performance, without failure, for a specified period, subject to specified environmental conditions. In mathematical terms, steady-state availability (A) may be defined as

$$A = \frac{\text{Uptime}}{\text{Total time}} = \frac{\text{MTBF}}{\text{MTBF} + \text{MTTR}}$$

where MTBF = mean time between failures and MTTR = mean time to repair. The failure rate is defined as

$$\lambda = \frac{1}{\text{MTBF}}$$

The unavailability is defined as

$$\overline{A} = \frac{\text{MTTR}}{\text{MTBF} + \text{MTTR}}$$

Thus, availability is complementary to unavailability:

$$A + \overline{A} = 1$$

For example, if a control system runs on average 182 days (6 months) before failing and it takes 10 days to repair the system to get it running again, then

$$A = \frac{182}{182 + 10} = 0.948$$

$$\lambda = \frac{1}{182} = 0.00549 \text{ failures/day}$$

and

$$\overline{A} = \frac{10}{182 + 10} = 0.052$$

Series Availability

If the elements in a system are configured so that the failure of any element causes the whole system to fail, then these elements are said to be in *series* (Fig. 13-18). Here, the overall availability can be expressed as the product of the availabilities of all those elements, provided that the failures are independent; that is, the failure of any element does not change the probability of failure of any other element. Thus,

$$A = A_1 \times A_2 \times A_3 \cdot \cdot \cdot \times A_n$$

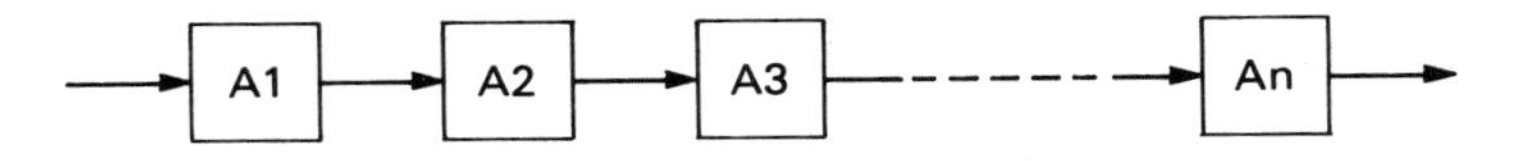

Figure 13-18. Functional diagram of controllers in series.

The equation for unavailability can be obtained by replacing each A by 1 − A. If the availability of individual elements is high, then multiples of $\overline{A}$ may be neglected, and the equation may be written as

$$\overline{A} = \overline{A}_1 + \overline{A}_2 + \overline{A}_3 + \cdots + \overline{A}_n$$

Parallel Availability

If the system elements work in *parallel*—that is, normally they all read the same inputs and generate the same outputs (Fig. 13-19)—then the system will work even if all elements except one fail, provided that the outputs are properly combined and the failures are to a predictable state. Then the unavailability of the system, that is, the failure of all the elements, can be expressed by

$$\overline{A} = \overline{A}_1 \times \overline{A}_2 \times \overline{A}_3 \times \cdots \times \overline{A}_n$$

By replacing $\overline{A}$ by 1 − A, we get

$$A = 1 - (1 - A_1)(1 - A_2)(1 - A_3) \cdots (1 - A_n)$$

Importance of Safety and Other Considerations

These definitions give a good idea of the general availability of systems, but they do not readily take into account factors such as

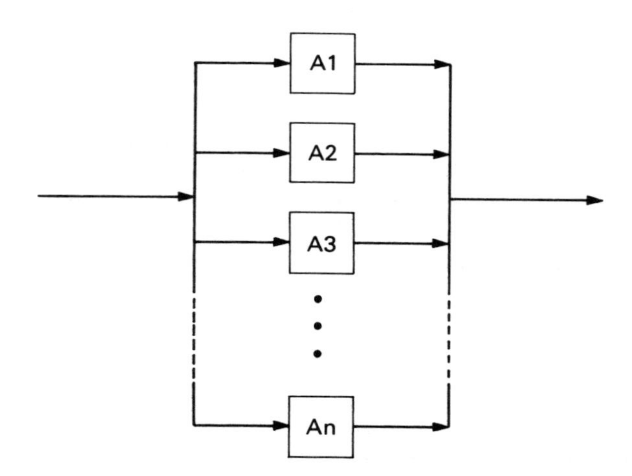

Figure 13-19. Functional diagram of controllers in parallel.

Whether on failure of the system the process is brought to a safe state automatically or not
Any degraded mode for running the process
Facilities provided to bring the process to a safe state manually
The ability to repair the failed portion while still operating

Safety considerations are most important in control system design. A system is considered safe when the personnel, the plant hardware, and process materials are protected from harm due to failure or malfunction of the control system. System reliability may be affected by the procedures chosen to enhance safety (Best, 1975*b*). The quenching of a reactor at the slightest sign of trouble, such as a sensor failure, would definitely improve safety but may so deteriorate reliability that the plant may well become uneconomical to operate. Thus, the control engineer is often confronted with the uncomfortable situation of trading safety for production.

The availability of a control system can be increased by increasing the mean time between failures and by decreasing the mean time to repair. Thus, the availability for a system may be increased considerably by only decreasing the mean time to repair the system. This may mean better diagnostics for detecting the faults and/or faster access by service personnel. In a nonredundant control system the availability of the system as a whole may be increased by increasing the reliability of the individual components in the system. This can be done by selecting components and packaging that suit the particular environments in which they will be operating and also by suitable burning-in to reduce infantile failures.

There is much literature on the reliability of continuous-control systems, a great deal of these are applicable to batch control systems; however, the discussions in this section are limited to those aspects that maximize the reliability of batch control systems exclusively. The discussions also focus on the reliability of a system as a whole rather than on the individual components in the control system.

Manual Backup

In a nonredundant batch control system the backup is typically manual. When a system failure is detected, plant operators, by using manual interfaces, may control the process to complete a batch or bring the process to a safe state. The use of manual backup as a viable system will depend on how critical is the process being controlled and also on the amount of information about the state of the batch readily available to the operators. In a computerized batch control system, the state of a batch may be displayed continuously on a dedicated monitor with its own storage of display information so that on failure of the computer, the monitor can retain the details of the state of the batch for operators to examine. This configuration, however, will be of little use if the failure of the computer control system is due to common power supply interruption. The way out in this situation may be to print out a hard copy of the latest state of a batch at the transition of each state. This has two drawbacks: It will generate large amounts of printed information, and it may not be easy to quickly extract critical data when required.

Distributed Control

The effectiveness of manual backup in a large centralized batch control system is marginal at best. Availability of sufficiently trained manpower to take over effectively an entire plant consisting of trains of parallel vessels is almost impossible. In this situation distributed batch control systems have some distinct advantages. In a distributed control system multiple computers and/or PCs or other hardware devices control the entire process. Because each piece of hardware system controls a part of the process, the failure of one of the systems may cause failure of only that part, so the manual back up may be more manageable. Proper care, however, is necessary in designing a distributed control system so as not to make it less reliable than a centralized control system.

A distributed control system should be designed to avoid multiple subsystems controlling either common process areas or areas that are tightly coupled. A common example of this is a batch reactor controlled by two controllers, one for the sequential control and another for the regulatory control. Here the failure of one of the controllers would make the whole reactor inoperative. These two controllers are in series, and if the availability of each of these controllers is 0.96, then the overall availability is

$$A = A_1 \times A_2 = 0.96 \times 0.96 = 0.92$$

In this case the availability of the system as a whole is decreased when two controllers are used instead of one. This is not so for two reactors where each is controlled by a separate controller and the failure of one will not affect the working of the other significantly. When a distributed control system is used to control process units in series, the isolation between the units may be increased by using intermediate storage facilities, thus reducing the effects of failure of one control subsystem on another. Isolation can increase facility and inventory costs, and it has to be weighed carefully against the increased availability obtained.

Hierarchical Control

So far we have considered distributed control subsystems essentially controlling process in the same level. It is possible and, indeed, desirable to decompose control system functionally in a hierarchical fashion where different subsystems may control different levels of batch control, as described in Chapters 7-11. Such a system will be reliable where functions are properly decomposed so that a failure at one level will not seriously affect the subsystem working at lower levels. The system may then still be running but in a degraded mode.

Dual Computer Systems

A dual computer system, where a primary computer is backed up by a similar computer, is a common method of increasing reliability. In a typical arrangement (Fig. 13-20) the control system, including operating system, regulatory control schemes, sequential logic, and data bases, is set up in one of the computers and then copied to the other; thus making the two computers functionally identical. One computer is then assigned the primary role; and the other, the backup.

The computer in the primary role carries on all the control actions and sends the status of the analog and switch inputs to the backup computer via the usual network channel(s) at regular intervals. Batch status parameters such as unit status, step and phase numbers, recipe assignment data, and so on, are also transmitted from the primary computer to the backup at regular intervals.

The backup computer assumes control only when the primary computer cannot communicate with it. When a failure in the communication is detected by the backup computer, there may be a number of retries, after which it is deemed that

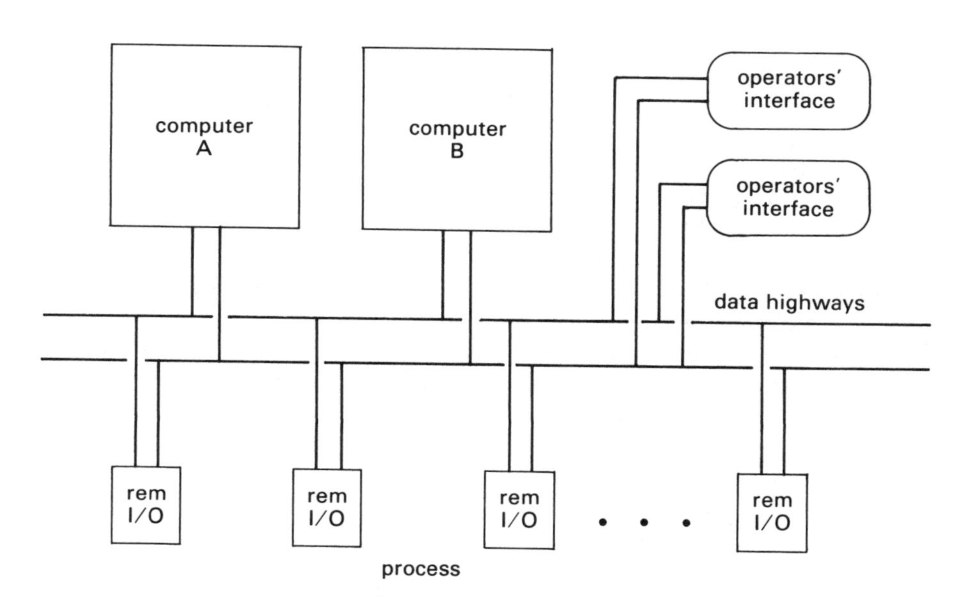

Figure 13-20. Dual computer backup configuration.

the primary computer has failed. The backup computer then takes over, deactivates the primary computer, and usually puts all the active units in safe states (hold), which allows the operator to decide where and how to restart the process. Optionally, the backup computer may start controlling the process from the point where, as perceived by it, the primary computer has failed. The main problem in allowing the backup to continue is in the windows between the last update of the status in the backup and the actual failure of the primary computer. Here the degree of criticality of the process would determine whether to continue control without manual intervention. Updating the backup computer more frequently reduces the window size but cannot eliminate it entirely.

The window is eliminated by an "update as you go" arrangement. Here, instead of updating the backup computer at fixed intervals, the primary computer updates a common buffer (or redundant buffers) as changes occur. The common buffer is made accessible to both computers, and the backup computer normally updates itself from this buffer at regular intervals. When the primary fails, the backup takes over only after making the necessary updates from this buffer. This arrangement is not commonly used because of the greatly increased load on the communication channel to the buffer, the increase in the load on the CPU of the primary computer, and the specialized design of the common buffer arrangement and functions to prevent the updating of the backup computer with erroneous data from the primary.

Another arrangement is to allow both computers to run at the same time and execute sequential and regulatory control functions. Here both computers scan analog and switch inputs independently, but the output of the backup computer is inhibited until the failure of the primary computer is detected. This type of arrangement may work well for regulatory control functions, but for the sequential control there is low likelihood that both computers will be in perfect synchronization just before the primary computer fails.

In a modified arrangement the primary computer supplies all the process input information to the backup. Both computers then work on the same data and normally generate the same outputs. The outputs are compared by the backup computer to check for synchronization. The output of the backup computer is inhibited from manipulating the process. If the outputs of the two computers disagree, then automatic transfer of all process tasks and data base, from the primary to the backup computer via dedicated high-speed data channels, takes place. This arrangement allows reasonable synchronization between the computers and has been successfully used in some systems.

For dual computer systems where one computer is backing up the other, the overall theoretical availability may be calculated from the equation for parallel availability:

$$A = 1 - (1 - A_1)(1 - A_2)$$

Thus if the availability of each computer is 0.96, then the overall availability is

$$A = 1 - (1 - 0.96)^2 = 0.998$$

which is a considerable improvement from the original. However, we are assuming a perfect detection of the failure of the primary computer and a perfect takeover. In reality this is not quite true, so the actual increase in availability by using a redundant computer, though significant, may not be as dramatic as the theoretical calculations suggest.

Modular Redundant System

In the dual computer systems discussed so far, the backup controller takes over when the primary controller fails. In such a system the proper and timely detection of the failure of the primary controller is very important. Additionally, the backup controller needs to know the process state at takeover. These problems may be avoided by using a *modular redundant system*. Here multiple controllers (three or more) are used in parallel, and all read the same process inputs. Their outputs are fed to a voting circuit that masks the effect of a faulty controller. In its simplest form three identical controllers are used (Fig. 13-21), and any malfunction of one of the controllers is detected immediately by the voting circuit. The voting circuit, on detection of a fault, alerts the operator, who may arrange for the servicing of the faulty controller as soon as possible. The availability of a modular redundant system may be calculated from the equations for parallel availability. In the example with three controllers the system will fail when any two controllers fail, with the third healthy, and this can happen in three ways. The system will also fail when all computers fail.

Now the unavailability of all three controllers at any one time is

$$\overline{A}_1 \cdot \overline{A}_2 \cdot \overline{A}_3 = \overline{A}_1^{\,3} \quad (\text{when } \overline{A}_1 = \overline{A}_2 = \overline{A}_3)$$

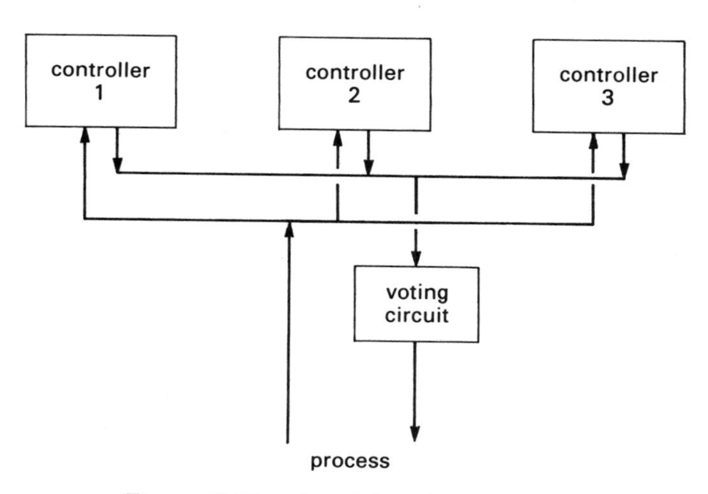

Figure 13-21. A modular redundant system.

The unavailability of any two controllers with the third running is

$$(\overline{A}_1 \cdot \overline{A}_2 - \overline{A}_1 \cdot \overline{A}_2 \cdot \overline{A}_3) + (\overline{A}_2 \cdot \overline{A}_3 - \overline{A}_1 \cdot \overline{A}_2 \cdot \overline{A}_3) + (\overline{A}_3 \cdot \overline{A}_1 - \overline{A}_1 \cdot \overline{A}_2 \cdot \overline{A}_3)$$
$$= 3(\overline{A}_1^2 - \overline{A}_1^3) \quad (\text{when } \overline{A}_1 = \overline{A}_2 = \overline{A}_3)$$

Thus, the total unavailability of either two or three controllers in a triple modular redundant system is

$$\overline{A} = 3(\overline{A}_1^2 - \overline{A}_1^3) + \overline{A}_1^3 = 3\overline{A}_1^2 - 2\overline{A}_1^3$$

Converting this to an availability equation, we get

$$A = 3A_1^2 - 2A_1^3$$

Let A_v be the availability of the voting circuit, which is in series. We then obtain the overall availability:

$$A = (3A_1^2 - 2A_1^3)A_v$$

Generally the voting circuit is much simpler than the controllers, so its availability is much higher; thus A_v can be taken as 1. Then if the availability of individual computers is 0.96, the overall availability is

$$A = 3(0.96)^2 - 2(0.96)^3 = 0.995$$

The triple modular redundant system offers higher reliability than its individual components, but theoretically this is lower than an ideal dual computer backup system. In practice it is claimed that the triple modular redundant system is more reliable than the dual computer backup system because of less than perfect failure detection and takeover mechanism of the backup computer (Wensley, 1982).

A modular redundant system is yet to be commonly accepted in the process industries. This system is expensive because it needs several computers and the voting circuit. The problem of synchronizing multiple processors in a batch environment needs to be looked at carefully. Multiple clocks, one for each processor can cause failure in synchronization, whereas a single clock makes it nonredundant.

Multitechnology Approach

In the redundant systems so far considered, the primary controller with backup or the modular redundant system, we have assumed that the individual controllers are identical. Here the anticipated benefits of redundancy are based on the assumption that the failures of individual controllers are independent of each other. This assumption is valid for many types of equipment—especially for devices that fail because of manufacturing uncertainty or random disturbances. Such elements typically have low mean time between failures. When controllers are made of relatively reliable components, but configurations are complex, software-based, or operated outside prescribed limits, failures tend to occur in groups. There the assumption of failure independence is invalid, and redundancy may not significantly improve overall availability. If similar controllers are used in primary and backup systems and are subjected to the same overstressed conditions (e.g., high ambient temperature or corrosive atmosphere), then there is considerable chance that they will fail at nearly the same time. Also, if they are supplied by the same power source, then any problem due to this source will equally affect both controllers.

The solution of this type of problem is to use different technologies for the primary and backup controllers (Fig. 13-22). Pneumatic controllers may back up electrical controllers, digital controllers may back up analog controllers, and so on. In hierarchical control systems this multitechnology approach may be used for the different levels in hierarchy, where the failure in one level may still allow the operations in other levels, thus controlling the process probably in a degraded mode.

The multitechnology approach does have drawbacks. For a reasonably reliable controller, failures are few and far between. This makes it increasingly difficult to keep operators fully conversant in working with the backup system. Dissimilar systems also have maintenance problems and need more tools and spare parts.

The multitechnology approach was once used extensively in process control, when the digital computer system first appeared. Engineers then had little trust in this new technology and always backed it up with the traditional pneumatic analog controllers and electrical relay circuits. This meant retaining large and expensive

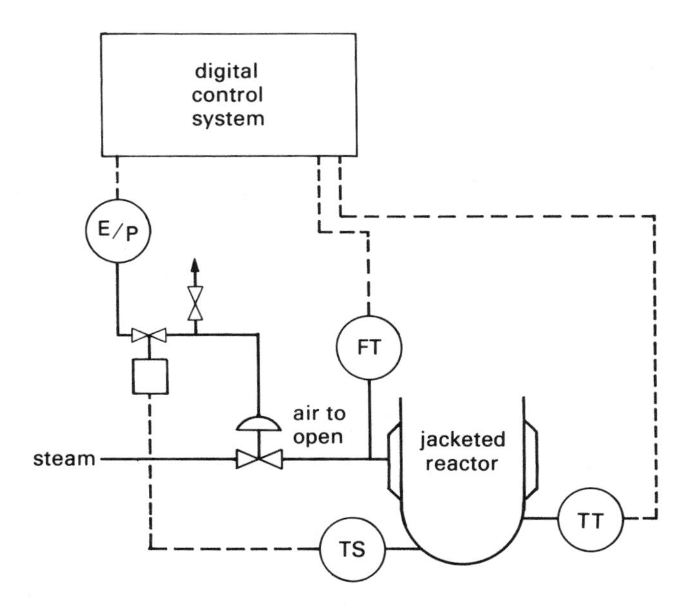

Figure 13-22. A multi-technology approach. Digital control is augmented by electrical and pneumatic technologies.

control panels. With the increased reliability of the digital computer system, this approach has become less common. However, the multitechnology approach is still worthy of consideration for the reasons stated earlier. The actual implementation may be altered from the traditional and expensive backup control panels to more modern, centralized human-machine communication systems. The design of standard human-machine communication consoles with the ability to communicate with dissimilar control systems is helping to realize this approach. This approach should be carefully considered for particular applications. Where such a need is established, a decision for complete backup or backup limited to certain critical areas, like safety interlocks, should be made.

Redundant Links and Interfaces

So far we have considered the redundancy of the controllers. The chances of failure of the links and interfaces between controllers and I/O multiplexers and between controllers are possibly higher than for controllers themselves. The chances of these failures can be reduced by using multiported controllers and multiplexers and redundant links (Figs. 13-7 and 13-8). Normally, one set of such links and interfaces is designated as primary, the other as backup. An automatic switchover mechanism on failure of link or interfaces is usually provided by the vendors.

Software Reliability

Over the years, hardware has tended to become more reliable and less expensive while software has become more complex and, therefore, error prone. Software failures then must be carefully considered along with hardware failures when evaluating a control system's reliability. The software reliability problem is somewhat dissimilar to that for the hardware. For one thing, software is not subject to degradation with use. Unreliability in software is the result of undetected errors or defects. They go undetected because not all possible conditions and paths in a piece of software are checked during its testing. The combinations of paths and inputs in a nontrivial software module can be enormous, which precludes its complete testing within any realistic time frame. The reliability of a piece of software depends on (1) the probability that it will operate without fault during a set time interval, and (2) how soon after a failure the cause may be detected and corrected.

When testing a software module, the error discovery rate can predict the mean time between failures (MTBF). An error detection curve may be drawn (Fig. 13-23) to estimate the number of remaining errors. This concept, however, is controversial: Some engineers maintain that measuring the MTBF makes no sense since software does not break or wear out but only contains undiscovered errors (Glass, 1979).

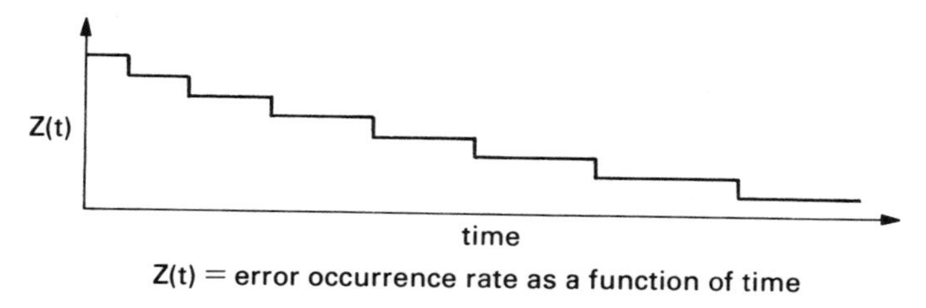

Figure 13-23. Software error detection curve.

Such a failure model at least provides some indication of reliability of a piece of software, allowing the user to decide whether the software module is ready to use.

The detection and correction of software errors after a module has been installed is generally the function of the software maintenance organization or plant engineering staff. To quickly pinpoint the area where a fault may be located requires proper tools and skills, including diagnostics, facilities for single-stepping the logic preferably in an off-line mode, and proper documentation of the software. Maintenance personnel also need extensive skill and experience to detect and fix software errors quickly. Unfortunately, maintenance is still one of the most neglected and least glamorous functions in programming. If software reliability is to be increased, this problem needs to be rectified quickly. One reason for the difficulty in maintenance is the wide variety of styles used by programmers. The management that imposes standards for style and documentation has gone quite a way toward making maintenance a more desirable occupation. For those organizations without a specialized software maintenance department, the vendors should be able to support this function.

Reliability of Standard Software

In considering reliability, one needs to take into account the two types of software usually present in a batch control system. The first type is the standard software usually supplied by the system vendor. Operating system, I/O handlers, regulatory and batch control packages, data base generators, utilities, and so on, fall into this category. The second type is the application-specific software, which is generated by the user, by the vendor's engineering group, or by a third party. Sequence logic, data base, graphic displays, and reports are in this category. Standard software is expected to be relatively error free, since it has been extensively tested by the vendor and field proven. The vendor should track the discovery of additional errors and send fixes to all installed systems. As stated before, the error discovery rate may follow the exponential decay curve (Fig. 13-23). By plotting the discovery rate, normalized for the number of installed systems, we might predict the future discovery rate for this type of software.

Reliability of Application-Specific Software

High reliability of the application-specific software should, in itself, be a major objective and ought to be considered early in the project. Ways of promoting the reliability of the software under development include

Top-down specification and design by levels
Proper design reviews
Modular programming
Walk-through and desk-checking the generated code
Testing the software using appropriate simulation

The importance of top-down specification and design by levels (preferably bottom-up) has been discussed extensively. Their role in generating more reliable software is emphasized here. The specification and design of a large batch control system are challenging tasks. A team of professionals is generally required. Any unresolved ambiguity in specification or design will cause similar ambiguity in coding, thus causing serious reliability problems. Conversely, unambiguous definitions leave little room for programming variations, and so they help generate more reliable software. This is easier to achieve when the total task is broken down into clearly defined levels with function modules in each level. The detailed designing at individual module levels should preferably be done with structured English or pseudocode (see Chapter 9). Proper design reviews among peers and between different parties, such as system developer and final user, help identify and resolve ambiguities.

Once the specification and design are complete, the coding should be done in a modular fashion using common subroutines for similar functions. Modular programs are easier to debug, simply because each module is (or should be) small enough to be easily understood.

Coding should preferably be done in a high-level language or by a *code generator.* Code generators use proven algorithms to generate large amounts of low-level code from high-level instructions (Brody, 1985). For example, custom-specified function blocks may be translated to relay ladder logic. Whenever possible, common functions should be coded only once, and only one copy should reside in the system—this will reduce the testing and maintenance time.

Walk-throughs and desk-checking of code help to maintain consistency in style and coding practices and to resolve misunderstandings about the functions of individual modules. A walk-through should be done with three to five persons closely associated with the project. Desk-checking is done by one individual at a time.

Testing is the traditional method of removing errors. Rigorous testing of application-specific software, using a representative data base, should be carried out after coding. Testing should be done with the specification and design documentation as the reference, not the coding. Proper test cases that simulate the process are necessary to do the required testing. This may be done by hardware (simulation equipment) or software. Software simulation is more flexible but requires more effort from the tester. Software simulation can be done internally with a controller or externally in another controller. Testing individual software modules by internal simulation poses no great problem. However, system level testing may not be possible internally, because this may introduce an excessive processing load to the system.

Conclusions

Reliability and safety are important considerations in designing a batch control system. Hardware reliability problems cannot always be solved by redundant systems. Redundant systems are expensive because they need more equipment. Also the mechanism for failure detection and control transfer (in case of backup equipment) is not always reliable. Thus, for many smaller batch control systems, manual backup with safety interlocks may suffice. For larger batch systems properly designed distributed control with sufficient isolation between the process areas can go a long way in insuring reliability. In such a system the failure of a single controller among many would be more manageable and may not seriously affect the safety or production of the process. Similar benefits are obtained by designing the control system in a hierarchical fashion, where the failure of one level in the hierarchy will not affect lower levels, and the control system may still run in a degraded mode. Then redundancy may only be required at the lowest level of control, the safety interlock. But here the need for redundancy is minimized by the simplicity and functional isolation provided by several available approaches. The multitechnology approach of backup may be used here advantageously. In a network system the use of multiported controllers and process I/O devices, along with redundant links, is recommended.

Software reliability may be increased by using field-proven standard packages as much as possible. For application-specific software detailed specification, modular design, and high-level languages reduce the chance of unexpected behavior. Rigorous testing of the code, first as individual modules and then at subsystem and system levels, further reduces errors in the software.

CHAPTER

14

TESTING

A batch control system designed in a modularized, structured fashion, as suggested in this book, can be tested in a way that matches the structure and facilitates the isolation and correction of errors. It is axiomatic that the most economical testing is that which identifies defects at the lowest integration level. A rule of thumb used for electronic hardware is that costs increase an order of magnitude for each integration level. In other words, if a defective part costs X dollars to replace before it has been installed on a printed circuit board, the part will cost 10X dollars to replace once installed. If the defect is not found until the various completed circuit boards are assembled into a piece of equipment, the cost associated with the defect rises to 100X dollars. Multiply the cost again if the defective equipment leaves the factory, and is only identified as such when the user attempts to operate it.

It is true then, though an oversimplification, to say that the purpose of testing is to find mistakes and problems. The majority of design errors should have been found during reviews, and syntactical errors should have been minimized through standard tools used in the program development process.

TEST CATEGORIES

There are two broad test categories: functional test and structural test. In a *functional* test the software is tested as a "black box." The tester presents sets of data (both legal and illegal) and checks the corresponding outputs (Fig. 14-1). The tester needs no knowledge of the internal programming structure. The tester concen-

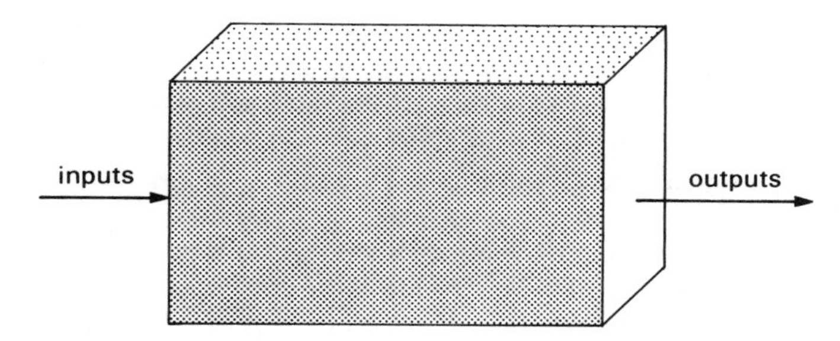

Figure 14-1. The black box test.

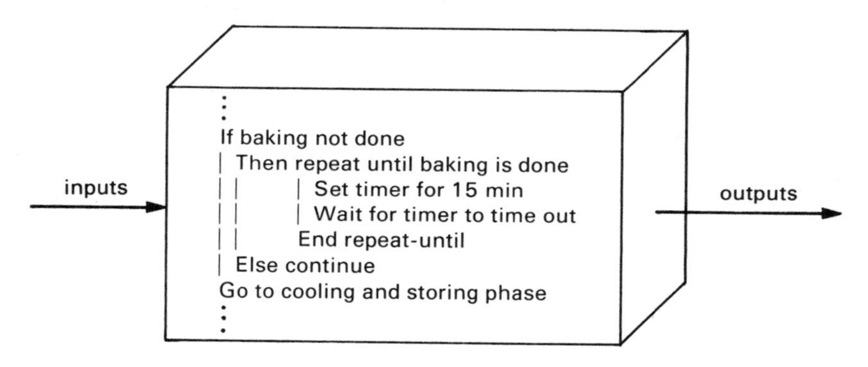

Figure 14-2. The white box test.

trates on the external behavior of the system and thus can detect functions that have not been implemented or are not operating properly. This is suitable for top-level tests and is oriented towards acceptance tests for the final user of the system. The functional tests may be performed both by the developer and the final user of the system either jointly or individually.

Structural tests, on the other hand, are the "white box" tests where detailed knowledge of the structure and coding of the software is required (Fig. 14-2). This type of test can find defects in isolation and concentrates on missing paths, wrong paths, and wrong actions within and between software modules. These tests are typically performed by engineers and programmers intimately involved in the design and implementation of the system and generally involve testing all possible paths in a set of sequence logic.

Functional and structural tests are both needed to prove that a control system is working correctly. Structural tests may be done as a software module is coded, whereas function tests are generally performed after a system or a major subsystem is integrated. Thus, individual software modules, such as a phase or an interlock function, require structural test. For each control level, both structural and functional tests are required, whereas the testing of the total system is largely functional.

A typical batch project requires involvement of the vendor, the system developer, and the user. The *vendor* is the organization that supplies the control system hardware and the standard software. The *user* is the organization that will use the control system to control its process. The *system developer* is an organization that designs and implements the system according to the user's needs. The system developer may be an engineering group within the user's or vendor's organization or a combination of personnel from both. In some cases it may be an independent third party—a system house. Testing plays a very important part here in proving that proper hardware and software are being developed and delivered from one party to another.

We will not discuss here the testing of standard hardware and software that are delivered by the vendor to the system developer, since this is a "generic" activity not uniquely associated with batch control. But testing of batch applications is a different matter. Typically, the testing of batch applications may be considered in three distinct stages: in-process tests (tests during the development of the system), factory acceptance test, and field acceptance test.

In-process tests are those conducted by system development organization members to prove to themselves that the individual software modules and subsystems are working properly. Most of these tests are structural, but they usually include some functional tests. They are generally not formal and do not have extensive test plans. However, it is our opinion that proper design and execution of these tests yield a considerable saving in the time and effort required to debug a system.

The *factory acceptance test* is the formal test to prove to the user organization that the system to be delivered meets its requirements. This test may be conducted by the system developer with user personnel witnessing it or jointly by the user and the system developer. These tests are conducted without the actual connection of inputs and outputs to the process to be controlled. The process is simulated by external hardware, such as manually operated switches, or by internal software. Formal test plans are usually agreed to before starting the test, and the test is generally functional. Sometimes the factory acceptance test is broken into a series of tests, each conducted separately to test an individual subsystem, and then a final test is conducted for the whole system.

After the system is installed at the site and connected to the process equipment, the field acceptance test may be carried out. This is a functional test where the process is run dry or with inert material, and it is generally conducted by the user's organization with help from the system developer.

TEST PLAN

It should be obvious that testing is best conducted according to a plan. A properly executed test plan provides appropriate demonstration of functions, features, performance, and reliability. It uncovers most defects with least effort by avoiding duplication and provides greatest coverage with fewer test cases. Plans are required for both structural and functional tests. The plans should specify the sequence in which a module or a function is to be tested, the required inputs, and the anticipated results. The plan for a structural test should cover all possible paths of logic in a module by specifying suitable inputs. All possible combination of paths in a module, however, are generally too numerous to be tested, so the test designer must exercise judgment to test an appropriate subset of combinations. The structural test plan for a module or a set of modules may be generated immediately after such a module, or modules, are designed.

The purpose of functional testing is to establish conformance to the specification, so the functional test plan can be prepared at any time after the specification is finished. In preparing a test plan, the engineer may discover omissions or errors in the specification, which may then be changed or rectified. The test plan may also be used for informal "desk checks" and for formal reviews of subsequent design activity.

IN-PROCESS TESTS

In-process tests should be started as soon as possible after modules, such as an interlock function or a phase, have been coded. In larger projects it is advantageous to perform desk-checking and walk-through for the code. In a walk-through programmers and engineers (usually three to five persons) in the project team go through the code line by line to check for bugs and inconsistencies. Doing so allows programming errors within the module and problems in intermodule communications to be quickly detected.

When the walk-through is finished, the module may be checked in its proper software and hardware environments (within practical limitations) by loading it in the machine along with the appropriate data base. Suitable hardware or software simulation is required to mimic the process and/or the environment.

In-process tests are primarily structural, and it is imperative that all logic paths within the module be checked at least once. However, as stated before, it is virtually impossible to check the function of a logic for all possible inputs and combinations of paths. A properly designed test plan, which requires doing all meaningful tests (within practical limits) without overburdening the tester, becomes very useful at this stage.

When all the modules in a system or a subsystem have been structurally tested, appropriate functional tests may be performed. Functional tests for the system may be performed as a rehearsal for the factory acceptance test.

FACTORY ACCEPTANCE TEST

The factory acceptance test is a set of formal tests performed when all the application software and data bases have been coded and installed and the structural tests of individual modules have been successfully completed. The testing is performed with simulation gear; the control system is not connected to the process. This is primarily a functional test, which is witnessed by, and sometimes carried out by, the final user of the system. A formal test plan, agreed to between the system developer and the user, is generally used. The specific goals of the factory acceptance test are as follows:

To confirm proper design and implementation, demonstrating to the user that the system meets user specifications.

To establish that the system performs properly under load and under conditions of stress, such as failure of a computer peripheral.

To confirm proper system tuning, where applicable. This involves setting program priorities (which program runs first when several should be run), for example.

To look for unanticipated interactions between modules, such as might be caused by two programs inadvertently sharing the same memory location for different data.

The term "factory" is used here as generic and means an area with proper space, power, simulation equipment, and meeting facilities. The factory setting is provided as a matter of course by control system vendors. When a user organization receives a system before the application work is done, a facility like this should be considered, especially where the control system is to be installed in an existing plant and ongoing operations might be disrupted.

When design is conducted according to the recommendations given in this book, then individual specifications exist for portions of the application small enough to be reasonably tested as application modules. Adequate access is provided to enable the user to direct the operation of the application modules. The design-by-levels philosophy provides for an operator interface at each level, but not necessarily the ability for the operator to view and manipulate internal signals, which may be required when testing part of a level. This ability should be provided as utility functions furnished with the ready-to-apply batch control system. If specific functions are not furnished, it may well be worthwhile for the user to construct them as application programs, and the sooner the better, to avoid missed opportunities for simplifying the work.

With these utilities in place, testing proceeds preferably from the lower levels up. First, the ability of the system to correctly read the field inputs must be confirmed. Then these inputs are manipulated, preferably by appropriate signal sources connected to the input terminals, and proper operation of the interlock levels is checked. The ability of the operator to access the system becomes important at the computational safety level (level 2). In this level interlocks may have multiple operating modes, normally selected by the higher levels, each of which must be separately tested. The tester replaces the higher levels for the purposes of this testing. Then proper operation of the regulatory level is checked, with the tester providing set points and other directives, through the operator interface.

The sequence levels typically represent the greatest part of a batch control system application and require the greatest testing time, especially for multiproduct or multisequence systems. First, the shared units should be tested with their local operator interfaces, if available. Once this operation is confirmed to be correct, testing proceeds outward through the envelopes (see Chapter 13) as they exist, until it is established that the shared units all respond correctly to simulated signals from the main process vessels.

Then the logic for the main process vessels themselves is tested. At this point the logic is supported by all of the functionality for the levels below and for the shared units. Outputs to files for logging and reporting should be tested here; the logging and reporting programs themselves may be tested later, independently, once it is confirmed that the data received from the logic are correct.

Software for production scheduling and other higher-level functions is then tested along with the proven batch control software. Here it is useful to prepare test cases and determine their solutions independently for comparison with results generated by the control computer.

This is not the only way to test a computer-based system; there are advocates of testing from the top down, using "stubs" to replace parts of the system not yet available. Testing from the bottom up, as preferred here, emphasizes safety and allows plant production to safely start when the batch logic is completed before, for example, production scheduling software is available.

There may be times when it is useful to test some functions to a higher level than others. For example, testing may stop when the sequence logic for shared units is tested, but logic for main process vessels is not. Then the plant operators can adjust set points, take samples, and so on, according to manual procedures, but still have shared unit logic to facilitate material charging.

All process and equipment failure conditions must be simulated and tested during this period, since many failure conditions are difficult and/or time consuming to simulate when connected to the actual process for the field acceptance test.

FIELD ACCEPTANCE TEST

After the factory acceptance test the system is installed and connected to the process equipment. For a new plant subsequent testing may involve both process equipment and the control system, and it can be lengthy. For an existing plant connections to the process must be checked, and assumptions made about plant dynamics (e.g., valve travel time) must be confirmed or corrected. Parts of the system not simulated earlier are tested here. This stage of testing may represent the first opportunity for members of the plant operating staff to become familiar with the system, so there is an opportunity for training as well as for incorporating suggestions for improvements.

The field acceptance test is functional and may be carried out in stages with procedures similar to the factory acceptance test. However, with the actual process input and output connected, it may not be possible to perform all the tests within a limited time (process dynamics comes into play). The process is usually run dry or with some inert material for one or more batches.

Testing of sequence logic for a multisequence or multiproduct batch plant is normally done with one test recipe, or at most a small number, designed to exercise all logic paths. Nevertheless, batch control systems are complex; discounting simple variations in ingredient amounts, temperature set points, and the like, there may still be a very large number of distinct recipes possible. Any one of them is capable of finding a previously hidden error. Therefore, the control system should be separately qualified to run each unique recipe before the recipe is released for normal production.

TEST ENVIRONMENT

Proper environment is required for in-process and factory acceptance tests. They include hardware and/or software simulations, utilities, and test data, which are briefly discussed here.

Simulation

Both hardware and software are used to establish an environment equivalent to that encountered in actual use to enable the tester to find the greatest number of problems prior to the system's placement in service. Contact inputs and outputs are commonly wired to switches and indicator lamps, respectively; test personnel observe the indicator lamps and change switch states accordingly.

A simple input-to-output connection, provided that the electrical requirements of both input and output circuits are met, can simulate a process plant response. Failure to respond can be simulated by opening a switch wired into the circuit or by simply disconnecting one of the wires (not recommended where potentially dangerous electrical conditions exist!). If timing is important, the output-to-input connection can be made through a time-delay relay with delay adjustable over the range of interest. More complex plant responses can be simulated by interconnecting standard and time-delay relays.

High-speed pulse trains, typical of the outputs of several types of flow transmitters, may be generated by inexpensive square-wave generators. Reed relays or optoisolators provide electrical isolation. Optoisolators consist of a light source and a light receiver separated usually by a transparent solid to allow signal transmission without electrical connection. Reed relays have been known to last for months at switching rates well above 1/s, but it is generally better to apply pulses to the coil only when necessary. Connections should be made carefully in light of the varying electrical requirements found among input circuits. The pulse generator may be set to generate an output similar to, or faster than, that expected from the plant to speed up testing. Where it is necessary to simulate the variation of flow rate with control valve position, the analog output to the control valve may be connected to a voltage-to-frequency converter.

Most testing of analog loops is performed with outputs connected back to inputs. Where it is necessary to simulate dynamics, lag units or laboratory-type process simulators may be used. Complex controller-process interactions, like those required for weight transfer sequences and heat-cool switching, are simulated within the limits of economic reason. The calculation of economic reason must consider the simulation's usefulness throughout system testing.

It is not always necessary to fully simulate a plant, especially where there are many identical process trains using the same software. Shared units should always be simulated, as should at least two process lines—the latter to confirm the ability of the shared units to resolve contentions, and the ability of the system as a whole to track multiple batches where required.

With varying degrees of success, attempts have been made to use spare capacity within the control system, or even additional programmable hardware, to provide simulation of the plant. The economics of program-based simulation have the same appeal as program-based control: easier changes and more complex structures without major cost increases. Contact I/O relationships may be simulated within the same programmable controller used for interlocks. Continuous-loop dynamics may be simulated with dynamic compensation and deadtime control blocks. Process sequences may be simulated by batch control language statements.

The attractiveness of this kind of simulation is limited, though, by several factors:

It is not necessarily easier to design a simulation function in software than in hardware. The availability of a less expensive implementation method may lead to the specification and design of excessively complex and expensive simulation.

Two separate software systems must be maintained—the one with only control software, and the one with software for both control and simulation. Software simulation often involves control system modifications, for example, to interact with internal (simulation) inputs and outputs rather than field I/O.

Since the details of the simulation are less visible in software than in hardware, demonstrations may lack full credibility.

Complex simulations may, themselves, contribute to perceived control system nonconformances.

Software-based simulation should therefore be implemented only when a reasonable likelihood of sufficient economic return can be demonstrated.

Utilities

Software *utilities* can greatly enhance the testing process by providing visibility and access for manipulation of the control system under test. However, they are generally not required for normal process operations. Valuable utilities include:

A means to "dump" the current contents of recipes, unit variables, status bits, I/O files, and so on, to the CRT screen or hard copy.

A means to manipulate unit variables.

A means to observe and manipulate data flow among software modules, including between process units. Ability to trend several of these signals on one CRT screen.

Capability of executing sequential logic one step at a time. Combined with the dumping capability mentioned, this allows the engineer to observe the changes caused by each step. An automatic comparison function, pointing out the differences between successive dumps, would be useful as well. Where single-stepping is not explicitly provided, it can usually be implemented as part of batch application programming.

Ability to force both contact and analog I/O.

Ability to disconnect output signals from the process to enable simulation testing to take place once the control system has been installed and wired.

The ease with which these utilities are used can vary widely; as an example, one manufacturer may allow signal manipulation by tag, whereas another may require knowledge of the signal's address in computer memory.

APPENDIXES

APPENDIX

1

COMPLEXITY IN ANALYZING THE FAILURE CONDITIONS

A batch application can quickly get extremely complex as an effort is made to analyze all possible alarm or failure conditions and define the actions needed for taking care of them. This frequently happens even when a plant or a process appears simple. The problem is the vast number of conditions that must be handled for even the simplest cases, which makes it very difficult to bound an application in advance. Usually there are no natural bounds that tend to limit an application.

Example

Consider a simplified batch application that has only four binary sensors, each of which may indicate an alarm or failure condition. All possible combinations of these alarm or failure conditions may be calculated by the equation

$$S = 2^N - 1$$

where S is the total number of combinations of alarm or failure states and N is the total number of binary elements, each of which may cause alarm or failure condition. Thus in this example a total of 15 states need to be considered for taking care of all possible alarm and failure conditions.

If the sequence in which the four binary elements go into alarm or failure condition is important (assuming that the time interval of occurrence is irrelevant), then all of the permutations of the states need to be calculated. This calculation is given by the equation

$$P = \sum_{K=1}^{N} \frac{N!}{(N-K)!}$$

where P is the total number of possible ways the failure states may be reached and N is the total number of binary elements, each of which may cause alarm on failure condition. Thus in this example a total of 64 cases need to be considered. If $N = 100$, all the combinations give 1.26×10^{30} possible states, and all permutations give much more.

Conceptually and physically such a large number of alarm or failure states is impossible to handle. Complexity is reduced, somewhat artificially, by carefully lumping multiple states together, to cope more effectively with the problem, and in most cases that suffices.

APPENDIX
2

A LANGUAGE FOR SPECIFYING SEQUENCE LOGIC

Following is a representative set of commands in a typical special-purpose language for the computer control of sequential processes. Functionally, the set is quite comprehensive but may vary considerably in detail from one computer to another. The parameters specified with these commands, in many cases, may be actual values or addresses that contain these values. For example, the command

CLOSE [contact,time]

closes a contact output or flag after the specified time. The time argument could either be an actual time interval (in seconds) or an address, the contents of which is the time interval.

Commands That Manipulate Contact Outputs, Contact Input Desired States, Devices, and Flags

CLOSE[contact,time]
Close contact output or flag after specified time.

OPEN[contact,time]
Open contact output or flag after specified time.

DESON[contact,time]
Set the desired state of contact input to closed and start testing after specified time. (Used for detection of off-normal conditions.)

DESOFF[contact,time]
Set the desired state of contact input to open and start testing after specified time. (Used for detection of off-normal conditions.)

ACT[device,time1,time2]
Drive the device to active state after time1 and start checking the state of the device after time2.

DEACT[device,time1,time2]
Drive the device to inactive state after time1 and start checking the state of the device after time2.

MSGON[contact]
Allow alarm message generation when there is discrepancy between the actual and the desired states for the contact input.

MSGOFF[contact]
Suppress alarm message for any discrepancy between the actual and the desired states for the contact input or contact output feedback.

SVCON[contact]
Allow calling service logic on discrepancy between the actual and the desired state for the contact input.

SVCOFF[contact]
Suppress calling service logic on discrepancy between the actual and the desired state for the contact input.

HLDON[contact]
Allow calling hold logic, to drive the unit to hold, on discrepancy between the actual and the desired state for the contact input.

HLDOFF[contact]
Suppress calling hold logic on discrepancy between the actual and the desired state for contact input.

DRIVE
Drive all contact outputs and contact input desired states for the unit to that specified for hold state in the data base.

REDRIVE
Restore all contact outputs and contact input desired states for the unit to the states they were in before the unit was driven to the hold state.

Commands That Interface with Regulatory Control Functions

BLKON[block]
Place the regulatory control block on scan or control.

BLKOFF[block]
Place the regulatory control block off scan or control.

AUTO[block]
Put the output of the regulatory control block on automatic control (as opposed to manually specified output value).

MANUAL[block]
Set the regulatory control block to receive manually specified output value.

OUTPUT[block,value]
Set the output of the regulatory control block to the specified value. The block has to be in manual.

TUNE[block,P,I,D]
Set the tuning parameters of the regulatory control block to the proportional (P), integral (I), and derivative (D) constants specified.

MEASURE[block,variable]
Get the output of the regulatory control block and store it in the specified variable.

SETPNT[block,value]
Change the set point of the regulatory control block to the specified value.

RAMP[block,value,rate]
Linearly increase or decrease the set point of the regulatory control block to the specified value at the rate specified.

HILIM[block,value]
Set the high absolute alarm limit of the output of the regulatory control block to the specified value.

LOLIM[block,value]
Set the low absolute alarm limit of the output of the regulatory control block to the specified value.

HIDEV[block,value]
Set the high deviation alarm limit of the regulatory control block to the specified value.

LODEV[block,value]
Set the low deviation alarm limit of the regulatory control block to the specified value.

ALMON[block]
Allow alarm message to be generated when absolute or deviation alarm limits are violated for the regulatory control block.

ALMOFF[block]
Suppress messages for absolute or deviation alarm limit violations for the regulatory control block.

CLSCAS[block]
Close the cascade control to the regulatory control block.

OPNCAS[block]
Open the cascade control to the regulatory control block.

Commands That Perform Arithmetic

ADD[value1,value2,variable]
Add the two numerical values, and store the sum in the specified variable.

SUB[value1,value2,variable]
Subtract value 2 from value 1 and store the difference in the specified variable.

MULDIV[value1,value2,value3,variable]
Multiply value 1 and value 2, divide the product by value 3, and store the result in the specified variable.

FIX[value,variable]
Convert the floating point value to integer and store it in the specified variable.

FLOAT[value,variable]
Convert the integer value to floating point and store it in the specified variable.

Commands for Communication with Personnel

ALARM
Set the alarm annunciator for the unit.

CRTMSG[message,value,flash]
Display the message with the optional numerical value on the CRT console and flash the message, if specified.

PRTMSG[message,value,highlight]
Print the message with the optional numerical value on the alarm printer and highlight the message (by color or otherwise), if specified.

CLOCK[variable1,variable2]
Fetch the current date and time and store in the variable 1 and variable 2 respectively.

LOG[record pointer,number,file name,filepointer]
Store the specified number of variables from the batch record starting at record pointer to the specified file, starting at file pointer and update the file pointer.

Commands That Make Decisions and Control Sequence Logic Steps

CHECK[contact,address1,address2]
Perform branching to address 1 or address 2, depending on the state (open or closed) of the contact.

GOTO[address]
Perform unconditional branch to the logic statement specified by the address.

GOSUB[name]
Branch to the subroutine specified.

GOBACK
Return from the subroutine called by GOSUB command.

GOHOLD[address]
Branch to the service logic for putting the unit in hold condition and store the return logic address.

RETURN
Return to the normal logic from service/hold condition. If returning from hold condition entered by GOHOLD command, then return to the address specified in that command. Otherwise, return to the normal logic at the point where it was interrupted.

IF[value1,value2,less,equal,great]
Branch to the appropriate address, depending on whether value 1 is less than, equal to, or greater than value 2.

WAIT[time,address]
Suspend processing the sequence logic for the specified time and then resume execution at logic address specified.

WAITFOR[contact,state,time,address]
Suspend processing of the sequence until the contact is in the specified state or the time has expired. Resume execution at the address if the time has expired or at the next statement if contact is in the specified state.

PAUSE[message,time,address1,address2]
Output the message on the CRT console and suspend processing of the sequence until operator responds from the console or the time has expired. Branch to address 1 if operator responds, else branch to address 2 on time-out.

SETTMR[timer,time,flag]
Set the timer to initial time value and set the flag when the time expires.

CLRTMR[timer]
Clear the timer (i.e., set it to 0).

READTMR[timer,variable]
Read the timer number and store its value in the specified variable.

REASON[address1,address2,address3,address4]
Branch to the appropriate logic address according to the reason for entering the service/hold logic.
Branch to

address1	if normal logic or operator has requested to put the unit to hold.
address2	if operator is requesting the service logic (not hold).
address3	if a switch input discrepancy has called for hold.
address4	if a switch input discrepancy has called for service logic (not hold).

PHASE[phase name,initialize]
Start at specified phase and optionally initialize the contact outputs, flags, and the input desired states.

APPENDIX

3

SYSTEM AND DATA BASE SPECIFICATION

For a computer-controlled batch system a set of application-specific data needs to be specified by the user. They include the configuration data, which allow the setting up of the data files of right sizes, and the data required by the control system, which typically include names and types of units, switches and devices, their initial states and their alarming requirements. These data are entered either interactively from a console or read from files. The formats and contents of these data vary considerably between different vendors' systems. Here is a typical example.

Configuration Data

One set of data required for each batch control system.

1. Number of units
2. Number of types of units
3. Number of phases
4. Number of procedures
5. Number of recipes
6. Number of contact inputs
   - Hardware switches
   - Analog alarm limits
   - Flags
7. Number of contact outputs
   - Hardware
   - Flags

8. Number of devices
9. Number of timers
10. Maximum number of phases per procedure
11. Maximum number of steps per phase
12. Maximum number of contact inputs per unit
13. Maximum number of contact outputs per unit
14. Maximum number of devices per unit
15. Maximum number of timers per unit
16. Maximum number of recipe variables per recipe
17. Contact input scan rate(s)
18. Batch cycle time (time between executions of the batch software)

Unit Specification Data

One set of data required for each unit in a system.

1. Unit name/ID
2. Unit type
3. Console number(s) (for entering and monitoring)
4. Annunciator number(s) (for alarming)
5. Alarm printer number
6. Integer variable identifiers and values
7. Floating-point variable identifiers and values
8. Continuous control block cross-reference data

Switch Input Specification Data

One set of data required for each switch input.

1. Switch number (local to the unit)
2. Switch name
3. Unit name (unit with which this switch is associated)
4. Description
5. Open state mnemonic
6. Closed state mnemonic
7. Switch type
   - Hardware
   - Analog alarm
   - Flag

Switch Output Specification Data

One set of data required for each switch output.

1. Switch number (local to the unit)
2. Switch name
3. Unit name (unit with which this switch is associated)
4. Description
5. Open state mnemonic
6. Closed state mnemonic
7. Switch type
   Hardware
   Flag

Device Specification Data

One set of data required for each device in the system.

1. Device name
2. Device number
3. Description
4. Input-switch name(s)
   Input 1
   Input 2
   .
   .
   .
   Input n
5. Output switch name(s)
   Output 1
   Output 2
   .
   .
   .
   Output n
6. State 1 mnemonic
7. State 2 mnemonic
8. State 1 definition
   Input states (1,2,...,n)
   Output states (1,2,...,n)
9. State 2 definition
   Input states (1,2,...n)
   Output states (1,2,...n)
10. Transition time allowed between states

Switch Input Initialization Data

One set of data required for each switch input for a particular unit type.

1. Unit type
2. Switch number
3. Initialization data

Phase	*Desired State*	*Alarm*	*Action on Discrepancy*
PH1	O/C	Y/N	S/H/N
PH2	O/C	Y/N	S/H/N
.	.	.	.
.	.	.	.
.	.	.	.
PHn	O/C	Y/N	S/H/N

Note: n = total number of phases for the unit type, O = open, C = closed, Y = yes, N = no/no action, S = call service, H = drive unit to hold.

Switch Output Initialization Data

One set of data required for each switch output for a particular unit type.

1. Unit type
2. Switch number
3. Initialization data

Phase	*Desired State*	*Desired State for Hold*
PH1	O/C	O/C
PH2	O/C	O/C
.	.	.
.	.	.
.	.	.
PHn	O/C	O/C

Note: For explanation, see notes under *Switch Input Initialization Data.*

APPENDIX
4

NUMBERING SYSTEMS

The *decimal* numbering system is the most familiar. Starting at the decimal point and proceeding left, digit positions represent 1's, 10's, 100's, and so on, or 10^0's, 10^1's, 10^2's, ... Computer circuits in practical use have two states, 1 and 0. (Commercial three-state logic has a "disconnected" state as well, allowing multiple chip outputs to be electrically connected. They have only two active output states, so this discussion applies to these devices as well.) When we wish to represent numbers within a computer, we must do so by combinations of 1's and 0's.

The *binary* numbering system assigns the weights 2^0, 2^1, 2^2, . . . or 1, 2, 4, . . . to successive positions to the left of the binary point. Thus, the decimal number 203 can be expressed as 128 + 64 + 8 + 2 + 1 and be represented in binary notation as 11001011.

Some devices used with industrial control systems offer binary number outputs; analog-to-digital converters are usually binary, for example. For operator communication, however, binary notation is clearly not ideal. A thumbwheel switch for numeric data entry has to have 10 positions per digit, not 8 or 16, which would enable binary notation numbers to be directly represented. The *binary coded decimal* (BCD) number system is used in these cases. BCD codes each digit separately as a binary number (i.e., in the range of 0000 to 1001 binary). Codes with higher numeric values, representing 10-15 in decimal, are not valid. BCD coding is usually accepted by programmable controllers and similar equipment, which, depending on the model and the application, either retain the BCD representation throughout or convert to binary by a standard algorithm.

There is no technical reason why simple decimal notation using 10 binary digits ("bits") per decimal digit cannot be used. Thus the number 4 decimal would be represented as 0000010000; the "1" is in the position corresponding to value 4, starting with value 0 at the right. For most applications it simply is not economical to do so; this requires 10 signals per decimal digit versus 4 for BCD. Some numeric keypads do use this method, since only 10 numeric signals, plus "clear" and "enter" can be used to key in a number of any length. (A strobe signal is sometimes used as well to indicate the presence of new data.) Similarly, before the availability of numeric readouts, such coding was commonly used for laboratory instrument readouts—one indicator lamp per digit value, 40 lamps for the range 0000-9999.

Numeric readouts commonly used today use seven segments to form any of the numerals 0-9 (as well as some letters, like "A," in some applications). The logic required to convert BCD to *seven-segment coding* is embodied within certain integrated circuits, which can then be packaged with the displays to form a display module capable of accepting BCD inputs. This same logic can also be installed in a programmable controller if, for example, the user wishes to drive a high-power readout.

Octal and *hexadecimal* (hex) *coding* designate base-8 and base-16 counting systems, respectively. They are used to represent binary numbers in a convenient and readable format by forming them into groups of three or four bits. Thus binary number 11001011 is shown as 313 octal (the leftmost grouping may have one, two, or three bits—in this case, two) or CB in hex. (For hex representations, "10" is shown as "A," "11" as "B," etc.)

Eight bits can be assigned in 2^8, or 256, combinations. They can represent values between 1 and 256, -127 to +128, .01 to 2.56, and so on. With proper scaling these *fixed-point* numbers of typically eight or more bits can be used to represent a wide range of process variables—just as signals of 4-20 mA dc, 0-10 Vdc, and 3-15 psig can be scaled to represent, say, 50-250°F. For communications with the operator the real-world values usually are displayed. For pneumatic and electronic analog controls, this is done by printed scales. Digital systems can use *floating-point* representations to allow the manipulation and display of values from very small to very large. Floating-point numbers are a form of scientific notation, which allows us to represent such numbers as 6.02×10^{23} without explicitly writing all of the zeroes. In this example, 6.02 is the *mantissa* and 23 is the *exponent*. Floating-point representations simply reserve part of the variable for the mantissa and part for the exponent, each of which may be positive or negative. A typical process control minicomputer uses 32 bits to achieve a range of roughly 10^{38}-10^{-39}, whereas with fixed-point 32 bits would typically achieve a range of from 0 to 4×10^6.

There are conventions for the representation of floating-point numbers, but they are not universally followed. The manufacturer of a distributed control system is responsible for insuring compatibility between all of its components; those assembling a system from products of multiple vendors have this responsibility themselves. Except for distributed control systems, floating-point numbers are usually contained within a central processor and its memory so that the coding method is not a major concern for the system designer.

Keyboards and printers generally use the well-known ASCII (American Standard Code for Information Interchange) coding, which uses groups of eight bits (seven active, one for error detection) to represent upper- and lowercase letters, numerals, punctuation, and other special characters. ASCII coding is popular for the interconnection of items of equipment from multiple vendors—although electrical, mechanical, and protocol compatibility must also be established. Transmission of numerical data by ASCII is usually less efficient than by integer or floating-point methods: The number 24,021 requires 8 bits × 5 characters or 40 bits in ASCII, but as few as 15 bits for binary coding; the exact number depends on the transmission protocol.

APPENDIX

5

TERMINOLOGY

We have selected terminology for this book to typify conventions now commonly used by U.S. users and vendors. Concurrently with the preparation of this book, the International Purdue Workshop on Industrial Computer Systems, Special Applications Programming Committee TC-4 (see Purdue, 1984) has been developing a proposed standard terminology for batch process control. References to "Purdue" in the glossary following, refer to that committee's proceedings, in the few cases where their terminology differs significantly from that used here. A summary of the committee's recommendations was published by Bristol (1985).

Auto (or Automatic) State: The state of a unit when it operates in accordance with control system outputs.

Batch: 1. A process that produces a product or products in discrete quantities. 2. A discrete quantity of a product, as produced by a batch process.

Batch Cycle Time: Time between completion of contiguous batches.

Cam Sequencer: A sequence control system using rotating notched cams.

Contact Input: A *contact* whose state is an input to a control system.

Contact Input Checking: Comparing the state of a *contact input* to its *desired state.*

Contact Output: A control system output that is always in one of two possible states.

Desired State (of a **Contact Input**): The state in which the contact input should be.

Device: An item of field equipment whose interface to the computer consists of more than a single contact input or output. A device is displayed and operated as a single entity instead of individual inputs and outputs. For example, a valve that has one output for operation and two limit switches for position feedback may be defined as a device.

Drum Sequencer: A sequence control sytem using a rotating drum with plugs.

Formula: (Purdue) Data defining a particular type of product. (Similar to our *recipe;* see also *procedure.*)

Hold Logic: A set of instructions for driving a *unit* to a *hold state.* This is a subset of *service logic.*

Hold State: The state to which a *unit* is driven in order to safely suspend normal sequencing.

Interlock: Enforced relationship between items of equipment. Interlock relationships are often used to ensure safety. They may be specified by Boolean equations, *ladder diagrams,* and so on.

Ladder Diagram: A logic diagram that uses electromechanical relay symbology. "Ladder" implies that the "power" wiring is shown as vertical lines, interconnected with logic circuits shown horizontally.

Logic Location: Recipe value that indicates where, within a process, a particular recipe item is to be used.

Manual State: The state of a unit when it is not being controlled by the control system.

Normal Logic: That part of a sequence that directs the normal unit operations. Normal logic assumes that the process will perform without contingencies (see *service logic*).

Phase: A major operational section in the manufacture of a batch. Phase boundaries define major milestones or points of safe intervention in the process. Startup, shutdown, discharge, reaction, and so on, may each be considered a phase. A phase is generally a collection of *steps.*

Phase Initialization Data: A set of data regarding *contact inputs, outputs* and/or *devices* required at the start of a *phase.* They may include:

Desired states of the *contacts* and/or *devices*
Alarming and servicing requirements or discrepancy
States to which the *contacts* or *devices* are to be driven, if the *unit* is put in *hold state,* and so on.

Phase Logic: Logic for the operation of a *phase.*

Procedure: A set of instructions specifying the order of execution of a set of *phases* to manufacture a specific product.

Recipe: A set of data defining the distinct control requirements of a particular type or grade of a product. Typically, a recipe is a list of parameters such as feed quantities, set points, ramp rates, and so on, but may also include logical operations or the order of execution of *steps* of logic. (Purdue: A recipe is the combination of a procedure and formula; it always specifies quantities, etc., and sequence.)

Scanner/Driver: A program that scans *inputs* and drives *outputs.*

Service Logic: That part of a sequence dealing with plant or process contingencies and other unexpected or random events, such as an operator's request to deviate from normal operations. (See *normal logic*)

Step: A set of instructions to complete a simple process activity, such as opening a valve and waiting for confirmation.

Unit: A group of interrelated items of process equipment that work in a somewhat autonomous fashion. A process plant is generally divided into a number of units. A batch reactor, along with its associated piping, pumps, agitators, and so on, would commonly be considered a unit. Utilities that are plant-wide resources, such as steam supply, may also be considered as units.

Unit Status Table (UST): A set of data for keeping track of the execution of *phase logic* for a *unit.*

REFERENCES

Armstrong, W. S., and B. F. Coe, 1983, Multi-product batch reactor control, *Chemical Engineering Progress,* January, pp. 56-61.

Barona, N., and S. Bacher, 1983, *Fundamentals of Batch Processing,* (course notes) Section II, American Institute of Chemical Engineers.

Best, R. E., 1975*a,* Batch Control Problems—Part I: Coping with Controllability, *Instrumentation Technology,* September, pp. 53-57.

Best, R. E., 1975*b,* Batch Control Problems—Part II: Optimizing Safety and Reliability, *Instrumentation Technology,* October, pp. 49-52.

Bowen, H. C., A. Ghosh, A. McGrath, and P. Rogerson, 1975, User-Oriented Languages on a Batch Plant, *Trends in On-Line Computer Control Systems,* Conference Publication No. 127, April, Institution of Electrical Engineers, Herts, England, pp. 154-161.

Bristol, E. H., 1985, A design tool kit for batch process control: Terminology and a structural model, *InTech* October, pp. 47-50.

Brody, K., 1985, Tolerating faults, *Instruments & Control Systems,* February, pp. 35-40.

California Raisin Advisory Board, 1983, California Golden Raisins—This Christmas Give Them Gold, *Ladies' Home Journal,* December, p. 17.

Canup, R. T., 1981, Process control system increases digester productivity at mill, *Pulp & Paper,* September, pp. 159-161.

Czech, R. S., 1982, The application of computer systems in refinery oil movements operations, *Advances in Instrumentation* **37**(part 2):797-821.

D'Angelo, P., 1981*a,* Structured analysis simplifies modern software design, *Electronic Design,* September, pp. 159-161.

D'Angelo, P. 1981*b*, Good software depends on proper testing and management, *Electronic Design*, October, pp. 187-190.

DeGraca, 1975, Graphic batch processing language simplified, *Instruments & Control Systems*, December, pp. 41-43.

De Marco, T., 1978, *Structured Analysis and System Specification*, Yourdon, New York.

Egli, U. M., and D. W. T. Rippin, 1981, Short-Term Scheduling for Multi-Product Chemical Batch Plants, American Institute of Chemical Engineers meeting, Houston, Texas, April.

Ellingsen, R. E., 1985, Operating Cost Optimization of a Batch Process, Fourth Annual Control Engineering Conference, Rosemount, IL, May 21-23.

Fadum, O. K., 1979, A computer-coordinated control system for batch digesters, *Advances in Instrumentation* **Vol. 34.**

Ghosh, A., 1982, Modular structuring of batch control logic, *Advances in Instrumentation* **37**(part 2):783-796.

Glass, R. L., 1979, *Software Reliability Guidebook*, Prentice-Hall, Englewood Cliffs, N.J., p. 204.

Ham, A. A., and J. M. Liemberg, 1976, Comparison of five temperature control schemes for an exothermic batch reactor, *Journal "A"* (Belgium) **16**(2).

Hochkamp, H. D., 1980, Computer Control Saves Energy in Dry Solids Batch Blending, *Instrumentation Technology*, May, pp. 57-58.

Instrument Society of America, Standard S18.1, "Annuciator Sequences and Specifications."

Kafer, P., 1982, Roasting Coffee: PC's Do It Better, *In Tech*, October.

King, J., 1984, Heuristic model for resource allocation in a contention environment, *Advances in Instrumentation* **39:**415-421.

Lackmeyer, R. J., and D. G. Kempfer, 1977, Computer Simulation of a Batch Reaction Which Utilizes an Inferential Conversion Monitor for Control, IEEE 1977 Joint Automatic Control Conference, San Francisco, CA, June 22-24.

Lemay, R., 1979, Batch digester computer cooking control, *Pulp & Paper Canada* **80**(6):109-113. (Paper also presented at 1978 Intl. Sulphite Conf. of the Technical Section, CPPA.)

Liptak, B. G., 1986, letter, Re: Use Three Reservoirs for Integrated Heating and Cooling Control, Oct. InTech, *InTech*, January, p. 38.

Lloyd, M., 1985, Graphical function chart programming for programmable controllers, *Control Engineering*, October, pp. 73-76.

Löhteemöki, E., E. Jutila, and M. Paasila, Profitability as a Criterion of Batch Process Design, Twelfth Symposium on Computer Applications in Chemical Engineering, Montreaux, Switzerland, April 8-11. (Also in *Computers and Chemical Engineering* **3**(1-4):197.)

McEvoy, L. D., 1978, Control systems for belt feeders, *Instrumentation Tech.*, February.

Mauderli, A., and D. W. T. Rippin, 1979, Production planning and scheduling for multi-purpose batch chemical plants, *Computers and Chemical Engineering* **3:**199-206.

Mauderli, A., and D. W. T. Rippin, 1980, Scheduling production in multi-purpose batch plants: The batchman program, *Chemical Engineering Progress,* April, pp. 37-45.

Mehta, G. A., 1983, Batch Process Control: Considerations and Benefits, American Institute of Chemical Engineers, Spring 1983 meeting, Houston, Texas, March 27-31.

Morris, R. C., 1983, Simulating batch processes, *Chemical Engineering,* May 16, pp. 77-81.

Morrison, R. V., 1984, Analyzing batch process cycles, *Chemical Engineering,* June, p. 175.

National Association of Relay Manufacturers, 1980, *Engineers' Relay Handbook,* 3rd ed., NARM, Elkhart, Indiana.

National Electrical Manufacturers Association, 1983, *Standards Publication ICS 3-1983,* Part ICS 3-304, NEMA, Washington, D.C.

Norris, R. G., 1980, Batch digester cooking control at Harmac, *Pulp & Paper Canada* **81**(12):88-90.

Perron, M., G. Chalaye, C. Foulard, and H. Schmith, 1975, A dynamic optimization of a batch cooking plant, in *Sixth Triennal World Congress of the International Federation of Automatic Control Proceedings,* Instrument Society of America, Research Triangle Park, N.C.

Purdue, 1984, *Minutes, Twelfth Annual Meeting, International Purdue Workshop on Industrial Computer Systems,* October 1-4, Purdue Laboratory for Industrial Control, W. Lafayette, IN.

Rosenof, H. P., 1980, Multiple PC's share scans, *Instruments and Control Systems,* March, pp. 118-119.

Rosenof, H. P., 1981, Diagramming method simplifies sequential control documentation, *Control Engineering,* March, pp. 73-77.

Rosenof, H. P., 1982*a,* Successful batch control planning: A path to plantwide automation, *Control Engineering,* September, pp. 107-109.

Rosenof, H. P., 1982*b,* Building batch control systems around recipes, *Chemical Engineering Progress,* September, pp. 59-62.

Rosenof, H. P., 1982*c,* Interunit synchronization in batch systems, *Advances in Instrumentation* **37**(part 2):863-877.

Rosenof, H. P., 1985, Data logging and reporting for effective batch control, *Instruments & Control Systems,* July, pp. 29-32.

Rothstein, M. B., and R. F. Sweeney, 1971, Sub-optimal control in instrumentation, *Advances in Instrumentation* **26**(part 4).

Rubenstein, E., 1979, The accident that shouldn't have happened, *IEEE Spectrum,* November, p. 38.

Saxon, M. J., and K. Glover, 1982, Adaptive Stochastic Control of an Industrial

Batch Process, IEE Conference on Applications of Adaptive and Multivariable Control, Hull, England, July.

Sicignano, A., J. D. McKeand, and S. F. LeMasters, 1984, IBM PC schedules batch processes, cuts inventories at Houston refinery, *Oil and Gas Journal,* February, pp. 63-67.

Shinskey, F. G., 1977, *Distillation Control for Productivity and Energy Conservation,* McGraw-Hill, New York.

Shinskey, F. G., 1979, *Process Control Systems* 2nd ed., McGraw-Hill, New York.

Shinskey, F. G., and Weinstein, 1965, A Dual-Mode Control System for a Batch Exothermic Reactor, Twentieth Annual ISA Conference, Los Angeles, California, October 4-7.

Spruytenberg, R., et al., 1976, Experience with a computer-coupled bioreactor, *The Chemical Engineer,* June, pp. 447-479.

Trüeb and Klanica, 1978, Development of an Algorithm for Temperature Control of Stirred Batch Reactors, Fourth IFAC/IFIP Conference on Digital Computer Applications to Process Control, March 19-22, Zurich, Switzerland (Springer-Verlag, Berlin, W. Germany).

Wensley, J. H., 1982, Reliability in batch control processes, *Advances in Instrumentation* **37**(part 3):1423-1436.

Wu, R. S. H., 1985, Dynamic thermal analyzer for monitoring batch processes, *Chemical Engineering Progress,* September, pp. 57-61.

Whitley, J. R., 1985, Use three reservoirs for integrated heating and cooling control, *InTech,* October, pp. 57-59. (See also Liptak's letter to the editor, *InTech,* January, 1986.)

Yamada, M., H. Fukui, Z. Hokanoso, and O. Adica, 1981, Automatic Start-up of Distillation Column by Computer, Eighth Triennial World Congress of the International Federation of Automatic Control, Kyoto, Japan.

INDEX

INDEX